Universitext

Werner Greub

Multilinear Algebra

2nd Edition

Springer-Verlag
New York Heidelberg Berlin

Werner Greub

Department of Mathematics
University of Toronto
Toronto M5S 1A1
Canada

AMS Subject Classifications: 15-01, 15A75, 15A72

Library of Congress Cataloging in Publication Data

Greub, Werner Hildbert, 1925-
Multilinear algebra,

(Universitext)
Includes index.
1. Algebras, Linear. I. Title.
QA184.G74 1978 512'.5 78-949
ISBN 0-387-90284-8

Printed in the United States of America

9 8 7 6 5 4 3 2 1

ISBN 0-387-90284-8 Springer-Verlag New York
ISBN 3-540-90284-8 Springer-Verlag Berlin Heidelberg

Preface

This book is a revised version of the first edition and is intended as a sequel and companion volume to the fourth edition of *Linear Algebra* (Graduate Texts in Mathematics 23).

As before, the terminology and basic results of *Linear Algebra* are frequently used without reference. In particular, the reader should be familiar with Chapters 1–5 and the first part of Chapter 6 of that book, although other sections are occasionally used.

In this new version of *Multilinear Algebra*, Chapters 1–5 remain essentially unchanged from the previous edition. Chapter 6 has been completely rewritten and split into three (Chapters 6, 7, and 8). Some of the proofs have been simplified and a substantial amount of new material has been added. This applies particularly to the study of characteristic coefficients and the Pfaffian.

The old Chapter 7 remains as it stood, except that it is now Chapter 9. The old Chapter 8 has been suppressed and the material which it contained (multilinear functions) has been relocated at the end of Chapters 3, 5, and 9.

The last two chapters on Clifford algebras and their representations are completely new. In view of the growing importance of Clifford algebras and the relatively few references available, it was felt that these chapters would be useful to both mathematicians and physicists.

In Chapter 10 Clifford algebras are introduced via universal properties and treated in a fashion analogous to exterior algebra. After the basic isomorphism theorems for these algebras (over an arbitrary inner product space) have been established the chapter proceeds to a discussion of finite-dimensional Clifford algebras. The treatment culminates in the complete classification of Clifford algebras over finite-dimensional complex and real inner product spaces.

The book concludes with Chapter 11 on representations of Clifford algebras. The twisted adjoint representation which leads to the definition of the spin-groups is an important example. A version of Wedderburn's theorem is the key to the classification of all representations of the Clifford algebra over an 8-dimensional real vector space with a negative definite inner product. The results are applied in the last section of this chapter to study orthogonal multiplications between Euclidean spaces and the existence of orthonormal frames on the sphere. In particular, it is shown that the $(n-1)$-sphere admits an orthonormal k-frame where k is the Radon–Hurwitz number corresponding to n. A deep theorem of F. Adams states that this result can not be improved.

The problems at the end of Chapter 11 include a basis-free definition of the Cayley algebra via the complex cross-product analogous to the definition of quaternions in Section 7.23 of the fourth edition of *Linear Algebra*. Finally, the Cayley multiplication is used to obtain concrete forms of some of the isomorphisms in the table at the end of Chapter 10.

I should like to express my deep thanks to Professor J. R. Vanstone who worked closely with me through each stage of this revision and who made numerous and valuable contributions to both content and presentation. I should also like to thank Mr. M. S. Swanson who assisted Professor Vanstone and myself with the proof reading.

Toronto, April 1978 W. H. Greub

Table of Contents

Tensor Products 1

Throughout this chapter except where noted otherwise all vector spaces will be defined over a fixed, but arbitrarily chosen, field Γ.

Multilinear Mappings

1.1. Bilinear Mappings

Suppose E, F and G are any three vector spaces, and consider a mapping

$$\varphi: E \times F \to G.$$

φ is called *bilinear* if it satisfies the conditions

$$\varphi(\lambda x_1 + \mu x_2, y) = \lambda\varphi(x_1, y) + \mu\varphi(x_2, y) \qquad x_1, x_2 \in E, y \in F, \lambda, \mu \in \Gamma,$$

$$\varphi(x, \lambda y_1 + \mu y_2) = \lambda\varphi(x, y_1) + \mu\varphi(x, y_2) \qquad x \in E, y_1, y_2 \in F.$$

Recall that if $G = \Gamma$, then φ is called a *bilinear function.*

The set S of all vectors in G of the form $\varphi(x, y)$, $x \in E$, $y \in F$ is not in general a vector subspace of G. As an example, let $E = F$ and G be respectively 2- and 4-dimensional vector spaces. Select a basis a_1, a_2 in E and a basis c_ν $(\nu = 1, \ldots, 4)$ in G and define the bilinear mapping φ by

$$\varphi(x, y) = \xi^1\eta^1 c_1 + \xi^1\eta^2 c_2 + \xi^2\eta^1 c_3 + \xi^2\eta^2 c_4$$

where $x = \xi^1 a_1 + \xi^2 a_2$ and $y = \eta^1 a_1 + \eta^2 a_2$. Then it is easy to see that a vector

$$z = \sum_\nu \lambda^\nu c_\nu$$

of G is contained in S if and only if the components satisfy the relation

$$\lambda^1\lambda^4 - \lambda^2\lambda^3 = 0.$$

Since the vectors $z_1 = 2c_1 + 2c_2 + c_3 + c_4$ and $z_2 = c_1 + c_3$ satisfy this condition, while the vector $z = z_1 - z_2 = c_1 + 2c_2 + c_4$ does not, it follows that S is not a subspace of G.

We shall denote by $\operatorname{Im}\varphi$ the subspace of G *generated* by S.

Now consider the set $B(E, F; G)$ of all bilinear mappings of $E \times F$ into G. By defining the sum of two bilinear mappings φ_1 and φ_2 by

$$(\varphi_1 + \varphi_2)(x, y) = \varphi_1(x, y) + \varphi_2(x, y)$$

and the mapping $(\lambda\varphi)$ by

$$(\lambda\varphi)(x, y) = \lambda\varphi(x, y) \qquad x \in E, y \in F, \lambda \varepsilon \Gamma,$$

we can introduce a vector space structure in the set $B(E, F; G)$. The space $B(E, F; \Gamma)$ of all bilinear functions in $E \times F$ will be denoted simply by $B(E, F)$.

1.2. Bilinear Mappings of Subspaces and Factor Spaces

Given a bilinear mapping $\varphi: E \times F \to G$ and a pair of subspaces $E_1 \subset E$, $F_1 \subset F$, a bilinear mapping $\varphi_1: E_1 \times F_1 \to G$ is induced by

$$\varphi_1(x_1, y_1) = \varphi(x_1, y_1) \qquad x_1 \in E_1, y_1 \in F_1.$$

φ_1 is called the *restriction* of φ to $E_1 \times F_1$.

Let $E = \sum_\alpha E_\alpha$ and $F = \sum_\beta F_\beta$ be two direct decompositions of E and F respectively and assume that for every pair (α, β) a bilinear mapping $\varphi_{\alpha\beta}: E_\alpha \times F_\beta \to G$ is given. Then there exists precisely one bilinear mapping $\varphi: E \times F \to G$ whose restriction to $E_\alpha \times F_\beta$ is $\varphi_{\alpha\beta}$. In fact, if $\pi_\alpha: E \to E_\alpha$ and $\rho_\beta: F \to F_\beta$ are the canonical projections, define φ by

$$\varphi(x, y) = \sum_{\alpha, \beta} \varphi_{\alpha\beta}(\pi_\alpha x, \rho_\beta y) \qquad x \in E, y \in F.$$

Then the restriction of φ to $E_\alpha \times F_\beta$ is $\varphi_{\alpha\beta}$.

Now let φ_1 and φ_2 be two bilinear mappings of $E \times F$ into G whose restrictions to $E_\alpha \times F_\beta$ are $\varphi_{\alpha\beta}$. Then it follows that

$$\begin{aligned}(\varphi_1 - \varphi_2)(x, y) &= \varphi_1(x, y) - \varphi_2(x, y) \\ &= \sum_{\alpha, \beta} \varphi_{\alpha\beta}(\pi_\alpha x, \rho_\beta y) - \sum_{\alpha, \beta} \varphi_{\alpha\beta}(\pi_\alpha x, \rho_\beta y) = 0\end{aligned}$$

whence $\varphi_1 = \varphi_2$.

If $E_1 \subset E$ and $F_1 \subset F$ are subspaces and $\varphi_1: E_1 \times F_1 \to G$ is a bilinear mapping then there exists a (not uniquely determined) bilinear mapping $\varphi: E \times F \to G$ whose restriction to $E_1 \times F_1$ is φ_1. To prove this choose two subspaces $E_2 \subset E$ and $F_2 \subset F$ such that

$$E = E_1 \oplus E_2, \qquad F = F_1 \oplus F_2$$

and define the bilinear mappings $\varphi_{ij}: E_i \times F_j \to G$ $(i, j = 1, 2)$ by $\varphi_{11} = \varphi_1$ and $\varphi_{ij} = 0$, $(i, j) \neq (1, 1)$. In view of the above remark there exists a bilinear mapping $\varphi: E \times F \to G$ whose restriction to $E_i \times F_j$ is φ_{ij}.

Now suppose that $\varphi: E \times F \to G$ is bilinear, and for some subspaces $E_1 \subset E$ and $G_1 \subset G$ $\varphi(x_1, y) \in G_1$ for every $x_1 \in E_1$, $y \in F$. Let $\rho: E \to E/E_1$ and $\pi: G \to G/G_1$ be the canonical projections, and define a bilinear mapping

$$\tilde{\varphi}: (E/E_1) \times F \to G/G_1$$

by

$$\tilde{\varphi}(\rho x, y) = \pi\varphi(x, y) \qquad \rho x \in E/E_1, y \in F.$$

It is clear that $\tilde{\varphi}$ is a well-defined bilinear mapping. We say that $\tilde{\varphi}$ is the bilinear mapping *induced by* φ.

If also for some subspace $F_1 \subset F$, $\varphi(x, y_1) \in G_1$ for each $x \in E$, $y_1 \in F_1$, then $\tilde{\varphi}(\rho x, y_1) = 0$ for each $\rho x \in E/E_1$, $y_1 \in F_1$. Denoting the canonical projection of F onto F/F_1 by σ we see that φ induces a bilinear mapping

$$\tilde{\varphi}: E/E_1 \times F/F_1 \to G/G_1$$

defined by

$$\tilde{\varphi}(\rho x, \sigma y) = \pi\varphi(x, y) \qquad \rho x \in E/E_1, \sigma y \in F/F_1.$$

1.3. Multilinear Mappings

Suppose we are given $p + 1$ vector spaces E_i $(i = 1, \ldots, p)$, G. A mapping $\varphi: E_1 \times \cdots \times E_p \to G$ is called *p-linear* if for every i $(1 \leq i \leq p)$

$$\varphi(x_1, \ldots, x_{i-1}, \lambda x_i + \mu y_i, x_{i+1}, \ldots, x_p) = \lambda\varphi(x_1, \ldots, x_i, \ldots, x_p) + \mu\varphi(x_1, \ldots, y_i, \ldots, x_p) \qquad x_i, y_i \in E_i, \lambda, \mu \in \Gamma.$$

If $G = \Gamma$, then φ is called a *p-linear function*.

As in the case $p = 2$ the subspace of G generated by the vectors $\varphi(x_1, \ldots, x_p)$, $x_i \in E_i$ will be denoted by $\operatorname{Im}\varphi$. Let $L(E_1, \ldots, E_p; G)$ be the set of all p-linear mappings $\varphi: E_1 \times \cdots \times E_p \to G$. Defining the linear operations by

$$(\varphi + \psi)(x_1, \ldots, x_p) = \varphi(x_1, \ldots, x_p) + \psi(x_1, \ldots, x_p)$$

and

$$(\lambda\varphi)(x_1, \ldots, x_p) = \lambda\varphi(x_1, \ldots, x_p)$$

we obtain a vector space structure in $L(E_1, \ldots, E_p; G)$. The space of all p-linear functions in $E_1 \times \cdots \times E_p$ will be denoted by $L(E_1, \ldots, E_p)$.

Problems

1. Establish natural isomorphisms

$$B(E, F; G) \cong L(E; L(F; G)) \cong L(F; L(E; G)).$$

2. Given a bilinear mapping $\varphi: E \times F \to G$, define a mapping $\psi: E \times F \to G$ by

$$\psi z = \varphi(\pi_1 z, \pi_2 z) \qquad z \in E \times F$$

where $\pi_1: E \times F \to E$ and $\pi_2: E \times F \to F$ are the canonical projections. Show that ψ satisfies the relations

$$\psi(z_1 + z_2) + \psi(z_1 - z_2) = 2\psi(z_1) + 2\psi(z_2)$$

and

$$\psi(\lambda z) = \lambda^2 \psi(z).$$

3. Let E and F be arbitrary. Show that the mapping $\beta: L(E; F) \times E \to F$ defined by $(\varphi, x) \to \varphi x$ is bilinear. Prove that $\operatorname{Im} \beta = F$.

4. Let E, F, G be finite-dimensional real vector spaces with the natural topology. Show that every bilinear mapping $\varphi: E \times F \to G$ is continuous. Conclude that the mapping $L(E; F) \times E \to F$ defined by $(\varphi, x) \to \varphi x$ is continuous.

5. Given a bilinear mapping $\varphi: E \times F \to G$ define the *null-spaces* $N_1(\varphi) \subset E$ and $N_2(\varphi) \subset F$ as follows:

$$N_1(\varphi): \{x; \varphi(x, y) = 0\} \quad \text{for every } y \in F$$

and

$$N_2(\varphi): \{y; \varphi(x, y) = 0\} \quad \text{for every } x \in E$$

(a) Consider the induced bilinear mapping

$$\tilde{\varphi}: E/N_1(\varphi) \times F/N_2(\varphi) \to G$$

(cf. Section 1.2). Show that $N_1(\tilde{\varphi}) = 0$ and $N_2(\tilde{\varphi}) = 0$.

Given a linear map $f: G \to H$ consider the bilinear mapping $\psi: E \times F \to H$ defined by $\psi(x, y) = f\varphi(x, y)$. Show that

$$N_1(\varphi) \subset N_1(\psi) \quad \text{and} \quad N_2(\varphi) \subset N_2(\psi)$$

(b) Conversely, let $\psi: E \times F \to H$ be a bilinear mapping such that $N_1(\varphi) \subset N_1(\psi)$ and $N_2(\varphi) \subset N_2(\psi)$. Prove that there exists a linear map $f: G \to H$ such that

$$\psi(x, y) = f\varphi(x, y).$$

Consider the space L of linear maps $f: G \to H$ satisfying this condition. Establish a linear isomorphism

$$L \xrightarrow{\cong} L(G/\operatorname{Im} \varphi; H).$$

Conclude that f is uniquely determined by ψ if and only if $\operatorname{Im} \varphi = G$.

6. Let E be a vector space and F be the space of all functions $h:E \to \Gamma$. Define a bilinear mapping $\varphi:L(E) \times L(E) \to F$ by

$$\varphi(f, g)(x) = f(x)g(x) \qquad x \in E.$$

Show that $N_1(\varphi) = 0$ and $N_2(\varphi) = 0$.

7. Let E, E^* be a pair of dual spaces and assume that $\Phi:E^* \times E \to \Gamma$ is a bilinear function such that

$$\Phi(\tau^{*-1}x^*, \tau x) = \Phi(x^*, x)$$

for every pair of dual automorphisms. Prove that $\Phi(x^*, x) = \lambda\langle x^*, x\rangle$ where λ is a scalar.

The Tensor Product

1.4. The Universal Property

Let E and F be vector spaces and let $\otimes$ be a bilinear mapping from $E \times F$ into a vector space T. We shall say that $\otimes$ has the *universal property*, if it satisfies the following conditions:

$\otimes_1$: The vectors $x \otimes y$ $(x \in E,\ y \in F)$ generate T, or equivalently, $\operatorname{Im} \otimes = T$.

$\otimes_2$: If φ is a bilinear mapping from $E \times F$ into any vector space H, then there exists a linear map $f:T \to H$ such that the diagram

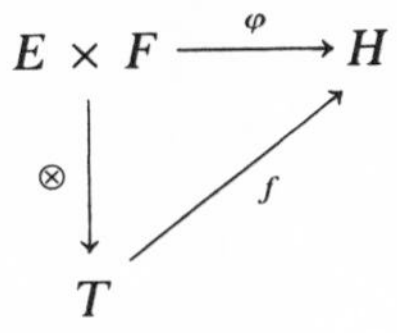

(1.1)

commutes.

The two conditions above are equivalent to the following single condition

$\otimes$: To every bilinear mapping $\varphi:E \times F \to H$ there exists a *unique* linear map $f:T \to H$ such that Diagram (1.1) commutes.

In fact, assume that $\otimes_1$ and $\otimes_2$ hold and let $f_1:T \to H$ and $f_2:T \to H$ be linear maps such that

$$\varphi(x, y) = f_1(x \otimes y) \quad \text{and} \quad \varphi(x, y) = f_2(x \otimes y).$$

Then we have

$$f_1(x \otimes y) = f_2(x \otimes y) \qquad x \in E, y \in F.$$

Now $\otimes_1$ implies that $f_1 = f_2$ and so f is uniquely determined by φ.

Conversely, assume that $\otimes$ holds. Then $\otimes_2$ is obviously satisfied. To prove $\otimes_1$, let T_1 be the subspace of T generated by the vectors $x \otimes y$ with $x \in E$ and $y \in F$. Then $\otimes$ determines a bilinear mapping $\varphi: E \times F \to T_1$ such that

$$i\varphi(x, y) = x \otimes y \qquad x \in E, y \in F,$$

where $i: T_1 \to T$ denotes the inclusion map. By $\otimes$ there is a linear map $f: T \to T_1$ such that

$$\varphi(x, y) = f(x \otimes y) \qquad x \in E, y \in F.$$

Applying i to this relation we obtain

$$(i \circ f)(x \otimes y) = (i \circ \varphi)(x, y) = x \otimes y.$$

On the other hand, clearly,

$$\iota(x \otimes y) = x \otimes y \qquad x \in E, y \in F$$

where ι is the identity map of T. Now the uniqueness part of $\otimes$ implies that $i \circ f = \iota$. Thus i is surjective and so $T_1 = T$. This proves $\otimes_1$.

Example. Consider the bilinear mapping $\Gamma \times F \to F$ given by $\lambda \otimes y = \lambda y$. Since $1 \otimes y = y$ this map satisfies condition $\otimes_1$. To verify $\otimes_2$, let $\varphi: \Gamma \times F \to H$ be any bilinear mapping and define a linear map $f: F \to H$ by setting

$$f(y) = \varphi(1, y).$$

Then we have for $\lambda \in \Gamma$ and $y \in F$

$$\varphi(\lambda, y) = \lambda\varphi(1, y) = \lambda f(y) = f(\lambda y) = f(\lambda \otimes y)$$

and so $\otimes_2$ is proved.

1.5. Elementary Properties

Before proving existence and uniqueness of bilinear mappings with the universal property we shall derive a few properties which follow directly from the definition.

Thus we assume that $\otimes: E \times F \to T$ is a bilinear mapping with the universal property.

Lemma 1.5.1. *Let a_i $(i = 1, \ldots, r)$ be linearly independent vectors in E and let b_i $(i = 1, \ldots, r)$ be arbitrary vectors in F. Then the relation*

$$\sum_i a_i \otimes b_i = 0$$

implies that $b_i = 0$ $(i = 1, \ldots, r)$.

PROOF. Since the a_i are linearly independent we can choose r linear functions f^i in E such that

$$f^i(a_j) = \delta^i_j \qquad (i, j = 1, \ldots, r).$$

Now consider the bilinear function

$$\Phi(x, y) = \sum_{i=1}^{r} f^i(x)g^i(y), \qquad x \in E,\ y \in F$$

where the g^i are arbitrary linear functions in F. In view of $\otimes_2$, there exists a linear function h in T such that

$$h(x \otimes y) = \sum_i f^i(x)g^i(y).$$

Then

$$h\left(\sum_j a_j \otimes b_j\right) = \sum_{i,j} f^i(a_j)g^i(b_j) = \sum_i g^i(b_i).$$

Since $\sum_j a_j \otimes b_j = 0$, we obtain

$$\sum_i g^i(b_i) = 0.$$

But the g^i are arbitrary and so it follows that

$$b_i = 0 \qquad (i = 1, \ldots, r).$$ □

Corollary. If $a \neq 0$ and $b \neq 0$, then $a \otimes b \neq 0$.

Lemma 1.5.2. *Let $\{e_\alpha\}_{\alpha \in A}$ be a basis of E. Then every vector $z \in T$ can be written in the form*

$$z = \sum_\alpha e_\alpha \otimes b_\alpha, \qquad b_\alpha \in F,$$

where only finitely many b_α are different from zero. Moreover, the b_α are uniquely determined by z.

PROOF. In view of $\otimes_1$, z is a finite sum

$$z = \sum_\nu x_\nu \otimes y_\nu \qquad x_\nu \in E,\ y_\nu \in F.$$

Now write

$$x_\nu = \sum_\alpha \lambda^\alpha_\nu e_\alpha \qquad \lambda^\nu_\alpha \in \Gamma.$$

Then we have

$$z = \sum_{\nu,\alpha} \lambda^\alpha_\nu e_\alpha \otimes y_\nu = \sum_{\nu,\alpha} e_\alpha \otimes \lambda^\alpha_\nu y_\nu = \sum_\alpha e_\alpha \otimes b_\alpha,$$

where

$$b_\alpha = \sum_\nu \lambda^\alpha_\nu y_\nu.$$

To prove uniqueness assume that

$$\sum_\alpha e_\alpha \otimes b_\alpha = \sum_\alpha e_\alpha \otimes b'_\alpha \qquad b_\alpha, b'_\alpha \in F.$$

Then

$$\sum_\alpha e_\alpha \otimes (b_\alpha - b'_\alpha) = 0$$

and so Lemma 1.5.1 implies that $b_\alpha = b'_\alpha$. □

Lemma 1.5.3. *Every nonzero vector $z \in T$ can be written in the form*

$$z = \sum_{i=1}^r x_i \otimes y_i \qquad x_i \in E,\ y_i \in F,$$

where the $x_i\,(1, \ldots, r)$ are linearly independent and the $y_i\,(i = 1, \ldots, r)$ are linearly independent.

Proof. Choose a representation $z = \sum_{i=1}^r x_i \otimes y_i$ where r is minimized. If $r = 1$, it follows from bilinearity that $x_1 \neq 0$ and $y_1 \neq 0$.

Now consider the case $r \geq 2$. If the vectors x_i are linearly dependent we may assume that

$$x_r = \sum_{i=1}^{r-1} \lambda^i x_i.$$

Then we have

$$z = \sum_{i=1}^{r-1} x_i \otimes y_i + \sum_{i=1}^{r-1} \lambda^i x_i \otimes y_r = \sum_{i=1}^{r-1} x_i \otimes (y_i + \lambda^i y_r) = \sum_{i=1}^{r-1} x_i \otimes y'_i$$

which contradicts the minimality of r. Thus the vectors x_i are linearly independent. In the same way it follows that the vectors y_i are linearly independent as well. □

1.6. Uniqueness

Suppose that $\otimes: E \times F \to T$ and $\tilde{\otimes}: E \times F \to \tilde{T}$ are bilinear mappings with the universal property. Then there exists a linear isomorphism $f: T \xrightarrow{\cong} \tilde{T}$ such that

$$f(x \otimes y) = x \mathbin{\tilde{\otimes}} y \qquad x \in E, y \in F.$$

In fact, in view of $\otimes_2$, we have linear maps

$$f: T \to \tilde{T} \quad \text{and} \quad g: \tilde{T} \to T$$

such that

$$f(x \otimes y) = x \mathbin{\tilde{\otimes}} y$$

and

$$g(x \mathbin{\tilde{\otimes}} y) = x \otimes y \qquad x \in E, y \in F.$$

These relations imply that

$$gf(x \otimes y) = x \otimes y \quad \text{and} \quad fg(x \mathbin{\tilde{\otimes}} y) = x \mathbin{\tilde{\otimes}} y.$$

Now $\otimes_1$ shows that

$$g \circ f = \iota \quad \text{and} \quad f \circ g = \iota.$$

Thus f and g are inverse linear isomorphisms.

1.7. Existence

To prove existence consider the free vector space $C(E \times F)$ generated by the set $E \times F$ (see Section 1.7 of *Linear Algebra*). Let $N(E, F)$ denote the subspace of $C(E \times F)$ generated by the vectors

$$(\lambda x_1 + \mu x_2, y) - \lambda(x_1, y) - \mu(x_2, y)$$

and

$$(x, \lambda y_1 + \mu y_2) - \lambda(x, y_1) - \mu(x, y_2).$$

Set

$$T = C(E \times F)/N(E, F)$$

and let $\pi: C(E \times F) \to T$ be the canonical projection.
Now define a set map $\otimes: E \times F \to T$ by

$$x \otimes y = \pi(x, y).$$

We shall show that $\otimes$ is a bilinear mapping and has the universal property. In fact, since

$$\pi(\lambda x_1 + \mu x_2, y) = \lambda\pi(x_1, y) + \mu\pi(x_2, y),$$

it follows that

$$\begin{aligned}(\lambda x_1 + \mu x_2) \otimes y &= \pi(\lambda x_1 + \mu x_2, y) \\ &= \lambda\pi(x_1, y) + \mu\pi(x_2, y) = \lambda x_1 \otimes y + \mu x_2 \otimes y.\end{aligned}$$

In the same way it is shown that $\otimes$ is linear in y.
To prove $\otimes_1$, observe that every vector $z \in T$ is a finite sum

$$z = \pi \sum_{\nu, \mu} \lambda^{\nu\mu}(x_\nu, y_\mu) \qquad x_\nu \in E, \ y_\mu \in F.$$

It follows that

$$\sum_{\nu, \mu} \lambda^{\nu\mu} x_\nu \otimes y_\mu = \sum_{\nu, \mu} \lambda^{\nu\mu}\pi(x_\nu, y_\mu) = \pi \sum_{\nu, \mu} \lambda^{\nu\mu}(x_\nu, y_\mu) = z.$$

To verify $\otimes_2$, consider a bilinear mapping ψ of $E \times F$ into a third vector space H. Since the pairs (x, y) $x \in E$, $y \in F$ form a basis for $C(E \times F)$ there is a uniquely determined linear map

$$g: C(E \times F) \to H$$

such that

$$g(x, y) = \psi(x, y).$$

The bilinearity of ψ implies that $N(E, F) \subset \ker g$. In fact, if

$$z = (\lambda x_1 + \mu x_2, y) - \lambda(x_1, y) - \mu(x_2, y)$$

is a generator of $N(E, F)$, then

$$\begin{aligned} g(z) &= g(\lambda x_1 + \mu x_2, y) - \lambda g(x_1, y) - \mu g(x_2, y) \\ &= \psi(\lambda x_1 + \mu x_2, y) - \lambda\psi(x_1, y) - \mu\psi(x_2, y) \\ &= 0. \end{aligned}$$

In a similar way it is shown that

$$g[(x, \lambda y_1 + \mu y_2) - \lambda(x, y_1) - \mu(x, y_2)] = 0$$

and it follows that $N(E, F) \subset \ker g$. Hence g induces a linear map

$$f: C(E \times F)/N(E, F) \to H$$

such that

$$f \circ \pi = g.$$

In particular, it follows that

$$(f \circ \otimes)(x, y) = f\pi(x, y) = g(x, y) = \psi(x, y).$$

This shows that the bilinear mapping $\otimes$ has the universal property.

Definition. The *tensor product* of two vector spaces E and F is a pair $(T, \otimes)$, where $\otimes: E \times F \to T$ is a bilinear mapping with the universal property. The space T, which is uniquely determined by E and F up to an isomorphism, is also called the tensor product of E and F and is denoted by $E \otimes F$.

Now we show that the tensor product is commutative in the sense that

$$E \otimes F \cong F \otimes E.$$

In fact, consider the bilinear mappings

$$\varphi: E \times F \to F \otimes E \quad \text{and} \quad \psi: F \times E \to E \otimes F$$

given by

$$\varphi(x, y) = y \otimes x \quad \text{and} \quad \psi(y, x) = x \otimes y.$$

In view of $\otimes_2$, they induce linear maps $f: E \otimes F \to F \otimes E$ and $g: F \otimes E \to E \otimes F$ such that

$$y \otimes x = f(x \otimes y) \quad \text{and} \quad x \otimes y = g(y \otimes x)$$

for all $x \in E$ and $y \in F$. These relations imply, in view of $\otimes_1$, that $g \circ f = \iota$ and $f \circ g = \iota$. Thus f and g are inverse isomorphisms.

1.8. Reduction of Bilinear Mappings to Linear Maps

Fix E and F and let G be a third vector space. Then a linear isomorphism

$$\Phi: L(E \otimes F; G) \xrightarrow{\cong} B(E, F; G)$$

is defined by

$$\Phi(f) = f \circ \otimes \qquad f \in L(E \otimes F; G).$$

In fact, $\otimes_2$ implies that Φ is surjective, since it states that any bilinear mapping $\varphi: E \times F \to G$ may be factored over the tensor product. To show that Φ is injective assume that $f \circ \otimes = 0$ for a certain linear map $f: E \otimes F \to G$. In view of $\otimes_1$ the space $E \otimes F$ is generated by the products $x \otimes y$ and hence it follows that $f = 0$.

The correspondence between the bilinear mappings $\varphi: E \times F \to G$ and the linear maps $f: E \otimes F \to G$ which is obtained by the above result is expressed by the following commutative diagram:

$$\begin{array}{ccc} E \times F & \xrightarrow{\varphi} & G \\ {\scriptstyle \otimes}\downarrow & \nearrow_{f} & \\ E \otimes F & & \end{array}$$

Proposition 1.8.1. *Let $\varphi: E \times F \to G$ be a bilinear mapping and $f: E \otimes F \to G$ be the induced linear map. Then f is surjective if and only if φ satisfies $\otimes_1$. Moreover f is injective if and only if φ satisfis $\otimes_2$.*

Proof. The first part follows immediately from the relation

$$\operatorname{Im} \varphi = \operatorname{Im} f.$$

To prove the second part assume that f is injective. Then the pair $(\operatorname{Im} \varphi, \varphi)$ is a tensor product for E and F. Hence every bilinear mapping $\psi: E \times F \to K$ induces a linear map $g: \operatorname{Im} \varphi \to K$ such that

$$\psi(x, y) = g\varphi(x, y).$$

If f is an extension of g to a linear map $f: G \to K$ it follows that

$$\psi(x, y) = f\varphi(x, y)$$

and hence φ has the property $\otimes_2$.

Conversely, assume that φ satisfies $\otimes_2$. Then the bilinear mapping $\otimes: E \times F \to E \otimes F$ induces a linear map $h: G \to E \otimes F$ such that

$$x \otimes y = h\varphi(x, y).$$

On the other hand, we have $\varphi(x, y) = f(x \otimes y)$ and it follows that

$$x \otimes y = hf(x \otimes y).$$

Hence $h \circ f = \iota$ and so f is injective. □

Problems

1. Consider the bilinear mapping $\beta: \Gamma^n \times \Gamma^m \to M^{n \times m}$ defined by

$$(\xi^1, \ldots, \xi^n) \times (\eta^1, \ldots, \eta^m) \to \begin{pmatrix} \xi^1\eta^1 & \cdots & \xi^1\eta^m \\ \vdots & & \vdots \\ \xi^n\eta^1 & \cdots & \xi^n\eta^m \end{pmatrix}$$

   Prove that the pair $(M^{n \times m}, \beta)$ is the tensor product of Γ^n and Γ^m.

2. Show that the bilinear mapping $\Gamma^n \times E \to \overset{n}{\oplus} E$ defined by

$$(\xi^1, \ldots, \xi^n) \otimes x = (\xi^1 x, \ldots, \xi^n x)$$

   is the tensor product.

3. Let S and T be two arbitrary sets and consider the vector spaces $C(S)$ and $C(T)$ generated respectively by S and T (cf. Section 1.7 of *Linear Algebra*). Show that $C(S \times T)$ is isomorphic to $C(S) \otimes C(T)$.

4. Assuming that $a \otimes b \neq 0$, $a \in E$, $b \in F$ prove that

$$a \otimes b = a' \otimes b'$$

   if and only if

$$a' = \lambda a \quad \text{and} \quad b' = \lambda^{-1} b \qquad \lambda \in \Gamma, \lambda \neq 0.$$

5. Let Δ be a subfield of Γ and consider a vector space E_Δ over Δ. Then $\Gamma \otimes E_\Delta$ is again a Δ-vector space. Define a scalar multiplication $\Gamma \times (\Gamma \otimes E_\Delta) \to \Gamma \otimes E_\Delta$ by

$$(\lambda, \alpha \otimes x) \to \lambda\alpha \otimes x \qquad \lambda, \alpha \in \Gamma, x \in E_\Delta$$

   (a) Prove that this multiplication makes $\Gamma \otimes E_\Delta$ into a Γ-vector space E_Γ.
   (b) Show that the restriction of this multiplication to $\Delta \times E_\Delta$ coincides with the scalar multiplication in E_Δ.
   (c) If $\{e_\alpha\}$ is a basis of E_Δ prove that $\{1 \otimes e_\alpha\}$ is a basis of E_Γ.
   (d) Let $\Gamma = \mathbb{C}$ and $\Delta = \mathbb{R}$. Prove that $E_\mathbb{C}$ is isomorphic to the vector space $E \times E$ constructed in Problem 5, §1, Chapter 1 of *Linear Algebra*.

6. With the notation of Problem 5 let φ_Δ be a linear transformation of E_Δ.
   (a) Prove that $\varphi_\Gamma = \iota \otimes \varphi_\Delta$ is a linear transformation of E_Γ.
   (b) For any polynomial $f \in \Gamma[t]$ prove that

$$(f(\varphi_\Delta))_\Gamma = f(\varphi_\Gamma).$$

   (c) Find the minimum polynomial of φ_Γ in terms of the minimum polynomial of φ_Δ.
   (d) Show that φ_Γ is semisimple (nilpotent) if φ_Δ is semisimple (nilpotent) and hence construct the decomposition of φ_Γ into semisimple and nilpotent parts.

Subspaces and Factor Spaces

1.9. Tensor Products of Subspaces

Suppose that the bilinear mapping $\otimes : E \times F \to T$ has the universal property and consider two subspaces $E_1 \subset E$ and $F_1 \subset F$. Let $\otimes'$ denote the restriction of $\otimes$ to $E_1 \times F_1$ and set $T_1 = \operatorname{Im} \otimes'$. We shall show that $(T_1, \otimes')$ is the tensor product of E_1 and F_1.

Property $\otimes_1$ is immediate from the definitions. To verify $\otimes_2$, let $\varphi_1 : E_1 \times F_1 \to H$ be a bilinear mapping. Extend φ_1 to a bilinear mapping $\varphi : E \times F \to H$. Since $\otimes$ has the universal property, there is a linear map

$$f : T \to H$$

such that

$$f(x \otimes y) = \varphi(x, y) \qquad x \in E, y \in F.$$

This relation implies that

$$f(x_1 \otimes' y_1) = \varphi(x_1, y_1) = \varphi_1(x_1, y_1) \qquad x_1 \in E_1, y_1 \in F_1,$$

and so φ_1 factors over $\otimes$.

1.10. Tensor Product of Factor Spaces

Again let $E_1 \subset E$ and $F_1 \subset F$ be subspaces and set

$$T(E_1, F_1) = E_1 \otimes F + E \otimes F_1.$$

Define a bilinear mapping $\beta : E \times F \to (E \otimes F)/T(E_1, F_1)$ by

$$\beta(x, y) = \pi(x \otimes y),$$

where π denotes the canonical projection.

Since $\beta(x_1, y) = 0$ if $x_1 \in E_1$, $y \in F$ and $\beta(x, y_1) = 0$, if $x \in E$, $y_1 \in F_1$, β induces a bilinear mapping

$$\bar{\beta} : E/E_1 \times F/F_1 \to (E \otimes F)/T(E_1, F_1)$$

such that

$$\bar{\beta}(\bar{x}, \bar{y}) = \beta(x, y) \qquad \bar{x} \in E/E_1, \bar{y} \in F/F_1.$$

To prove that $\bar{\beta}$ satisfies $\otimes$ first notice that

$$\operatorname{Im} \bar{\beta} = \operatorname{Im} \beta = \operatorname{Im} \pi = (E \otimes F)/T(E_1, F_1)$$

and so property $\otimes_1$ follows. To check $\otimes_2$, let

$$\psi : E/E_1 \times F/F_1 \to H$$

be any bilinear mapping. Define a bilinear mapping $\varphi : E \times F \to H$ by setting

$$\varphi(x, y) = \psi(\bar{x}, \bar{y})$$

Then there is a linear map $f: E \otimes F \to H$ such that

$$\varphi(x, y) = f(x \otimes y) \qquad x \in E, y \in F.$$

Moreover,

$$f(x_1 \otimes y) = \varphi(x_1, y) = \psi(0, \bar{y}) = 0 \qquad x_1 \in E_1, y \in F$$

and similarly,

$$f(x \otimes y_1) = 0 \qquad x \in E, y \in F_1.$$

Hence $T(E_1, F_1) \subset \ker f$, and so f induces a linear map

$$\bar{f}:(E \otimes F)/T(E_1, F_1) \to H$$

such that

$$\bar{f} \circ \pi = f.$$

It follows that

$$\begin{aligned}\psi(\bar{x}, \bar{y}) &= \varphi(x, y) = f(x \otimes y) = \bar{f}\pi(x \otimes y)\\ &= \bar{f}\beta(x, y) = \bar{f}\bar{\beta}(\bar{x}, \bar{y}) \qquad \bar{x} \in E/E_1, \bar{y} \in F/F_1\end{aligned}$$

whence $\psi = \bar{f} \circ \bar{\beta}$. Thus $\bar{\beta}$ satisfies condition $\otimes_2$ and so the proof is complete.

The result obtained above shows that there is a canonical isomorphism

$$E/E_1 \otimes F/F_1 \xrightarrow{\cong} (E \otimes F)/(E_1 \otimes F + E \otimes F_1).$$

Problem

Let E_1 and E_2 be subspaces of E such that $E = E_1 + E_2$ and set $E_1 \cap E_2 = F$. Establish an isomorphism

$$E/E_1 \otimes E/E_2 \xrightarrow{\cong} (E_1 \otimes E_2)/(F \otimes E_2 + E_1 \otimes F).$$

Direct Decompositions

1.11. Tensor Product of Direct Sums

Assume that two families of linear spaces $E_\alpha, \alpha \in I$ and $F_\beta, \beta \in J$ are given and that for every pair (α, β), $(E_\alpha \otimes F_\beta, \otimes)$ is the tensor product of E_α and F_β. Then a bilinear mapping φ of $\tilde{E} = \oplus_\alpha E_\alpha$ and $\tilde{F} = \oplus_\beta F_\beta$ into the direct sum $\tilde{G} = \oplus_{\alpha, \beta}(E_\alpha \otimes F_\beta)$ is defined by

$$\varphi(\tilde{x}, \tilde{y}) = \sum_{\alpha, \beta} i_{\alpha\beta}(\pi_\alpha \tilde{x} \otimes \rho_\beta \tilde{y}),$$

where

$$\pi_\alpha: E \to E_\alpha, \qquad \rho_\beta: \tilde{F} \to F_\beta$$

are the canonical projections and

$$i_{\alpha\beta}:E_\alpha \otimes F_\beta \to \tilde{G}$$

are the canonical injections. It will be shown that the pair $(\tilde{G}, \varphi)$ is the tensor product of $\tilde{E}$ and $\tilde{F}$.

Condition $\otimes_1$ is trivially fulfilled. To verify $\otimes_2$ let $\tilde{\psi}:\tilde{E} \times \tilde{F} \to H$ be an arbitrary bilinear mapping. Define $\psi_{\alpha\beta}:E_\alpha \times F_\beta \to H$ by

$$\psi_{\alpha\beta}(x, y) = \tilde{\psi}(i_\alpha x, j_\beta y),$$

where

$$i_\alpha:E_\alpha \to \tilde{E} \quad \text{and} \quad j_\beta:F_\beta \to \tilde{F}$$

are the canonical injections. Then $\psi_{\alpha\beta}$ induces a linear map $f_{\alpha\beta}:E_\alpha \otimes F_\beta \to H$ such that

$$\psi_{\alpha\beta}(x, y) = f_{\alpha\beta}(x \otimes y) \qquad x \in E_\alpha,\ y \in F_\beta.$$

Define a linear map $f:\tilde{G} \to H$ by

$$f = \sum_{\alpha, \beta} f_{\alpha\beta} \circ \pi_{\alpha\beta},$$

where

$$\pi_{\alpha\beta}:\tilde{G} \to E_\alpha \otimes F_\beta$$

are the canonical projections. Then it follows that

$$\begin{aligned}(f \circ \varphi)(\tilde{x}, \tilde{y}) &= f \sum_{\alpha, \beta} i_{\alpha\beta}(\pi_\alpha \tilde{x} \otimes \rho_\beta \tilde{y})\\ &= \sum_{\alpha, \beta} f_{\alpha\beta}(\pi_\alpha \tilde{x} \otimes \rho_\beta \tilde{y})\\ &= \sum_{\alpha, \beta} \psi_{\alpha\beta}(\pi_\alpha \tilde{x}, \rho_\beta \tilde{y})\\ &= \tilde{\psi}\left(\sum_\alpha i_\alpha \pi_\alpha \tilde{x}, \sum_\beta j_\beta \rho_\beta \tilde{y}\right)\\ &= \tilde{\psi}(\tilde{x}, \tilde{y}).\end{aligned}$$

Hence $f \circ \varphi = \tilde{\psi}$ and so $\otimes_2$ is satisfied.

1.12. Direct Decompositions

Assume that the pair $(E \otimes F, \otimes)$ is the tensor product of the vector spaces E and F and that two direct decompositions $E = \sum_\alpha E_\alpha$ and $F = \sum_\beta F_\beta$ are given. It will be shown that $E \otimes F$ is the direct sum of the subspaces $E_\alpha \otimes F_\beta$,

$$E \otimes F = \sum_{\alpha, \beta} E_\alpha \otimes F_\beta. \tag{1.2}$$

In view of $\otimes_1$ the space $E \otimes F$ is generated by the products $x \otimes y$; $x \in E$, $y \in F$.

Since $x = \sum_\alpha x_\alpha$, $x_\alpha \in E_\alpha$ and $y = \sum_\beta y_\beta$, $y_\beta \in F_\beta$ it follows that

$$x \otimes y = \sum_{\alpha, \beta} x_\alpha \otimes y_\beta.$$

This equation shows that the space $E \otimes F$ is the sum of the subspaces $E_\alpha \otimes F_\beta$.

To prove that the decomposition (1.2) is direct consider the direct sums $\tilde{E} = \oplus_\alpha E_\alpha$, $\tilde{F} = \oplus_\beta F_\beta$ and $\tilde{G} = \oplus_{\alpha,\beta} E_\alpha \otimes F_\beta$ and let the injections i_α, j_β, $i_{\alpha\beta}$ and the projections π_α, ρ_β, $\pi_{\alpha\beta}$ be defined as in the previous section. Then if $\varphi: \tilde{E} \times \tilde{F} \to \tilde{G}$ is the bilinear mapping given by

$$\varphi(\tilde{x}, \tilde{y}) = \sum_{\alpha, \beta} i_{\alpha\beta}(\pi_\alpha \tilde{x} \otimes \rho_\beta \tilde{y})$$

we have shown (in the previous section) that the pair $(\tilde{G}, \varphi)$ is the tensor product of $\tilde{E}$ and $\tilde{F}$.

Now consider the linear isomorphisms

$$f: E \to \tilde{E} \quad \text{and} \quad g: F \to \tilde{F}$$

defined by

$$fx = \sum_\alpha i_\alpha x_\alpha \qquad \text{and} \quad gy = \sum_\beta j_\beta y_\beta,$$

where

$$x = \sum_\alpha x_\alpha,\ x_\alpha \in E_\alpha \quad \text{and} \quad y = \sum_\beta y_\beta,\ y_\beta \in F_\beta.$$

Define a bilinear mapping $\psi: E \times F \to \tilde{G}$ by

$$\psi(x, y) = \varphi(fx, gy).$$

In view of the factorization property there exists a linear map $h: E \otimes F \to \tilde{G}$ such that

$$h(x \otimes y) = \psi(x, y)$$

and hence

$$h(x \otimes y) = \varphi(fx, gy).$$

If $x \in E_\tau$ and $y \in F_\sigma$ it follows from the definition of f, g, and φ that

$$\begin{aligned} h(x \otimes y) &= \varphi(fx, gy) = \varphi(i_\tau x, j_\sigma y) \\ &= \sum_{\alpha, \beta} i_{\alpha\beta}(\pi_\alpha i_\tau x \otimes \rho_\beta j_\sigma y) = i_{\tau\sigma}(x \otimes y). \end{aligned}$$

But this equation shows that h maps every subspace $E_\alpha \otimes F_\beta$ of $E \otimes F$ into the subspace $i_{\alpha\beta}(E_\alpha \otimes F_\beta)$ of $\tilde{G}$. Since the decomposition

$$\tilde{G} = \sum_{\alpha, \beta} i_{\alpha\beta}(E_\alpha \otimes F_\beta)$$

is direct, the decomposition

$$E \otimes F = \sum_{\alpha, \beta} E_\alpha \otimes F_\beta$$

must be direct. This completes our proof.

Conversely, suppose that direct decompositions

$$E = \sum_\alpha E_\alpha, \qquad F = \sum_\beta F_\beta, \qquad G = \sum_{\alpha, \beta} G_{\alpha\beta}$$

and bilinear mappings $\otimes : E_\alpha \times F_\beta \to G_{\alpha\beta}$ are given such that the pair $(G_{\alpha\beta}, \otimes)$ is the tensor product of E_α and F_β. Define a bilinear mapping $\varphi : E \times F \to G$ by

$$\varphi(x, y) = \sum_{\alpha, \beta} x_\alpha \otimes y_\beta,$$

where

$$x = \sum_\alpha x_\alpha \quad \text{and} \quad y = \sum_\beta y_\beta.$$

Then the pair (G, φ) is the tensor product of E and F.

The condition $\otimes_1$ is obviously satisfied. To prove $\otimes_2$ let $\psi : E \times F \to H$ be an arbitrary bilinear mapping and consider the restriction $\psi_{\alpha\beta}$ of ψ to $E_\alpha \times F_\beta$. Then there exists a linear map $f_{\alpha\beta} : G_{\alpha\beta} \to H$ such that $f_{\alpha\beta}(x_\alpha \otimes y_\beta) = \psi_{\alpha\beta}(x_\alpha, y_\beta)$. Define a linear map $f : G \to H$ by

$$f(z) = \sum_{\alpha, \beta} f_{\alpha\beta}(z_{\alpha\beta})$$

where $z = \sum_{\alpha, \beta} z_{\alpha\beta}$, $z_{\alpha\beta} \in G_{\alpha\beta}$.

Then

$$\begin{aligned}(f \circ \varphi)(x, y) &= f\varphi(x, y) \\ &= \sum_{\alpha, \beta} f_{\alpha\beta}(x_\alpha \otimes y_\beta) \\ &= \sum_{\alpha, \beta} \psi_{\alpha\beta}(x_\alpha, y_\beta) = \psi(x, y)\end{aligned}$$

whence $f \circ \varphi = \psi$. Thus $\otimes_2$ is satisfied and the proof is complete.

1.13. Tensor Product of Basis Vectors

Suppose that $(a_\alpha)_{\alpha \in I}$ and $(b_\beta)_{\beta \in J}$ are, respectively, bases of vector spaces E and F. Then the products $(a_\alpha \otimes b_\beta)_{\alpha \in I, \beta \in J}$ form a basis of $E \otimes F$. To prove this, let E_α, F_β denote the one-dimensional subspaces of E and F generated by a_α and b_β respectively. Then $E = \sum_\alpha E_\alpha$, $F = \sum_\beta F_\beta$; in view of the result of the previous section it follows that

$$E \otimes F = \sum_{\alpha, \beta} E_\alpha \otimes F_\beta.$$

Now it was shown in Section 1.5 that $a_\alpha \neq 0$, $b_\beta \neq 0$ implies $a_\alpha \otimes b_\beta \neq 0$. On the other hand, $\otimes_1$ applied to E_α, F_β and $E_\alpha \otimes F_\beta$ gives that $E_\alpha \otimes F_\beta$ is spanned by the single element $a_\alpha \otimes b_\beta$. Thus $E \otimes F$ is the direct sum of the one-dimensional subspaces generated by the products $a_\alpha \otimes b_\beta$, and hence these products form a basis of $E \otimes F$ (see Lemma 1.5.1).

In particular, it follows from these remarks that if E and F have finite dimensions, then $E \otimes F$ has finite dimension, and

$$\dim(E \otimes F) = \dim E \cdot \dim F. \tag{1.3}$$

1.14. Application to Bilinear Mappings

Let E and F be vector spaces with bases $(x_\alpha)_{\alpha \in I}$ and $(y_\beta)_{\beta \in J}$ respectively. Then since $x_\alpha \otimes y_\beta$ is a basis of $E \otimes F$, it follows that every set map of $(x_\alpha \otimes y_\beta)$ into a third vector space G can be extended in a unique way to a linear map

$$f: E \otimes F \to G$$

and every linear map $f: E \otimes F \to G$ is obtained in this way. In view of the isomorphism

$$L(E \otimes F; G) \cong B(E, F; G),$$

it follows that every set map

$$(x_\alpha, y_\beta) \to G$$

can be extended in a unique way to a bilinear mapping $\varphi: E \times F \to G$ and every bilinear mapping φ is obtained in this way. In particular, the space $\operatorname{Im} \varphi$ is generated by the vectors $\varphi(x_\alpha, y_\beta)$. This result implies that if E and F have finite dimension, then

$$\dim \operatorname{Im} \varphi \leqq \dim E \cdot \dim F.$$

It is now easy to construct a basis of the space $B(E, F; G)$ provided that the dimensions of E and F are finite. Let x_i $(i = 1, \ldots, n)$, y_j $(j = 1, \ldots, m)$ and $(z_\gamma)_{\gamma \in K}$ be bases of E, F, and G respectively. Then the products $x_i \otimes y_j$ form a basis of $E \otimes F$ and hence the linear maps f_γ^{kl} $(k = 1, \ldots, n;\ l = 1, \ldots, m;\ \gamma \in K)$ defined by

$$f_\gamma^{kl}(x_i \otimes y_j) = \delta_i^k \delta_j^l z_\gamma$$

form a basis of $L(E \otimes F; G)$. Consequently, the bilinear mappings φ_γ^{kl} given by

$$\varphi_\gamma^{kl}(x_i, y_j) = \delta_i^k \delta_j^l z_\gamma$$

form a basis of $B(E, F; G)$.

If G has finite dimension as well, it follows that

$$\dim B(E, F; G) = \dim E \cdot \dim F \cdot \dim G$$

and so in particular

$$\dim B(E, F) = \dim E \cdot \dim F.$$

1.15. Intersection of Tensor Products

Let E_1 and E_2 be two subspaces of a vector space E. Then if F is a second vector space,

$$\boxed{(E_1 \otimes F) \cap (E_2 \otimes F) = (E_1 \cap E_2) \otimes F.} \tag{1.4}$$

Clearly,

$$(E_1 \cap E_2) \otimes F \subset (E_1 \otimes F) \cap (E_2 \otimes F).$$

To prove the inclusion in the other direction, let z be an arbitrary vector of $(E_1 \otimes F) \cap (E_2 \otimes F)$. If $(b_\beta)_{\beta \in J}$ is a basis of F we can write

$$z = \sum_\beta u_\beta \otimes b_\beta, \qquad u_\beta \in E_1 \quad \text{and} \quad z = \sum_\beta v_\beta \otimes b_\beta, \qquad v_\beta \in E_2.$$

This yields

$$\sum_\beta (u_\beta - v_\beta) \otimes b_\beta = 0$$

and since the b_β are linearly independent we obtain $u_\beta = v_\beta$. Hence $u_\beta \in E_1 \cap E_2$ and $z \in (E_1 \cap E_2) \otimes F$. This completes the proof of (1.4).

Next consider two subspaces $E_1 \subset E$ and $F_1 \subset F$. Then

$$\boxed{(E_1 \otimes F) \cap (E \otimes F_1) = E_1 \otimes F_1.} \tag{1.5}$$

To prove (1.5), choose a subspace $F' \subset F$ such that

$$F = F_1 \oplus F'.$$

Then we have in view of Section 1.12 that

$$E_1 \otimes F = E_1 \otimes F_1 \oplus E_1 \otimes F'.$$

Intersecting with $E \otimes F_1$ and observing that $E_1 \otimes F_1 \subset E \otimes F_1$ we obtain

$$(E_1 \otimes F) \cap (E \otimes F_1) = (E_1 \otimes F_1) \oplus [(E_1 \otimes F') \cap (E_1 \otimes F_1)]. \tag{1.6}$$

Now Formula (1.4) yields

$$(E_1 \otimes F') \cap (E_1 \otimes F_1) = E_1 \otimes (F' \cap F_1) = E_1 \otimes 0 = 0$$

and thus (1.5) follows from (1.6).

Finally let E_1, E_2 and F_1, F_2 be subspaces respectively of E and F. Then

$$\boxed{(E_1 \otimes F_1) \cap (E_2 \otimes F_2) = (E_1 \cap E_2) \otimes (F_1 \cap F_2).} \tag{1.7}$$

Clearly,

$$(E_1 \cap E_2) \otimes (F_1 \cap F_2) \subset (E_1 \otimes F_1) \cap (E_2 \otimes F_2).$$

Moreover, we have in view of (1.4) that

$$(E_1 \otimes F_1) \cap (E_2 \otimes F_2) \subset (E_1 \otimes F) \cap (E_2 \otimes F) = (E_1 \cap E_2) \otimes F.$$

In the same way it follows that

$$(E_1 \otimes F_1) \cap (E_2 \otimes F_2) \subset (E \otimes F_1) \cap (E \otimes F_2) = E \otimes (F_1 \cap F_2).$$

Now formula (1.5) yields

$$(E_1 \otimes F_1) \cap (E_2 \otimes F_2) \subset (E_1 \cap E_2) \otimes (F_1 \cap F_2).$$

This completes the proof of (1.7).

Problems

1. Let $z \in E \otimes F, z \neq 0$, be any vector. Show by an explicit example that in general there exist several representations of the form
$$z = \sum_i x_i \otimes y_i \qquad x_i \in E,\ y_i \in F$$
where the x_i, and the y_i, are respectively linearly independent. Given two such representations
$$z = \sum_{i=1}^{r} x_i \otimes y_i \quad \text{and} \quad z = \sum_{j=1}^{s} x'_j \otimes y'_j$$
prove that $r = s$.

2. Let $\varphi: E \times F \to G$ be a bilinear mapping. Show that the following property is equivalent to $\otimes_2$. Whenever the vectors $x_\alpha \in E$ and $y_\beta \in F$ are linearly independent then so are the vectors $\varphi(x_\alpha, y_\beta)$.

3. Let $A \neq 0$ be an algebra of finite dimension and assume that the pair (A, β) is a tensor product for A and A where β denotes the multiplication. Prove that $\dim A = 1$.

4. Let E be an arbitrary vector space and F be a vector space of dimension m. Establish a (noncanonical) isomorphism
$$\underbrace{E \oplus \cdots \oplus E}_{m} \xrightarrow{\cong} E \otimes F.$$

5. Let E, E^* and F, F^* be two pairs of dual spaces of finite dimension. Consider the bilinear mapping
$$\beta: E \times F \to B(E^*, F^*)$$

given by

$$\beta_{x,y}(x^*, y^*) = \langle x^*, x\rangle\langle y^*, y\rangle.$$

Prove that the pair $(B(E^*, F^*), \beta)$ is the tensor product of E and F.

Linear Maps

1.16. Tensor Product of Linear Maps

Given four vector spaces E, E', F, F' consider two linear maps

$$\varphi : E \to E' \quad \text{and} \quad \psi : F \to F'.$$

Then a bilinear mapping $E \times F \to E' \otimes F'$ is defined by

$$(x, y) \to \varphi x \otimes \psi y.$$

In view of the factorization property there exists a linear map

$$\chi : E \otimes F \to E' \otimes F'$$

such that

$$\chi(x \otimes y) = \varphi x \otimes \psi y \tag{1.8}$$

and this map is uniquely determined by (1.8). The correspondence $(\varphi, \psi) \to \chi$ defines a bilinear mapping

$$\beta : L(E; E') \times L(F; F') \to L(E \otimes F; E' \otimes F').$$

Proposition 1.16.1. *Let $L(E; E') \otimes L(F; F')$ be the tensor product of $L(E; E')$ and $L(F; F')$. Then the linear map*

$$f : L(E; E') \otimes L(F; F') \to L(E \otimes F; E' \otimes F')$$

induced by the bilinear mapping β is injective.

PROOF. Let ω be an element such that $f(\omega) = 0$. If $\omega \neq 0$ we can write

$$\omega = \sum_{i=1}^{r} \varphi_i \otimes \psi_i \qquad \varphi_i \in L(E, E'), \psi_i \in L(F, F'),$$

where the φ_i and ψ_i are linearly independent.

Then

$$f(\omega) = \sum_i \beta(\varphi_i, \psi_i)$$

and hence $f(\omega) = 0$ implies that

$$\sum_{i=1}^{r} \varphi_i(x) \otimes \psi_i(y) = 0 \tag{1.9}$$

for every pair $x \in E$, $y \in F$. Now choose a vector $a \in E$ such that $\varphi_1(a) \neq 0$.

Let $p \geq 1$ be the maximal number of linearly independent vectors in the set $\varphi_1(a), \ldots, \varphi_r(a)$. Rearranging the φ_i we can achieve that the vectors $\varphi_1(a), \ldots, \varphi_p(a)$ are linearly independent. Then we have

$$\varphi_j(a) = \sum_{i=1}^{p} \lambda_{ji} \varphi_i(a) \qquad j = p+1, \ldots, r$$

and Relation (1.9) yields

$$\sum_{i=1}^{p} \varphi_i(a) \otimes \psi_i(y) + \sum_{j=p+1}^{r} \left(\sum_{i=1}^{p} \lambda_{ji} \varphi_i(a) \right) \otimes \psi_j(y) = 0 \qquad y \in F;$$

i.e.,

$$\sum_{i=1}^{p} \varphi_i(a) \otimes \left(\sum_{j=p+1}^{r} \lambda_{ji} \psi_j(y) + \psi_i(y) \right) = 0 \qquad y \in F.$$

Since the vectors $\varphi_i(a)$ $(i = 1, \ldots, p)$ are linearly independent it follows that

$$\psi_i(y) + \sum_{j=p+1}^{r} \lambda_{ji} \psi_j(y) = 0 \qquad i = 1, \ldots, p$$

for every $y \in F$, i.e. $\psi_i + \sum_{j=p+1}^{r} \lambda_{ji} \psi_j = 0$. This is in contradiction to our hypothesis that the maps ψ_j are linearly independent and hence f is injective. □

Corollary I. *The pair* (Im β, β) *is the tensor product of* $L(E; E')$ *and* $L(F; F')$.

Corollary II. *The bilinear mapping* $\beta: L(E) \times L(F) \to L(E \otimes F)$ *given by*

$$\beta(f, g)(x \otimes y) = f(x)g(y)$$

is such that (Im β, β) *is the tensor product of* $L(E)$ *and* $L(F)$.

Corollary III. *If E and F have finite dimension the elements* $\beta(\varphi, \psi)$ *generate the space*

$$L(E \otimes F; E' \otimes F)$$

as will be shown in Section 1.27 *and so the pair* $(L(E \otimes F; E' \otimes F'), \beta)$ *is the tensor product of* $L(E; E')$ *and* $L(F; F')$.

In general, however, this is not the case as the example below will show. Nevertheless, by an abuse of language, we call $\beta(\varphi, \psi)$ the tensor product of the linear maps φ and ψ and write $\beta(\varphi, \psi) = \varphi \otimes \psi$. Then formula (1.8) reads

$$(\varphi \otimes \psi)(x \otimes y) = \varphi x \otimes \psi y.$$

In particular, if $E = F = E' = F'$ and $\varphi = \psi = \iota$, then $\iota \otimes \iota = \iota$.

1.17. Example

Let $E = F$ be a vector space with a countable basis and put $E' = F' = \Gamma$. Then we have

$$L(E; E') = L(F; F') = L(E), \qquad L(E \otimes F; E' \otimes F') = L(E \otimes E)$$

and the bilinear mapping β is given by

$$\beta:(f, g) \to f \cdot g,$$

where

$$(f \cdot g)(x \otimes y) = f(x)g(y).$$

It will be shown that β does not satisfy the Condition $\otimes_1$.

We associate with every linear function h in $E \otimes E$ a subspace $E_h \subset E$ (called the nullspace of the corresponding function) in the following way: A vector $x_0 \in E$ is contained in E_h if and only if $h(x_0 \otimes y) = 0$ for every $y \in E$. Now assume that $h \in \operatorname{Im} \beta$. Then the factor space E/E_h has finite dimension. In fact, h can be written as a finite sum of the form

$$h(x \otimes y) = \sum_{i=1}^{r} f_i(x)g_i(y).$$

It follows that for any $x \in \bigcap_{i=1}^{r} \ker f_i$ we have

$$h(x \otimes y) = 0 \qquad y \in F;$$

i.e.,

$$E_h \supset \bigcap_{i=1}^{r} \ker f_i.$$

Now a result of Section 2.4 of *Linear Algebra* implies that $\dim E/E_h \leq r$.

On the other hand, consider the linear function ω on $E \otimes E$ given by

$$\omega(x \otimes y) = \sum_{\nu} \xi^\nu \eta^\nu$$

where ξ^ν and η^ν $(\nu = 1, 2, \dots)$ are the components of x and y with respect to a basis of E. It is easy to see that the nullspace $E_\omega = 0$ and hence $\dim E/E_\omega = \dim E = \infty$. Consequently, ω is not contained in $\operatorname{Im} \beta$.

1.18. Compositions

Consider four linear maps

$$\varphi: E \to E' \qquad \varphi': E' \to E''$$
$$\psi: F \to F' \qquad \psi': F' \to F''.$$

Then it is clear from the definition that

$$(\varphi' \otimes \psi') \circ (\varphi \otimes \psi) = (\varphi' \circ \varphi) \otimes (\psi' \circ \psi). \tag{1.10}$$

Now assume that φ and ψ are injective. Then there exist linear maps $\tilde{\varphi}: E' \to E$ and $\tilde{\psi}: F' \to F$ such that

$$\tilde{\varphi} \circ \varphi = \iota \quad \text{and} \quad \tilde{\psi} \circ \psi = \iota.$$

Hence, formula (1.10) yields

$$(\tilde{\varphi} \otimes \tilde{\psi}) \circ (\varphi \otimes \psi) = (\tilde{\varphi} \circ \varphi) \otimes (\tilde{\psi} \otimes \psi) = \iota \otimes \iota = \iota$$

showing that $\varphi \otimes \psi$ is injective.

1.19. Image Space and Kernel

It follows immediately from the definition of $\varphi \otimes \psi$ that

$$\boxed{\operatorname{Im}(\varphi \otimes \psi) = \operatorname{Im} \varphi \otimes \operatorname{Im} \psi.} \tag{1.11}$$

In particular, if φ and ψ are surjective, then so is $\varphi \otimes \psi$. Now the formula

$$\boxed{\ker(\varphi \otimes \psi) = \ker \varphi \otimes F + E \otimes \ker \psi} \tag{1.12}$$

will be established. Consider the induced injective linear maps

$$\bar{\varphi}: E/\ker \varphi \to E' \quad \text{and} \quad \bar{\psi}: F/\ker \psi \to F'.$$

Then

$$\bar{\varphi} \otimes \bar{\psi}: E/\ker \varphi \otimes F/\ker \psi \to E' \otimes F'$$

is injective as well (Section 1.18). Let

$$\pi_1: E \to E/\ker \varphi, \qquad \pi_2: F \to F/\ker \psi,$$

and

$$\pi: E \otimes F \to (E \otimes F)/T(\ker \varphi, \ker \psi)$$

$$T(\ker \varphi, \ker \psi) = \ker \varphi \otimes F + E \otimes \ker \psi$$

be the canonical projections. According to Section 1.10 there exists a linear isomorphism

$$g: E/\ker \varphi \otimes F/\ker \psi \xrightarrow{\cong} E \otimes F/T(\ker \varphi, \ker \psi) \tag{1.13}$$

such that

$$g(\pi_1 x \otimes \pi_2 y) = \pi(x \otimes y).$$

Now define a linear map

$$\chi: (E \otimes F)/T(\ker \varphi, \ker \psi) \to E' \otimes F'$$

by

$$\chi = (\bar{\varphi} \otimes \bar{\psi}) \circ g^{-1}.$$

Clearly χ is injective. Moreover, if $x \in E$ and $y \in F$ are arbitrary we obtain

$$\begin{aligned}(\chi \circ \pi)(x \otimes y) &= (\bar{\varphi} \otimes \bar{\psi})g^{-1}g(\pi_1 x \otimes \pi_2 y)\\ &= \bar{\varphi}\pi_1 x \otimes \bar{\psi}\pi_2 y\\ &= \varphi x \otimes \psi y\end{aligned}$$

whence

$$\chi \circ \pi = \varphi \otimes \psi.$$

Since χ is injective, it follows that

$$\ker(\varphi \otimes \psi) = \ker \pi = T(\ker \varphi, \ker \psi) = \ker \varphi \otimes F + E \otimes \ker \psi.$$

Problems

1. Consider two linear maps $\varphi: E \to E'$ and $\psi: F \to F'$.
   (a) Prove that $\varphi \otimes \psi$ is injective if and only if both mappings φ and ψ are injective.
   (b) Assume that E and F have finite dimension. Prove that

   $$r(\varphi \otimes \psi) = r(\varphi)r(\psi)$$

   where r denotes rank (see Section 2.34 of *Linear Algebra*).

2. Let E, F be two vector spaces of dimension n and m respectively and let $\varphi: E \to E$, $\psi: F \to F$ be two linear transformations. Prove that

   $$\operatorname{tr}(\varphi \otimes \psi) = \operatorname{tr} \varphi \operatorname{tr} \psi$$

   and

   $$\det(\varphi \otimes \psi) = (\det \varphi)^m(\det \psi)^n.$$

3. Consider a vector space E of dimension n. Given two linear transformations α and β of E let

   $$\Phi: L(E; E) \to L(E; E)$$

   be the linear transformation defined by

   $$\Phi\sigma = \alpha \circ \sigma \circ \beta \qquad \sigma \in L(E; E).$$

   Show that

   $$\operatorname{tr} \Phi = \operatorname{tr} \alpha \operatorname{tr} \beta$$

   and

   $$\det \Phi = \det(\alpha \circ \beta)^n.$$

4. Let E and F be two finite-dimensional vector spaces and E', F' be arbitrary vector spaces. Prove that the bilinear mapping

   $$\beta: L(E; E') \times L(F; F') \to L(E \otimes F; E' \otimes F')$$

   defined by $\beta(\varphi, \psi) = \varphi \otimes \psi$ is a tensor product.

5. Let E be a finite-dimensional vector space and consider two commuting linear transformations φ, ψ of E. Prove that

$$(\varphi \otimes \psi)_S = \varphi_S \otimes \psi_S$$

and

$$(\varphi \otimes \psi)_N = \varphi_S \otimes \psi_N + \varphi_N \otimes \psi_S + \varphi_N \otimes \psi_N$$

(cf. Section 13.24 of *Linear Algebra*). Conclude that the tensor product of commuting transformations is semisimple if and only if both transformations are semisimple.

Tensor Product of Several Vector Spaces

1.20. The Universal Property

Let E_i $(i = 1, \ldots, p)$ be any p vector spaces and let

$$\otimes : E_1 \times \cdots \times E_p \to T$$

be a p-linear mapping. This mapping is said to have the *universal property* if it satisfies the following conditions:

$\otimes_1$: The vectors $x_1 \otimes \cdots \otimes x_p$, $(x_i \in E_i)$ generate T.
$\otimes_2$: Every p-linear mapping $\varphi : E_1 \times \cdots \times E_p \to H$ (H any vector space) can be written in the form

$$\varphi(x_1, \ldots, x_p) = f(x_1 \otimes \cdots \otimes x_p)$$

where $f: T \to H$ is a linear map.

The existence and uniqueness theorems are proved in the same way as in the case $p = 2$ and we shall not repeat them.

Definition. The *tensor product of the spaces* E_i $(i = 1, \ldots, p)$ is a pair $(T, \otimes)$ where $\otimes : E_1 \times \cdots \times E_p \to T$ is a p-linear mapping with the universal property. T is also called the tensor product of the spaces E_i and is denoted by $E_1 \otimes \cdots \otimes E_p$.

If H is any vector space, then the correspondence $\varphi \leftrightarrow f$ expressed by the commutative diagram

$$\begin{array}{ccc} E_1 \times \cdots \times E_p & \xrightarrow{\varphi} & H \\ \downarrow \otimes & \nearrow f & \\ E_1 \otimes \cdots \otimes E_p & & \end{array}$$

determines a linear isomorphism

$$\Phi : L((E_1 \otimes \cdots \otimes E_p); H) \xrightarrow{\cong} L(E_1, \ldots, E_p; H).$$

Proposition 1.20.1. *Given three arbitrary spaces E_1, E_2, E_3 there exists a linear isomorphism*

$$f: E_1 \otimes E_2 \otimes E_3 \xrightarrow{\cong} (E_1 \otimes E_2) \otimes E_3$$

such that

$$f(x \otimes y \otimes z) = (x \otimes y) \otimes z.$$

PROOF. Consider the trilinear mapping

$$E_1 \times E_2 \times E_3 \to (E_1 \otimes E_2) \otimes E_3$$

defined by

$$(x, y, z) \to (x \otimes y) \otimes z.$$

In view of the factorization property, there is induced a linear map

$$f: E_1 \otimes E_2 \otimes E_3 \to (E_1 \otimes E_2) \otimes E_3$$

such that

$$f(x \otimes y \otimes z) = (x \otimes y) \otimes z. \tag{1.14}$$

On the other hand, to each fixed $z \in E_3$ there corresponds a bilinear mapping $\beta_z: E_1 \times E_2 \to E_1 \otimes E_2 \otimes E_3$ defined by

$$\beta_z(x, y) = x \otimes y \otimes z.$$

The mapping β_z induces a linear map

$$g_z: E_1 \otimes E_2 \to E_1 \otimes E_2 \otimes E_3$$

such that

$$g_z(x \otimes y) = x \otimes y \otimes z. \tag{1.15}$$

Define a bilinear mapping

$$\psi: (E_1 \otimes E_2) \times E_3 \to E_1 \otimes E_2 \otimes E_3$$

by

$$\psi(u, z) = g_z(u) \qquad u \in E_1 \otimes E_2, z \in E_3. \tag{1.16}$$

Then ψ induces a linear map

$$g: (E_1 \otimes E_2) \otimes E_3 \to E_1 \otimes E_2 \otimes E_3$$

such that

$$\psi(u, z) = g(u \otimes z) \qquad u \in E_1 \otimes E_2, z \in E_3. \tag{1.17}$$

Combining (1.17), (1.16), and (1.15) we find

$$g((x \otimes y) \otimes z) = \psi(x \otimes y, z) = g_z(x \otimes y) = x \otimes y \otimes z. \tag{1.18}$$

Equations (1.14) and (1.18) yield $gf(x \otimes y \otimes z) = x \otimes y \otimes z$ and $fg((x \otimes y) \otimes z) = (x \otimes y) \otimes z$ showing that f is a linear isomorphism of $E_1 \otimes E_2 \otimes E_3$ onto $(E_1 \otimes E_2) \otimes E_3$ and g is the inverse isomorphism. □

In the same way a linear isomorphism $h: E_1 \otimes E_2 \otimes E_3 \xrightarrow{\cong} E_1 \otimes (E_2 \otimes E_3)$ can be constructed such that

$$h(x \otimes y \otimes z) = x \otimes (y \otimes z).$$

Hence, $h \circ f^{-1}$ is an isomorphism of $(E_1 \otimes E_2) \otimes E_3$ onto $E_1 \otimes (E_2 \otimes E_3)$ such that $(x \otimes y) \otimes z \mapsto x \otimes (y \otimes z)$.

More generally, if E_i $(i = 1, \ldots, p + q)$ are $p + q$ vector spaces then there exists precisely one vector space isomorphism

$$f:(E_1 \otimes \cdots \otimes E_p) \otimes (E_{p+1} \otimes \cdots \otimes E_{p+q}) \to E_1 \otimes \cdots \otimes E_{p+q}$$

such that

$$f((x_1 \otimes \cdots \otimes x_p) \otimes (x_{p+1} \otimes \cdots \otimes x_{p+q})) = x_1 \otimes \cdots \otimes x_{p+q}.$$

It follows that there is a uniquely determined bilinear mapping

$$\beta:(E_1 \otimes \cdots \otimes E_p) \times (E_{p+1} \otimes \cdots \otimes E_{p+q}) \to E_1 \otimes \cdots \otimes E_{p+q}$$

such that

$$\beta(x_1 \otimes \cdots \otimes x_p, x_{p+1} \otimes \cdots \otimes x_{p+q}) = x_1 \otimes \cdots \otimes x_{p+q}$$

and that $(E_1 \otimes \cdots \otimes E_{p+q}, \beta)$ is the tensor product of $E_1 \otimes \cdots \otimes E_p$ and $E_{p+1} \otimes \cdots \otimes E_{p+q}$.

The theory developed for the case $p = 2$ carries over to the general case in an obvious way, and the reader will have no difficulty in making (and proving) the generalization himself. In particular, he should verify that if a_ν^i is a basis for E_i $(i = 1, \ldots, p)$ then the products $a_{\nu_1}^1 \otimes \cdots \otimes a_{\nu_p}^p$ form a basis for $E_1 \otimes \cdots \otimes E_p$, and then generalize the results of Section 1.14, obtaining in the finite-dimensional case that

$$\dim L(E_1, \ldots, E_p; G) = \dim E_1 \cdots \dim G$$

$$\dim L(E_1, \ldots, E_p) = \dim E_1 \cdots \dim E_p.$$

The tensor product of several linear maps can be defined in the same way as in the case $p = 2$. If $\varphi_i: E_i \to F_i$ $(i = 1, \ldots, p)$ are linear maps then there exists precisely one linear map

$$\chi: E_1 \otimes \cdots \otimes E_p \to F_1 \otimes \cdots \otimes F_p$$

such that

$$\chi(x_1 \otimes \cdots \otimes x_p) = \varphi_1 x_1 \otimes \cdots \otimes \varphi_p x_p \qquad x_i \in E_i.$$

As in the case $p = 2$ we shall write $\chi = \varphi_1 \otimes \cdots \otimes \varphi_p$. As in that case, the mapping $(\varphi_1, \ldots, \varphi_p) \mapsto \varphi_1 \otimes \cdots \otimes \varphi_p$ induces an injection

$$L(E_1; F_1) \otimes \cdots \otimes L(E_p; F_p) \to L(E_1 \otimes \cdots \otimes E_p; F_1 \otimes \cdots \otimes F_p).$$

If

$$\psi_i: F_i \to G_i \qquad (i = 1, \ldots, p)$$

is another system of linear maps it follows from the above definition that

$$(\psi_1 \otimes \cdots \otimes \psi_p) \circ (\varphi_1 \otimes \cdots \otimes \varphi_p) = (\psi_1 \circ \varphi_1) \otimes \cdots \otimes (\psi_p \circ \varphi_p).$$

An argument similar to the one given for $p = 2$ shows that

$$\operatorname{Im}(\varphi_1 \otimes \cdots \otimes \varphi_p) = \operatorname{Im} \varphi_1 \otimes \cdots \otimes \operatorname{Im} \varphi_p$$

and

$$\ker(\varphi_1 \otimes \cdots \otimes \varphi_p) = \sum_{i=1}^{p} E_1 \otimes \cdots \ker \varphi_i \otimes \cdots \otimes E_p.$$

Problem

Let E_i be p vector spaces.
(a) Prove that $x_1 \otimes \cdots \otimes x_p = 0$ if and only if at least one $x_i = 0$.
(b) Assuming that $x_1 \otimes \cdots \otimes x_p \neq 0$ prove that

$$x_1 \otimes \cdots \otimes x_p = y_1 \otimes \cdots \otimes y_p$$

if and only if $y_i = \lambda_i x_i$ $(i = 1, \ldots, p)$ and $\lambda_1 \cdots \lambda_p = 1$.

Dual Spaces

1.21. Bilinear Mappings

Let two triples of vector spaces E, E', E'' and F, F', F'' be given and consider two bilinear mappings

$$\varphi: E \times E' \to E'' \quad \text{and} \quad \psi: F \times F' \to F''.$$

Then there exists precisely one bilinear mapping

$$\chi:(E \otimes F) \times (E' \otimes F') \to E'' \otimes F''$$

such that

$$\chi(x \otimes y, x' \otimes y') = \varphi(x, x') \otimes \psi(y, y') \qquad x \in E, x' \in E', y \in F, y' \in F'. \quad (1.19)$$

Since the spaces $E \otimes F$ and $E' \otimes F'$ are generated by the products $x \otimes y$ and $x' \otimes y'$ respectively it is clear that if χ exists it is uniquely determined by φ and ψ. To prove the existence of χ consider the linear maps

$$f: E \otimes E' \to E'' \quad \text{and} \quad g: F \otimes F' \to F''$$

induced by φ and ψ respectively. Then $f \otimes g$ is a linear map of $(E \otimes E') \otimes (F \otimes F')$ into $E'' \otimes F''$. Now let

$$S:(E \otimes F) \otimes (E' \otimes F') \xrightarrow{\cong} (E \otimes E') \otimes (F \otimes F')$$

be the linear isomorphism defined by

$$S:(x \otimes y) \otimes (x' \otimes y') \to (x \otimes x') \otimes (y \otimes y')$$

and define a bilinear mapping χ by

$$\chi(u, v) = (f \otimes g)S(u \otimes v) \qquad u \in E \otimes F, v \in E' \otimes F'.$$

Then it follows that

$$\begin{aligned}\chi(x \otimes y, x' \otimes y') &= (f \otimes g)((x \otimes x') \otimes (y \otimes y')) \\ &= f(x \otimes x') \otimes g(y \otimes y') \\ &= \varphi(x, x') \otimes \psi(y, y').\end{aligned}$$

We shall denote χ by $\varphi \otimes \psi$, the justification for this notation being given in problem 1 at the end of Section 1.26.

1.22. Bilinear Functions

In particular, every pair of bilinear functions Φ and Ψ in $E \times E'$ and $F \times F'$ induces a bilinear function $\Phi \otimes \Psi$ in $(E \otimes F) \times (E' \otimes F')$ such that

$$(\Phi \otimes \Psi)(x \otimes y, x' \otimes y') = \Phi(x, x')\Psi(y, y').$$

We shall show that $\Phi \otimes \Psi$ is nondegenerate if and only if Φ and Ψ are both nondegenerate.

Consider the linear maps

$$\varphi:E \to L(E') \qquad \psi:F \to L(F') \qquad \chi:E \otimes F \to L(E' \otimes F')$$

which are determined by

$$\varphi_a x' = \Phi(a, x') \qquad \psi_b y' = \Psi(b, y') \qquad \chi_c z' = (\Phi \otimes \Psi)(c, z') \quad (1.20)$$

(Here $\varphi(a)$, $\psi(b)$, and $\chi(c)$ are denoted by φ_a, ψ_b, and χ_c). Then we have $\varphi \otimes \psi:E \otimes F \to L(E') \otimes L(F')$. On the other hand, $L(E') \otimes L(F')$ may be considered as a subspace of $L(E' \otimes F')$ (Corollary III to Proposition 1.16.1).

It will be shown that

$$\chi = i \circ (\varphi \otimes \psi), \qquad (1.21)$$

where i denotes the injection of $L(E') \otimes L(F')$ into $L(E' \otimes F')$. By definition we have

$$(\varphi \otimes \psi)_{a \otimes b} = \varphi_a \otimes \psi_b \qquad a \in E, b \in F$$

and thus

$$i(\varphi \otimes \psi)_{a \otimes b}(x' \otimes y') = \varphi_a x' \cdot \psi_b y' = \Phi(a, x')\Psi(b, y').$$

On the other hand it follows from (1.20) that

$$\chi_{a \otimes b}(x' \otimes y') = (\Phi \otimes \Psi)(a \otimes b, x' \otimes y') = \Phi(a, x')\Psi(b, y')$$

and so we obtain

$$i(\varphi \otimes \psi)_{a \otimes b} = \chi_{a \otimes b} \qquad a \in E, b \in F$$

whence (1.21).

Since i is an injection, it follows from (1.12) that

$$\ker \chi = \ker(\varphi \otimes \psi) = \ker \varphi \otimes F + E \otimes \ker \psi.$$

Hence, the nullspaces $N_E(\Phi)$, $N_F(\Psi)$, and $N_{E \otimes F}(\Phi \otimes \Psi)$ (see Section 2.21 of *Linear Algebra*) are connected by the formula

$$N_{E \otimes F}(\Phi \otimes \Psi) = N_E(\Phi) \otimes F + E \otimes N_F(\Psi). \tag{1.22}$$

In the same way it is shown that

$$N_{E' \otimes F'}(\Phi \otimes \Psi) = N_{E'}(\Phi) \otimes F' + E' \otimes N_{F'}(\Psi). \tag{1.23}$$

Formulas (1.22) and (1.23) imply that $\Phi \otimes \Psi$ is nondegenerate if and only if Φ and Ψ are nondegenerate.

Suppose now that E^*, E and F^*, F are two dual pairs and let both scalar products be denoted by $\langle , \rangle$. Then the above results show that there exists precisely one bilinear function $\langle , \rangle$ in $E^* \otimes F^*$, $E \otimes F$ such that

$$\langle x^* \otimes y^*, x \otimes y \rangle = \langle x^*, x \rangle \langle y^*, y \rangle \tag{1.24}$$

and this bilinear function is again nondegenerate. In other words, if E^*, E and F^*, F are dual pairs, then a duality between $E^* \otimes F^*$ and $E \otimes F$ is induced.

Next, consider the special case $F = E^*$ and $F^* = E$. Then we obtain a scalar product in the pair $E^* \otimes E$, $E \otimes E^*$ defined by

$$\langle x^* \otimes x, y \otimes y^* \rangle = \langle x^*, y \rangle \langle y^*, x \rangle.$$

Since the mapping $x^* \otimes x \mapsto x \otimes x^*$ is an isomorphism of $E^* \otimes E$ onto $E \otimes E^*$, this scalar product determines a scalar product in the pair $E \otimes E^*$, $E \otimes E^*$ given by

$$\langle x \otimes x^*, y \otimes y^* \rangle = \langle x^*, y \rangle \langle y^*, x \rangle. \tag{1.25}$$

Hence the space $E \otimes E^*$ can be considered as dual to itself. It is clear, moreover, that this scalar product is symmetric.

Now suppose that E_i^*, E_i $(i = 1, \ldots, p)$ are pairs of dual vector spaces and let all the scalar products be denoted by $\langle , \rangle$. As in the case $p = 2$ a scalar product is induced between $E_1^* \otimes \cdots \otimes E_p^*$ and $E_1 \otimes \cdots \otimes E_p$ such that

$$\langle x^{*1} \otimes \cdots \otimes x^{*p}, x_1 \otimes \cdots \otimes x_p \rangle = \langle x^{*1}, x_1 \rangle \cdots \langle x^{*p}, x_p \rangle.$$

1.23. Dual Mappings

Let E_i, E_i^* and F_i, F_i^* $(i = 1, 2)$ be four pairs of dual vector spaces, and let

$$\varphi: E_1 \to E_2 \qquad \varphi^*: E_1^* \leftarrow E_2^*$$

and

$$\psi: F_1 \to F_2 \qquad \psi^*: F_1^* \leftarrow F_2^*$$

be two pairs of dual mappings. Then the mappings

$$\varphi \otimes \psi: E_1 \otimes F_1 \to E_2 \otimes F_2$$

and

$$\varphi^* \otimes \psi^*: E_1^* \otimes F_1^* \leftarrow E_2^* \otimes F_2^*$$

are dual with respect to the induced scalar products.

In fact, let $x_1 \in E_1$, $y_1 \in F_1$, $x_2^* \in E_2^*$, and $y_2^* \in F_2^*$ be arbitrarily chosen vectors. Then we have

$$\begin{aligned}\langle x_2^* \otimes y_2^*, \varphi x_1 \otimes \psi y_1 \rangle &= \langle x_2^*, \varphi x_1 \rangle \langle y_2^*, \psi y_1 \rangle \\ &= \langle \varphi^* x_2^*, x_1 \rangle \langle \psi^* y_2^*, y_1 \rangle \\ &= \langle \varphi^* x_2^* \otimes \psi^* y_2^*, x_1 \otimes y_1 \rangle\end{aligned}$$

whence

$$(\varphi \otimes \psi)^* = \varphi^* \otimes \psi^*.$$

1.24. Example

Consider the dual spaces $E^* = L(E)$ and $F^* = L(F)$. Then the induced scalar product in $L(E) \otimes L(F)$ and $E \otimes F$ is given by

$$\langle f \otimes g, x \otimes y \rangle = f(x)g(y).$$

On the other hand, the space $L(E \otimes F)$ is dual to $E \otimes F$ with respect to the scalar product

$$\langle h, x \otimes y \rangle = h(x \otimes y) \qquad h \in L(E \otimes F). \tag{1.26}$$

Now consider the injection

$$i: L(E) \otimes L(F) \to L(E \otimes F)$$

defined by

$$i(f \otimes g)(x \otimes y) = f(x)g(y). \tag{1.27}$$

Then formulas (1.26), (1.27), and (1.24) yield the relation

$$\langle i(f \otimes g), (x \otimes y) \rangle = f(x)g(y) = \langle f \otimes g, x \otimes y \rangle$$

showing that the injection i preserves the scalar products.

1.25. Inner Product Spaces

An *inner product* in a vector space E is a nondegenerate symmetric bilinear function $(,)$ in E. So in particular, every Euclidean space is an inner product space.

Now let E and F be inner product spaces and denote both inner products by $(,)$. In view of Section 1.22, there is precisely one bilinear function $(,)$ in $E \otimes F$ satisfying

$$(x_1 \otimes y_1, x_2 \otimes y_2) = (x_1, x_2) \cdot (y_1, y_2).$$

Clearly, this bilinear function is again symmetric. Moreover, it is nondegenerate as follows from Section 1.22. The inner product space $E \otimes F$ so obtained is called the *tensor product of the inner product spaces E and F*.

Now assume that E and F are Euclidean spaces of dimensions n and m respectively. Choose orthonormal bases a_ν $(\nu = 1, \ldots, n)$ and b_μ $(\mu = 1, \ldots, m)$ in E and F. Then we have

$$(a_\nu \otimes b_\mu, a_\lambda \otimes b_\kappa) = \delta_{\nu\lambda}\, \delta_{\mu\kappa}.$$

Thus the products $a_\nu \otimes b_\mu$ form an orthonormal basis of $E \otimes F$. In particular, $E \otimes F$ is again a Euclidean space.

1.26. The Composition Algebra

Let E^*, E be a pair of dual vector spaces. Define a multiplication in the space $E^* \otimes E$ by setting

$$(x^* \otimes x) \circ (y^* \otimes y) = \langle x^*, y\rangle (y^* \otimes x).$$

It is easy to verify that this multiplication makes $E^* \otimes E$ into an associative (noncommutative) algebra called the *composition algebra*.

Next, consider the linear map $T: E^* \otimes E \to L(E; E)$ given by

$$T(a^* \otimes b)x = \langle a^*, x\rangle b \qquad x \in E.$$

Since

$$T[(a_1^* \otimes b_1) \circ (a_2^* \otimes b_2)] = T(a_1^* \otimes b_1) \circ T(a_2^* \otimes b_2),$$

T is an algebra homomorphism. We show that T is injective. In fact, assume that $T(z) = 0$, $z \in E^* \otimes E$. Choose a basis $\{e_\alpha\}$ of E. Then z is a finite sum

$$z = \sum_{\nu=1}^{r} a_\nu^* \otimes e_\nu \quad \text{where } a_\nu^* \in E^*$$

(see Lemma 1.5.2). Then, for every $x \in E$,

$$\sum_{\nu=1}^{r} \langle a_\nu^*, x\rangle e_\nu = 0$$

whence

$$\langle a_\nu^*, x\rangle = 0 \qquad x \in E.$$

This implies that $a_\nu^* = 0$ ($\nu = 1, \ldots, r$) and so $z = 0$. Hence T is injective.

For the further discussion we distinguish two cases:

Case 1. $\dim E < \infty$. Set $\dim E = n$. Then $\dim(E^* \otimes E) = n^2 = \dim L(E; E)$ and so T is a linear isomorphism (since it is injective). Thus, $\ell = T^{-1}(\iota)$ is the unit element of the composition algebra. It is called the *unit tensor* of E.

To obtain a more explicit expression for the unit tensor let $\{e^{*\nu}\}$, $\{e_\nu\}$ ($\nu = 1, \ldots, n$) be a pair of dual bases for E^* and E and consider the element $\sum_{\nu=1}^n e^{*\nu} \otimes e_\nu$. Then we have

$$T\left(\sum_{\nu=1}^n e^{*\nu} \otimes e_\nu\right)(x) = \sum_{\nu=1}^n \langle e^{*\nu}, x\rangle e_\nu = x \qquad x \in E,$$

whence

$$\sum_{\nu=1}^n e^{*\nu} \otimes e_\nu = \ell.$$

In particular, the sum $\sum_{\nu=1}^n e^{*\nu} \otimes e_\nu$ is independent of the choice of the dual bases $\{e^{*\nu}\}$, $\{e_\nu\}$.

Next observe that the spaces $E^* \otimes E$ and $L(E; E)$ are self-dual with respect to the scalar products given by

$$\langle x^* \otimes x, y^* \otimes y\rangle = \langle x^*, y\rangle\langle y^*, x\rangle$$

and

$$\langle \alpha, \beta\rangle = \operatorname{tr}(\alpha \circ \beta) \qquad \alpha, \beta \in L(E; E)$$

respectively. A simple calculation shows that

$$\langle T(x^* \otimes x), T(y^* \otimes y)\rangle = \langle x^* \otimes x, y^* \otimes y\rangle$$

and so T preserves the scalar products.

Case 2. $\dim E = \infty$. Then the composition algebra does not have a unit element. To prove this, assume that e is a unit element of $E^* \otimes E$. Let $\{e_\alpha\}$ be a basis of E. Then e is a *finite* sum

$$e = \sum_{\nu=1}^r a_\nu^* \otimes e_\nu \quad \text{with } a_\nu^* \in E^*.$$

For $x^* \in E^*$ and $x \in E$,

$$\begin{aligned} e \circ (x^* \otimes x) &= \sum_{\nu=1}^r (a_\nu^* \otimes e_\nu) \circ (x^* \otimes x) \\ &= \sum_{\nu=1}^r \langle a_\nu^*, x\rangle (x^* \otimes e_\nu) = \sum_{\nu=1}^r \lambda^\nu (x^* \otimes e_\nu) \end{aligned}$$

where $\lambda^\nu = \langle a_\nu^*, x\rangle$. Since e is a unit element, it follows that

$$\sum_{\nu=1}^{r} \lambda^\nu(x^* \otimes e_\nu) = x^* \otimes x,$$

whence

$$x = \sum_{\nu=1}^{r} \lambda^\nu e_\nu.$$

Thus the vectors $e_1, \ldots, e_r$ generate E and so E has finite dimension.

It follows from the result above that, whenever $\dim E = \infty$, the linear map T is not an isomorphism and hence not surjective.

Problems

1. Given six vector spaces E, E', E'' and F, F', F'' consider the bilinear mapping

$$\gamma: B(E, E'; E'') \times B(F, F'; F'') \to B(E \otimes F, E' \otimes F'; E'' \otimes F'')$$

defined by

$$\gamma(\varphi, \psi):(x \otimes y, x' \otimes y') \mapsto \varphi(x, x') \otimes \psi(y, y').$$

Show that the pair $(\operatorname{Im} \gamma, \gamma)$ is the tensor product of $B(E, E'; E'')$ and $B(F, F'; F'')$.

2. Let E, E^* and F, F^* be two pairs of dual vectors spaces and consider subspaces $E_1 \subset E$ and $F_1 \subset F$.

(a) Show that a nondegenerate bilinear function is induced in

$$(E^* \otimes F^*)/T(E_1^\perp, F_1^\perp) \quad \text{and} \quad E_1 \otimes F_1,$$

where

$$T(E_1^\perp, F_1^\perp) = E_1^\perp \otimes F^* + E^* \otimes F_1^\perp$$

by the scalar product in $E^* \otimes F^*, E \otimes F$.

(b) Prove the relation

$$(E_1 \otimes F_1)^\perp = E_1^\perp \otimes F^* + E^* \otimes F_1^\perp.$$

3. Given a pair of dual transformations $\varphi: E \to E$, $\varphi^*: E^* \leftarrow E^*$ prove that the linear transformation $\varphi \otimes \varphi^*$ is self-dual.

Finite-Dimensional Vector Spaces

For the remainder of this chapter all vector spaces will be assumed to have finite dimension.

1.27.

Let E and F be vector spaces of dimension n and m respectively. Then $E \otimes F$ has dimension nm (see Section 1.13).

Proposition 1.27.1. *Let $\varphi: E \times F \to T$ be a bilinear mapping where* $\dim T = nm$. *Then conditions $\otimes_1$ and $\otimes_2$ for φ are equivalent.*

PROOF. Consider the induced linear map

$$f: E \otimes F \to T.$$

Then, if φ satisfies $\otimes_1$, f is surjective, (Proposition 1.8.1). Since $\dim T = nm = \dim(E \otimes F)$, it follows that f is an isomorphism and so φ satisfies $\otimes_2$.

On the other hand, if φ satisfies $\otimes_2$, then f is injective (Proposition 1.8.1), and hence it must be a linear isomorphism. Thus φ satisfies $\otimes_1$. □

Next, let $\varphi: E \to E'$ and $\psi: F \to F'$ be linear maps where $\dim E' = n'$ and $\dim F' = m'$. It has been shown in Section 1.16 that the bilinear mapping

$$\beta: L(E; E') \times L(F; F') \to L(E \otimes F; E' \otimes F')$$

given by $\varphi \times \psi \to \varphi \otimes \psi$ satisfies $\otimes_2$. In the finite-dimensional case we have

$$\begin{aligned} \dim L(E \otimes F; E' \otimes F') &= (nm) \cdot (n'm') = (nn') \cdot (mm') \\ &= \dim L(E; E') \cdot \dim(F; F'). \end{aligned}$$

Hence, by the proposition above, β satisfies $\otimes_1$ as well. Thus β has the universal property and we may write

$$L(E \otimes F; E' \otimes F') = L(E; E') \otimes L(F; F').$$

This yields for $E' = F' = \Gamma$

$$L(E \otimes F; \Gamma) = L(E; \Gamma) \otimes L(F; \Gamma),$$

that is,

$$(E \otimes F)^* = E^* \otimes F^*.$$

Thus the tensor product of linear functions f and g in E and F is the linear function in $E \otimes F$ given by

$$(f \otimes g)(x \otimes y) = f(x) \cdot g(y) \qquad x \in E, y \in F.$$

1.28. The Isomorphism T

Let E^* be a dual space of E and consider the linear map

$$T: E^* \otimes F \to L(E; F)$$

given by

$$T(a^* \otimes b)x = \langle a^*, x \rangle b \qquad x \in E.$$

We show that T has the universal property. In fact, we have the commutative diagram

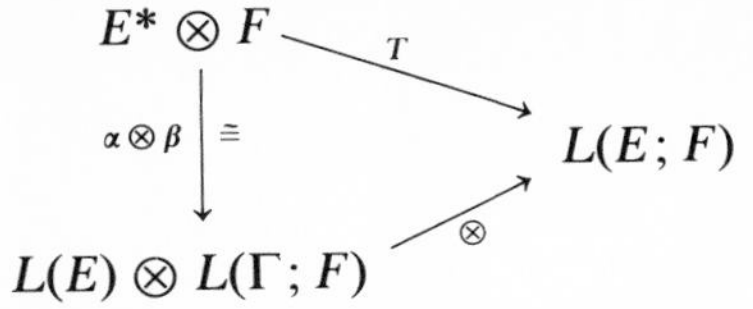

where $\alpha : E^* \xrightarrow{\cong} L(E)$ is the canonical isomorphism, $\beta : F \to L(\Gamma; F)$ is the isomorphism given by $\beta_y(\lambda) = \lambda y$, $y \in F$, $\lambda \in \Gamma$, and $\otimes$ is the map defined in Section 1.16. Since the bilinear mapping $\otimes$ has the universal property (Proposition 1.27.1), the same is true for T. Thus we may identify $E^* \otimes F$ with $L(E; F)$ under T.

A straightforward computation shows that

$$\psi \circ T(a^* \otimes b) = T(a^* \otimes \psi b) \qquad \psi \in L(F; E)$$

and

$$T(a^* \otimes b) \circ \psi = T(\psi^* a^* \otimes b) \qquad \psi \in L(F; E).$$

In particular, if $a \in E$, $a^* \in E^*$, $b \in F$, $b^* \in F^*$, then

$$T(b^* \otimes a) \circ T(a^* \otimes b) = \langle b^*, b \rangle T(a^* \otimes a). \tag{1.28}$$

Finally, consider the *trace form* $\operatorname{tr} : L(E, F) \times L(F, E) \to \Gamma$ given by $(\varphi \times \psi) \to \operatorname{tr}(\varphi \circ \psi)$. We show that the operator T satisfies

$$\operatorname{tr}(T(b^* \otimes a) \circ T(a^* \otimes b)) = \langle a^* \otimes b, b^* \otimes a \rangle = \langle a^*, a \rangle \langle b^*, b \rangle. \tag{1.29}$$

In fact, Formula (1.28) yields

$$\operatorname{tr}(T(b^* \otimes a) \circ T(a^* \otimes b)) = \langle b^*, b \rangle \operatorname{tr} T(a^* \otimes a).$$

But since the linear map $T(a^* \otimes a)$ is given by $T(a^* \otimes a)x = \langle a^*, x \rangle a$, we have

$$\operatorname{tr} T(a^* \otimes a) = \langle a^*, a \rangle \tag{1.30}$$

and so (1.29) follows.

Formula (1.29) implies in particular, that the trace form is nondegenerate (cf. Section 1.22).

1.29 The Algebra of Linear Transformations

To simplify notation we use the isomorphism T of the last section to identify $a^* \otimes a$ with the corresponding linear transformation.

Consider the associative algebra $A = A(E; E)$ of the linear transformations $\varphi : E \to E$. A linear map $\Omega : A \otimes A \to L(A; A)$ is defined by $(\alpha, \beta) \mapsto \Omega(\alpha \otimes \beta)$ where $\Omega(\alpha \otimes \beta)$ is the transformation defined by

$$\Omega(\alpha \otimes \beta)\varphi = \alpha \circ \varphi \circ \beta. \tag{1.31}$$

Proposition 1.29.1. *Ω is an isomorphism.*

PROOF. We recall that the space A is dual to itself with respect to the trace form $\langle \psi, \varphi \rangle = \operatorname{tr}(\psi \circ \varphi)$. Now let $F: A \times A \to L(A; A)$ be the linear map defined by

$$F(\alpha \otimes \beta)\varphi = \langle \alpha, \varphi \rangle \beta.$$

Then the mappings F and Ω are connected by the relation

$$\Omega = F \circ Q, \tag{1.32}$$

where Q is the linear automorphism of $L(A \otimes A)$ defined by

$$Q((a^* \otimes a) \otimes (b^* \otimes b)) = (a^* \otimes b) \otimes (b^* \otimes a).$$

To prove (1.32) it is sufficient to show that

$$\Omega((a^* \otimes a) \otimes (b^* \otimes b)) = F((a^* \otimes b) \otimes (b^* \otimes a)). \tag{1.33}$$

Let $\varphi: E \to E$ be an arbitrary linear transformation. Then we have, in view of the results of Section 1.28, that

$$\begin{aligned} \Omega((a^* \otimes a) \otimes (b^* \otimes b))\varphi &= (a^* \otimes a) \circ \varphi \circ (b^* \otimes b) = (a^* \otimes a) \circ (b^* \otimes \varphi b) \\ &= \langle a^*, \varphi b \rangle b^* \otimes a \\ &= \langle \varphi^* a^*, b \rangle b^* \otimes a \end{aligned}$$

and

$$\begin{aligned} F((a^* \otimes b) \otimes (b^* \otimes a))\varphi &= \langle a^* \otimes b, \varphi \rangle b^* \otimes a \\ &= \langle \varphi^* a^*, b \rangle b^* \otimes a \end{aligned}$$

whence (1.33).

In view of Section 1.28, F is a linear isomorphism. Since Q is a linear automorphism of $A \otimes A$, relation (1.32) implies that Ω is an isomorphism. □

Corollary. *Suppose α_i and β_i $(i = 1, \ldots, r)$ are elements of A such that the α_i are linearly independent. Then the relation*

$$\sum_i \alpha_i \circ \varphi \circ \beta_i = 0 \quad \textit{for every } \varphi \in A$$

implies that $\beta_i = 0$ $(i = 1, \ldots, r)$.

1.30. The Endomorphisms of A

Every linear automorphism α of E induces an endomorphism $h_\alpha \neq 0$ of the algebra A given by

$$h_\alpha \varphi = \alpha \circ \varphi \circ \alpha^{-1}.$$

It will now be shown that conversely, *every nonzero endomorphism of the algebra A is obtained in this way.* In other words, every endomorphism

$h \neq 0$ of the algebra A can be written in the form $h\varphi = \alpha \circ \varphi \circ \alpha^{-1}$ where α is a regular linear transformation of E.

Since the pair $(L(A; A), \Omega)$ is the tensor product of A and A, we can write

$$h\varphi = \sum_{i=1}^{r} \alpha_i \circ \varphi \circ \beta_i \qquad \alpha_i, \beta_i \in A,$$

where the α_i and the β_i are linearly independent.

Now the relation

$$h(\varphi \circ \psi) = h\varphi \circ h\psi$$

implies that

$$\sum_i \alpha_i \circ \varphi \circ (\psi \circ \beta_i - \beta_i \circ \sum_j \alpha_j \circ \psi \circ \beta_j) = 0.$$

Since the α_i are linearly independent and φ is an arbitrary element of A it follows that (see the Corollary to Proposition 1.29.1)

$$\psi \circ \beta_i - \beta_i \circ \sum_j \alpha_j \circ \psi \circ \beta_j = 0 \qquad (i = 1, \ldots, r).$$

This can be written in the form

$$\sum_j (\delta_{ij}\iota - \beta_i \circ \alpha_j) \circ \psi \circ \beta_j = 0 \qquad (i = 1, \ldots, r).$$

Now the linear independence of the elements β_j implies that

$$\beta_i \circ \alpha_j = \delta_{ij}\iota \qquad (i, j = 1, \ldots, r). \tag{1.34}$$

For $j \neq i$ we obtain from (1.34)

$$\beta_i \circ \alpha_j = 0$$

and for $j = i$

$$\beta_i \circ \alpha_i = \iota.$$

These relations are compatible only if $r = 1$. Denoting α_1 by α we obtain

$$h\varphi = \alpha \circ \varphi \circ \alpha^{-1}.$$

It is easy to show that the element α is uniquely determined by h up to a constant factor.

Our result shows, in particular, that every endomorphism $h \neq 0$ of the algebra A preserves the scalar product $\langle \psi, \varphi \rangle = \operatorname{tr}(\psi \circ \varphi)$. In fact, for every two elements $\varphi, \psi \in A$ we have

$$\begin{aligned} \langle h\varphi, h\psi \rangle &= \operatorname{tr}(h\varphi \circ h\psi) \\ &= \operatorname{tr} h(\varphi \circ \psi) = \operatorname{tr}(\alpha \circ \varphi \circ \psi \circ \alpha^{-1}) \\ &= \operatorname{tr}(\varphi \circ \psi) = \langle \varphi, \psi \rangle \end{aligned}$$

whence

$$\langle h\varphi, h\psi \rangle = \langle \varphi, \psi \rangle \qquad \varphi, \psi \in A.$$

PROBLEMS

1. Let $a^* \in E^*$ and $b \in F$ be two fixed vectors and consider the linear maps $a^* \otimes b: E \to F$ and $b \otimes a^*: E^* \leftarrow F^*$. Prove that

$$b \otimes a^* = (a^* \otimes b)^*.$$

2. Let E, F be Euclidean spaces and consider the induced inner product in $E \otimes F$. Given two linear transformations $\varphi: E \to E$, $\psi: F \to F$; prove that
(a) $\varphi \otimes \psi$ is a rotation if and only if $\varphi = \lambda\tau_E$ and $\psi = \lambda^{-1}\tau_F$ where τ_E and τ_F are rotations of E, F and $\lambda \neq 0$ is a real number.
(b) $\varphi \otimes \psi$ is selfadjoint if and only if both transformations φ and ψ are selfadjoint or skew.
(c) $\varphi \otimes \psi$ is skew if and only if precisely one of the transformations is selfadjoint and the other one is skew.
(d) $\varphi \otimes \psi$ is normal if and only if both transformations are normal.

3. Let E be a real n-dimensional vector space and consider two regular transformations φ and ψ of E. Given an orientation in $E \otimes E$ prove that
(a) if n is even then $\varphi \otimes \psi$ preserves the orientation.
(b) if n is odd, then $\varphi \otimes \psi$ preserves the orientation if and only if both mappings φ and ψ are orientation preserving or orientation reversing.

Tensor Products of Vector Spaces with Additional Structure 2

Tensor Products of Algebras

2.1. The Structure Map

If A is an algebra, then the multiplication $A \times A \to A$ determines a linear map $\mu_A : A \otimes A \to A$ such that

$$\mu_A(x \otimes y) = xy. \tag{2.1}$$

μ_A is called the *structure map* of the algebra A. (In this chapter the symbol μ_A will be reserved exclusively for structure maps. Since such a notation appears in no other chapters, there is no possibility of confusion.) Conversely, if A is a vector space and $\mu_A : A \otimes A \to A$ is a linear map, a multiplication is induced in A by

$$xy = \mu_A(x \otimes y) \tag{2.2}$$

and so A becomes an algebra. The above remark shows that there is a 1–1 correspondence between the multiplications in A and the linear maps $\mu_A : A \otimes A \to A$.

Now let B be a second algebra and $\mu_B : B \otimes B \to B$ be the corresponding structure map. If $\varphi : A \to B$ is a homomorphism we have

$$\varphi \mu_A(x \otimes y) = \mu_B(\varphi x \otimes \varphi y) = \mu_B(\varphi \otimes \varphi)(x \otimes y)$$

whence

$$\varphi \circ \mu_A = \mu_B \circ (\varphi \otimes \varphi). \tag{2.3}$$

Conversely, every linear map $\varphi : A \to B$ which satisfies this relation is a homomorphism.

2.2. The Canonical Tensor Product of Algebras

Let A and B be two algebras with structure maps μ_A and μ_B respectively. Consider the *flip-operator*

$$S:(A \otimes B) \otimes (A \otimes B) \to (A \otimes A) \otimes (B \otimes B)$$

defined by

$$S(x_1 \otimes y_1 \otimes x_2 \otimes y_2) = x_1 \otimes x_2 \otimes y_1 \otimes y_2.$$

Then a linear map

$$\mu_{A \otimes B}:(A \otimes B) \otimes (A \otimes B) \to A \otimes B$$

defined by

$$\mu_{A \otimes B} = (\mu_A \otimes \mu_B) \circ S \tag{2.4}$$

determines an algebra structure in $A \otimes B$. The algebra $A \otimes B$ is called the *canonical tensor product* of the algebras A and B. It is easily checked that the multiplication in $A \otimes B$ satisfies

$$(x_1 \otimes y_1)(x_2 \otimes y_2) = x_1 x_2 \otimes y_1 y_2. \tag{2.5}$$

This formula shows that a canonical tensor product of two associative (commutative) algebras is again associative (commutative). If A and B have unit elements I_A and I_B respectively then $I_A \otimes I_B$ is the unit element of $A \otimes B$. If B has a unit element I_B we can define an injective linear mapping $\varphi: A \to A \otimes B$ by

$$\varphi x = x \otimes I_B \qquad x \in A.$$

It follows from (2.5) that

$$\varphi(xx') = xx' \otimes I_B = (x \otimes I_B)(x' \otimes I_B) = \varphi x \cdot \varphi x', \qquad x, x' \in A,$$

i.e., φ preserves products and so it is a monomorphism.

2.3. Tensor Product of Homomorphisms

Let A_1, B_1, A_2, B_2 be algebras and suppose that

$$\varphi_1: A_1 \to B_1 \qquad \varphi_2: A_2 \to B_2$$

are homomorphisms. Then $\varphi_1 \otimes \varphi_2$ is a homomorphism of the algebra $A_1 \otimes A_2$ into the algebra $B_1 \otimes B_2$. In fact, since

$$(\varphi_1 \otimes \varphi_1 \otimes \varphi_2 \otimes \varphi_2) \circ S = S \circ (\varphi_1 \otimes \varphi_2 \otimes \varphi_1 \otimes \varphi_2)$$

it follows that

$$\begin{aligned}(\varphi_1 \otimes \varphi_2) \circ \mu_{A_1 \otimes A_2} &= [(\varphi_1 \circ \mu_{A_1}) \otimes (\varphi_2 \circ \mu_{A_2})] \circ S \\ &= [(\mu_{B_1} \circ (\varphi_1 \otimes \varphi_1)) \otimes \mu_{B_2} \circ (\varphi_2 \otimes \varphi_2)] \circ S \\ &= [(\mu_{B_1} \otimes \mu_{B_2}) \circ S] \circ (\varphi_1 \otimes \varphi_2 \otimes \varphi_1 \otimes \varphi_2) \\ &= \mu_{B_1 \otimes B_2} \circ [(\varphi_1 \otimes \varphi_2) \otimes (\varphi_1 \otimes \varphi_2)].\end{aligned}$$

This equation shows that $\varphi_1 \otimes \varphi_2$ is a homomorphism.

Now consider two involutions ω_A and ω_B in A and B respectively. Then the mapping

$$\omega_{A \otimes B} = \omega_A \otimes \omega_B$$

is an involution in the algebra $A \otimes B$.

2.4. Antiderivations

Let ω_A be an involution in the algebra A and Ω_A be an antiderivation with respect to ω_A; i.e.,

$$\Omega_A(x \cdot y) = \Omega_A x \cdot y + \omega_A x \cdot \Omega_A y. \tag{2.6}$$

In terms of the structure map, (2.6) can be rewritten as

$$\Omega_A \circ \mu_A = \mu_A \circ (\Omega_A \otimes \iota + \omega_A \otimes \Omega_A). \tag{2.7}$$

To simplify notation we write

$$\Omega_A \otimes \iota + \omega_A \otimes \Omega_A = \Omega_{A \otimes A}$$

and then (2.7) reads

$$\boxed{\Omega_A \circ \mu_A = \mu_A \circ \Omega_{A \otimes A}.} \tag{2.8}$$

Now let B be a second algebra and Ω_B be an antiderivation of B with respect to an involution ω_B,

$$\Omega_B \circ \mu_B = \mu_B \circ \Omega_{B \otimes B}. \tag{2.9}$$

Consider the linear maps

$$\Omega_{A \otimes B} = \Omega_A \otimes \iota_B + \omega_A \otimes \Omega_B, \tag{2.10}$$

$$\Omega_{(A \otimes B) \otimes (A \otimes B)} = \Omega_{A \otimes B} \otimes \iota_{A \otimes B} + \omega_{A \otimes B} \otimes \Omega_{A \otimes B}, \tag{2.11}$$

and

$$\Omega_{(A \otimes A) \otimes (B \otimes B)} = \Omega_{A \otimes B} \otimes \iota_{B \otimes B} + \omega_{A \otimes A} \otimes \Omega_{B \otimes B}. \tag{2.12}$$

On the assumption that

$$S \circ \Omega_{(A \otimes B) \otimes (A \otimes B)} = \Omega_{(A \otimes A) \otimes (B \otimes B)} \circ S \tag{2.13}$$

(this always holds if $\omega_A = \iota_A$, $\omega_B = \iota_B$), Equations (2.10), (2.11), (2.12), and (2.13) imply that

$$\begin{aligned}
\mu_{A \otimes B} \circ \Omega_{(A \otimes B) \otimes (A \otimes B)} &= (\mu_A \otimes \mu_B) \circ S \circ \Omega_{(A \otimes B) \otimes (A \otimes B)} \\
&= (\mu_A \otimes \mu_B) \circ \Omega_{(A \otimes A) \otimes (B \otimes B)} \circ S \\
&= (\mu_A \otimes \mu_B) \circ (\Omega_{A \otimes A} \otimes \iota_{B \otimes B} + \omega_{A \otimes A} \otimes \Omega_{B \otimes B}) \circ S \\
&= (\mu_A \circ \Omega_{A \otimes A} \otimes \mu_B + \mu_A \circ \omega_{A \otimes A} \otimes \mu_B \circ \Omega_{B \otimes B}) \circ S \\
&= (\Omega_A \circ \mu_A \otimes \mu_B + \omega_A \circ \mu_A \otimes \Omega_B \circ \mu_B) \circ S \\
&= (\Omega_A \otimes \iota_B + \omega_A \otimes \Omega_B) \circ (\mu_A \otimes \mu_B) \circ S \\
&= \Omega_{A \otimes B} \circ \mu_{A \otimes B}.
\end{aligned}$$

This relation shows, in view of (2.8), that $\Omega_{A \otimes B}$ is an antiderivation in the algebra $A \otimes B$ with respect to the involution $\omega_{A \otimes B}$.

Tensor Products of *G*-Graded Vector Spaces

2.5. Poincaré Series

Let $E = \sum_\alpha E_\alpha$, $\alpha \in G$ and $F = \sum_\beta F_\beta$, $\beta \in H$ be respectively G- and H-graded vector spaces. Then a $(G \oplus H)$-gradation is induced in the space $E \otimes F$ by

$$E \otimes F = \sum_{\alpha, \beta} E_\alpha \otimes F_\beta. \tag{2.14}$$

If $H = G$, then (2.14) is a G-bigradation. The corresponding (simple) G-gradation is given by

$$E \otimes F = \sum_\gamma (E \otimes F)_\gamma, \qquad (E \otimes F)_\gamma = \sum_{\alpha + \beta = \gamma} E_\alpha \otimes F_\beta. \tag{2.15}$$

The space $E \otimes F$, together with its G-gradation is called the *tensor product of the G-graded spaces E and F*.

It follows from (2.15) that for every two homogeneous elements $x \in E_\alpha$, $y \in F_\beta$ the element $x \otimes y$ is homogeneous and

$$\deg(x \otimes y) = \deg x + \deg y.$$

In particular, the linear isomorphism $f: E \otimes F \to F \otimes E$ given by

$$f(x \otimes y) = y \otimes x$$

is homogeneous of degree zero. Moreover, we have as well that each $(E \otimes F)_\gamma$ is linearly generated by homogeneous decomposable elements of the form $x \otimes y$, $x \in E_\alpha$, $y \in F_\beta$, $\alpha + \beta = \gamma$.

Let E, E', F, F' be G-graded vector spaces, and consider homogeneous linear maps

$$\varphi: E \to E' \quad \text{and} \quad \psi: F \to F'$$

of degrees k and l respectively. Then $\varphi \otimes \psi: E \otimes F \to E' \otimes F'$ is homogeneous of degree $k + l$. In fact, if x and y are homogeneous elements of degree α and β respectively it follows that

$$\begin{aligned}\deg(\varphi \otimes \psi)(x \otimes y) &= \deg(\varphi x \otimes \psi y)\\ &= \deg \varphi x + \deg \psi y\\ &= \alpha + k + \beta + l\\ &= (\alpha + \beta) + (k + l)\end{aligned}$$

and hence $\varphi \otimes \psi$ is homogeneous of degree $k + l$.

Now assume that $G = \mathbb{Z}$ and that the gradations of E and F are positive and almost finite. Then the *Poincaré series* of $E \otimes F$ is given by

$$P_{E \otimes F}(t) = \sum_k \dim(E \otimes F)_k t^k.$$

Since

$$\dim(E \otimes F)_k = \dim \sum_{i+j=k} E_i \otimes F_j = \sum_{i+j=k} \dim E_i \dim F_j,$$

the above formula reads

$$P_{E \otimes F}(t) = \sum_k \sum_{i+j=k} \dim E_i t^i \dim E_j t^j = P_E(t) \cdot P_F(t)$$

showing that the Poincaré series of $E \otimes F$ is the product of the Poincaré series of E and F.

2.6. Tensor Products of Several G-Graded Vector Spaces

Let $E_i = \sum_\alpha E_i^\alpha$, $\alpha \in G$ be G-graded vector spaces. Then a G p-gradation is induced in the space $E = E_1 \otimes \cdots \otimes E_p$ by assigning the degree $(\alpha_1, \ldots, \alpha_p)$ to the elements of $E_1^{\alpha_1} \otimes \cdots \otimes E_p^{\alpha_p}$. The corresponding simple G-gradation is given by $E = \sum_{\alpha \in G} E_\alpha$ where

$$E_\alpha = \sum E_1^{\alpha_1} \otimes \cdots \otimes E_p^{\alpha_p},$$

the sum being extended over all p-tuples $(\alpha_1, \ldots, \alpha_p)$ such that $\alpha_1 + \cdots + \alpha_p = \alpha$. The space E together with this gradation is called a *tensor product for the G-graded spaces E_i*. It follows from the definitions that

$$\deg(x_1 \otimes \cdots \otimes x_p) = \deg x_1 + \cdots + \deg x_p$$

for every p-tuple of homogeneous elements x_i. As another immediate consequence of the definition we note that the isomorphism $f:(E_1 \otimes E_2) \otimes E_3 \to E_1 \otimes (E_2 \otimes E_3)$ defined by

$$f:(x_1 \otimes x_2) \otimes x_3 \to x_1 \otimes (x_2 \otimes x_3)$$

is homogeneous of degree zero.

It is easy to verify that if E_i, F_i $(i = 1, \ldots, p)$ are graded spaces, and if $\varphi_i: E_i \to F_i$ are homogeneous of degree k_i then the map $\varphi_1 \otimes \cdots \otimes \varphi_p$ is homogeneous of degree $\sum_{i=1}^p k_i$.

Suppose now that $G = \mathbb{Z}$ and that all the gradations of the E_i are positive and almost finite. Then clearly the induced gradation in E is again positive and almost finite. Moreover, the Poincaré series of E is given by

$$P_E(t) = P_{E_1}(t) \cdots P_{E_p}(t).$$

The proof is similar to that given for $p = 2$.

2.7. Dual G-Graded Spaces

Let $E = \sum_{\alpha \in G} E_\alpha$, $E^* = \sum_{\alpha \in G} E_\alpha^*$ and $F = \sum_{\beta \in G} F_\beta$, $F^* = \sum_{\beta \in G} F_\beta^*$ be two pairs of dual G-graded vector spaces and consider the spaces

$$E \otimes F = \sum_{\alpha, \beta} E_\alpha \otimes F_\beta$$

$$E^* \otimes F^* = \sum_{\alpha, \beta} E_\alpha^* \otimes F_\beta^*$$

as G-bigraded vector spaces. Then the induced scalar product between $E \otimes F$ and $E^* \otimes F^*$ respects the G-bigradations. In fact, for any vectors $x \in E_{\alpha_1}$, $x^* \in E_{\alpha_2}^*$, $y \in F_{\beta_1}$, $y^* \in F_{\beta_2}^*$ we have

$$\langle x^* \otimes y^*, x \otimes y\rangle = \langle x^*, x\rangle\langle y^*, y\rangle = 0$$

unless $\alpha_1 = \alpha_2$ and $\beta_1 = \beta_2$. As an immediate consequence we have that the G-graded spaces $E \otimes F$, $E^* \otimes F^*$ are dual G-graded spaces.

2.8. Anticommutative Tensor Products of Graded Algebras

Let $A = \sum_p A_p$ and $B = \sum_q B_q$ be two graded algebras. Consider the *anticommutative flip operator*

$$Q:(A \otimes B) \otimes (A \otimes B) \to (A \otimes A) \otimes (B \otimes B)$$

defined by

$$Q(x \otimes y \otimes x' \otimes y') = (-1)^{p'q} x \otimes x' \otimes y \otimes y',$$

where $\deg x' = p'$ and $\deg y = q$. Then the linear map

$$\mu_{A \hat{\otimes} B}:(A \otimes B) \otimes (A \otimes B) \to A \otimes B$$

defined by

$$\mu_{A \hat{\otimes} B} = (\mu_A \otimes \mu_B) \circ Q$$

determines an algebra structure in the graded vector space $A \otimes B$. The resulting algebra, $A \hat{\otimes} B$, is called the *anticommutative tensor* product or the *skew tensor product* of A and B. The multiplication in $A \hat{\otimes} B$ is given by

$$(x \otimes y)(x' \otimes y') = (-1)^{p'q} xx' \otimes yy' \qquad p' = \deg x',\, q = \deg y. \tag{2.16}$$

If A and B are algebras without gradation, then by the tensor product of A and B we shall mean the canonical tensor product (Section 2.2). If, in this chapter, A and B are graded algebras, then by the tensor product of A and B we shall mean the anticommutative tensor product. Observe that the underlying vector spaces of the algebras $A \otimes B$ and $A \hat{\otimes} B$ coincide.

Now it will be shown that $A \hat{\otimes} B$ is a graded algebra. In fact, if $x_1 \in A_{p_1}$, $x_2 \in A_{p_2}$, $y_1 \in B_{q_1}$, $y_2 \in B_{q_2}$ are arbitrary we have

$$(x_1 \otimes y_1)(x_2 \otimes y_2) = (-1)^{p_2 q_1} x_1 x_2 \otimes y_1 y_2.$$

Since A and B are graded algebras, it follows that

$$\deg(x_1 x_2) = p_1 + p_2 \qquad \deg(y_1 y_2) = q_1 + q_2.$$

In view of the definition of the gradation in $A \hat{\otimes} B$ (Section 2.5) we obtain that $(x_1 \otimes y_1)(x_2 \otimes y_2)$ is homogeneous of degree $p_1 + p_2 + q_1 + q_2$ and hence $A \hat{\otimes} B$ is a graded algebra.

It is easy to verify that if C is a third graded algebra then the linear map $f:(A \hat{\otimes} B) \hat{\otimes} C \to A \hat{\otimes} (B \hat{\otimes} C)$ given by

$$f:(x \otimes y) \otimes z \to x \otimes (y \otimes z)$$

preserves products and hence is an isomorphism.

The anticommutative tensor product of two anticommutative graded algebras is again an anticommutative algebra. In fact, let $x \in A_p$, $y \in B_q$, $x' \in A_{p'}$, and $y' \in B_{q'}$ be homogeneous elements. Then we have

$$\begin{aligned}(x \otimes y)(x' \otimes y') &= (-1)^{p'q} xx' \otimes yy' \\ &= (-1)^{p'q + p'p + qq'} x'x \otimes y'y \\ &= (-1)^{p'q + pp' + qq' + pq'} (x' \otimes y')(x \otimes y) \\ &= (-1)^{(p+q)(p'+q')} (x' \otimes y')(x \otimes y).\end{aligned}$$

The reader should observe that the canonical tensor product of anticommutative graded algebras is not in general an anticommutative algebra.

2.9. Homomorphisms and Antiderivations

Let $C = \sum_r C_r$ and $D = \sum_s D_s$ be two more graded algebras and assume that

$$\varphi: A \to C, \qquad \psi: B \to D$$

are homomorphisms homogeneous of even degree k and l respectively. Then $\varphi \otimes \psi$ is a linear map homogeneous of degree $k + l$. Moreover, we have

$$(\varphi \otimes \varphi \otimes \psi \otimes \psi) \circ Q = Q \circ (\varphi \otimes \psi \otimes \varphi \otimes \psi)$$

as is easily checked. By the same argument as that used in Section 2.3 it follows that $\varphi \otimes \psi$ is a homomorphism of $A \mathbin{\hat{\otimes}} B$ into $C \mathbin{\hat{\otimes}} D$.

In particular, if ω_A and ω_B are involutions in A and B, homogeneous of degree zero, then $\omega_{A \otimes B} = \omega_A \otimes \omega_B$ is an involution of degree zero in $A \mathbin{\hat{\otimes}} B$. As a special case, suppose that ω_A and ω_B are the canonical involutions in A and B. Then $\omega_{A \otimes B}$ is the canonical involution in $A \mathbin{\hat{\otimes}} B$ (see Section 6.6 of *Linear Algebra*).

Proposition 2.9.1. *Let Ω_A and Ω_B be homogeneous antiderivations of odd degree k. Then the mapping*

$$\Omega_{A \hat{\otimes} B} = \Omega_A \otimes \iota_B + \omega_A \otimes \Omega_B$$

(where ω_A is the canonical involution) is an antiderivation, homogeneous of degree k, in the graded algebra $A \mathbin{\hat{\otimes}} B$. If θ_A and θ_B are derivations of even degree k then

$$\theta_{A \hat{\otimes} B} = \theta_A \otimes \iota_B + \iota_A \otimes \theta_B$$

is a homogeneous derivation of degree k in $A \mathbin{\hat{\otimes}} B$.

Proof. Clearly, $\Omega_{A \hat{\otimes} B}$ is a homogeneous linear map of degree k. To show that $\Omega_{A \otimes B}$ is an antiderivation we verify that

$$Q \circ \Omega_{(A \hat{\otimes} B) \hat{\otimes} (A \hat{\otimes} B)} = \Omega_{(A \hat{\otimes} A) \hat{\otimes} (B \hat{\otimes} B)} \circ Q \tag{2.17}$$

(see Section 2.4 for the notation). Then the same argument as that used in Section 2.4 proves that $\Omega_{A \hat{\otimes} B}$ is an antiderivation in $A \mathbin{\hat{\otimes}} B$. Similarly, to prove the second part of the proposition we need only establish the formula

$$Q \circ \theta_{(A \hat{\otimes} B) \hat{\otimes} (A \hat{\otimes} B)} = \theta_{(A \hat{\otimes} A) \hat{\otimes} (B \hat{\otimes} B)} \circ Q. \tag{2.18}$$

Now we proceed to the verification of (2.17) and (2.18).

Let

$$x \in A_p, \qquad y \in B_q, \qquad x' \in A_{p'}, \qquad y' \in B_{q'}$$

be homogeneous elements. Then we obtain

$$
\begin{aligned}
&Q\,\Omega_{(A\hat{\otimes}B)\hat{\otimes}(A\hat{\otimes}B)}(x\otimes y\otimes x'\otimes y')\\
&\quad= Q[\Omega_A x\otimes y\otimes x'\otimes y' + (-1)^p x\otimes \Omega_B y\otimes x'\otimes y'\\
&\qquad + (-1)^{p+q} x\otimes y\otimes \Omega_A x'\otimes y' + (-1)^{p+q+p'} x\otimes y\otimes x'\otimes \Omega_B y']\\
&\quad= (-1)^{qp'}\Omega_A x\otimes x'\otimes y\otimes y' + (-1)^{p+q+(k+p')q} x\otimes \Omega_A x'\otimes y\otimes y'\\
&\qquad + (-1)^{p+p'(q+k)} x\otimes x'\otimes \Omega_B y\otimes y'\\
&\qquad + (-1)^{p+q+p'+p'q} x\otimes x'\otimes y\otimes \Omega_B y'\\
&\quad= (-1)^{qp'}\{\Omega_A x\otimes x' + (-1)^{p+(k+1)q} x\otimes \Omega_A x'\}\otimes y\otimes y'\\
&\qquad + (-1)^{qp'}(x\otimes x')\otimes\{(-1)^{p+kp'}\Omega_B y\otimes y' + (-1)^{p+q+p'} y\otimes \Omega_B y'\}.
\end{aligned}
$$

Since k is odd we have

$$
\begin{aligned}
(\Omega_A x\otimes x' + (-1)^{p+(k+1)q} x\otimes \Omega_A x') &= (\Omega_A x\otimes x' + (-1)^p x\otimes \Omega_A x')\\
&= \Omega_{A\hat{\otimes}A}(x\otimes x')
\end{aligned}
$$

and

$$
\begin{aligned}
&(-1)^{p+kp'}\Omega_B y\otimes y' + (-1)^{p+q+p'} y\otimes \Omega_B y'\\
&\quad= (-1)^{p+p'}[\Omega_B y\otimes y' + (-1)^q y\otimes \Omega_B y']\\
&\quad= (-1)^{p+p'}\Omega_{B\hat{\otimes}B}(y\otimes y').
\end{aligned}
$$

Hence it follows that

$$
\begin{aligned}
&Q\Omega_{(A\hat{\otimes}B)\hat{\otimes}(A\hat{\otimes}B)}(x\otimes y\otimes x'\otimes y')\\
&\quad= (-1)^{qp'}[\Omega_{A\hat{\otimes}A}(x\otimes x')\otimes y\otimes y' + (-1)^{p+q'}(x\otimes x')\otimes \Omega_{B\hat{\otimes}B}(y\otimes y')]\\
&\quad= (-1)^{qp'}\Omega_{(A\hat{\otimes}A)\hat{\otimes}(B\hat{\otimes}B)}(x\otimes x'\otimes y\otimes y')\\
&\quad= \Omega_{(A\hat{\otimes}A)\hat{\otimes}(B\hat{\otimes}B)}Q(x\otimes y\otimes x'\otimes y')
\end{aligned}
$$

whence (2.17).

Now let θ_A and θ_B be homogeneous derivations of even degree k. Then we have

$$
\begin{aligned}
&Q\,\Omega_{(A\hat{\otimes}B)\hat{\otimes}(A\hat{\otimes}B)}(x\otimes y\otimes x'\otimes y')\\
&\quad= Q(\theta_A x\otimes y\otimes x'\otimes y' + x\otimes \theta_B y\otimes x'\otimes y'\\
&\qquad + x\otimes y\otimes \theta_A x'\otimes y' + x\otimes y\otimes x'\otimes \theta_B y')\\
&\quad= (-1)^{p'q}[\theta_A x\otimes x'\otimes y\otimes y' + x\otimes x'\otimes y\otimes \theta_B y']\\
&\qquad + (-1)^{q(p'+k)} x\otimes \theta_A x'\otimes y\otimes y' + (-1)^{p'(q+k)} x\otimes x'\otimes \theta_B y\otimes y'\\
&\quad= (-1)^{p'q}[\theta_A x\otimes x'\otimes y\otimes y' + x\otimes \theta_A x'\otimes y\otimes y'\\
&\qquad + x\otimes x'\otimes \theta_B y\otimes y' + x\otimes x'\otimes y\otimes \theta_B y']\\
&\quad= (-1)^{p'q}\theta_{(A\hat{\otimes}A)\hat{\otimes}(B\hat{\otimes}B)}(x\otimes x'\otimes y\otimes y')\\
&\quad= \theta_{(A\hat{\otimes}A)\hat{\otimes}(B\hat{\otimes}B)}Q(x\otimes y\otimes x'\otimes y')
\end{aligned}
$$

whence (2.18). □

Tensor Products of Differential Spaces

In Sections 2.10–2.17 the notations B_E, $B(E)$; Z_E, $Z(E)$; and H_E, $H(E)$ for the boundary, cycle, and homology spaces of a differential space will be used interchangeably.

2.10. Tensor Products of Differential Spaces

Suppose that (E, ∂_E) and (F, ∂_F) are differential spaces. We wish to make $E \otimes F$ into a differential space. In order to do so we shall need an involution ω of E such that

$$\partial_E \circ \omega + \omega \circ \partial_E = 0. \tag{2.19}$$

Suppose we are given such an involution. Define $D = \partial_{E \otimes F}$ by

$$\partial_{E \otimes F} = \partial_E \otimes \iota + \omega \otimes \partial_F.$$

Then we have

$$\partial_{E \otimes F}(x \otimes y) = \partial_E x \otimes y + \omega x \otimes \partial_F y. \tag{2.20}$$

From (2.19) we obtain that

$$\begin{aligned} \partial_{E \otimes F}^2 &= \partial_E^2 \otimes \iota + \omega\partial_E \otimes \partial_F + \partial_E\omega \otimes \partial_F + \iota \otimes \partial_F^2 \\ &= (\omega\partial_E + \partial_E\omega) \otimes \partial_F \\ &= 0 \end{aligned}$$

and so $(E \otimes F, \partial_{E \otimes F})$ is a differential space. Formula (2.20) implies that

$$Z_E \otimes Z_F \subset Z_{E \otimes F}. \tag{2.21}$$

Moreover

$$B_E \otimes Z_F \subset B_{E \otimes F} \quad \text{and} \quad Z_E \otimes B_F \subset B_{E \otimes F}. \tag{2.22}$$

In fact, if $\partial_E x \in B_E$ and $y \in Z_F$ are arbitrary elements, then

$$\partial_E x \otimes y = \partial_{E \otimes F}(x \otimes y).$$

Similarly, if $x \in Z_E$ and $\partial_F y \in B_F$, then

$$\partial_{E \otimes F}(\omega x \otimes y) = \omega^2 x \otimes \partial_F y = x \otimes \partial_F y.$$

It follows from Relations (2.21) and (2.22) that the bilinear mapping $(E \times F) \to E \otimes F$ induces a bilinear mapping $\varphi: H_E \times H_F \to H_{E \otimes F}$ such that

$$\varphi(\pi_E z_1, \pi_F z_2) = \pi_{E \otimes F}(z_1 \otimes z_2) \qquad z_1 \in Z_E,\ z_2 \in Z_F, \tag{2.23}$$

where π_E, π_F, and $\pi_{E \otimes F}$ are the canonical projections of the cycle spaces onto the homology spaces.

It is the purpose of this section to show that the pair $(H_{E\otimes F}, \varphi)$ is the tensor product of H_E and H_F. We first establish the formulas

$$Z_{E\otimes F} = Z_E \otimes Z_F + B_{E\otimes F} \tag{2.24}$$

$$(Z_E \otimes Z_F) \cap B_{E\otimes F} = B_E \otimes Z_F + Z_E \otimes B_F. \tag{2.25}$$

Consider the linear operators $D_1 : E \otimes F \to E \otimes F$ and $D_2 : E \otimes F \to E \otimes F$ given by

$$D_1 = \partial_E \otimes \iota \quad \text{and} \quad D_2 = \omega \otimes \partial_F. \tag{2.26}$$

Then

$$D_1^2 = D_2^2 = 0, \qquad D_1 D_2 + D_2 D_1 = 0, \tag{2.27}$$

and

$$D_1 + D_2 = D.$$

It follows from (2.26) and (1.11) that

$$\operatorname{Im} D_1 = B_E \otimes F, \qquad \operatorname{Im} D_2 = E \otimes B_F,$$

and

$$\operatorname{Im}(D_1 D_2) = B_E \otimes B_F$$

whence, in view of (1.7),

$$\boxed{\operatorname{Im}(D_1 D_2) = \operatorname{Im} D_1 \cap \operatorname{Im} D_2.}$$

The kernels of D_1 and D_2 are given by

$$\ker D_1 = Z_E \otimes F \qquad \ker D_2 = E \otimes Z_F$$

(cf. (1.12)) and so we obtain

$$\boxed{\ker D_1 \cap \ker D_2 = Z_E \otimes Z_F.} \tag{2.28}$$

Suppose now that $z \in Z_{E\otimes F}$ is arbitrary. Then $D_1 z = -D_2 z$ and so

$$D_1 z \in \operatorname{Im} D_1 \cap \operatorname{Im} D_2 = \operatorname{Im} D_1 D_2.$$

Let $x \in E \otimes F$ be a vector such that $D_1 D_2 x = D_1 z$. Then setting

$$y = z - (D_1 x + D_2 x)$$

we obtain

$$D_1 y = D_1 z - D_1 D_2 x = 0$$

and

$$D_2 y = D_2 z - D_2 D_1 x = -D_1 z + D_1 D_2 x = 0.$$

Thus

$$y \in \ker D_1 \cap \ker D_2 = Z_E \otimes Z_F.$$

It follows that

$$z = Dx + y \in B_{E \otimes F} + Z_E \otimes Z_F$$

whence

$$Z_{E \otimes F} \subset Z_E \otimes Z_F + B_{E \otimes F}.$$

Inclusion in the other direction is a consequence of (2.21) and so (2.24) is proved.

Next, note that every element of $B_{E \otimes F} \cap (Z_E \otimes Z_F)$ can be written in the form

$$Dx = D_1 x + D_2 x.$$

Then we obtain from (2.26) and (2.27)

$$D_2(D_1 x) = D_2 Dx = 0$$

and

$$D_1(D_2 x) = D_1 Dx = 0.$$

Hence

$$D_1 x \in \ker D_2 \cap \operatorname{Im} D_1 = B_E \otimes Z_F$$

and

$$D_2 x \in \ker D_1 \cap \operatorname{Im} D_2 = Z_E \otimes B_F.$$

It follows that $Dx \in B_E \otimes Z_F + Z_E \otimes B_F$; i.e.,

$$B_{E \otimes F} \cap (Z_E \otimes Z_F) \subset B_E \otimes Z_F + Z_E \otimes B_F.$$

The inclusion in the other direction follows from (2.22) and hence (2.25) is proved as well.

Now we are ready to prove that the pair $(H_{E \otimes F}, \varphi)$ is the tensor product of H_E and H_F.

Let σ be the restriction of the canonical projection $\pi_{E \otimes F}: Z_{E \otimes F} \to H_{E \otimes F}$ to the subspace $Z_E \otimes Z_F$. Then we have, in view of (2.24) and (2.25),

$$\operatorname{Im} \sigma = H_{E \otimes F} \tag{2.29}$$

and

$$\ker \sigma = B_E \otimes Z_F + Z_E \otimes B_F. \tag{2.30}$$

Consequently σ induces a linear isomorphism

$$\bar{\sigma}: (Z_E \otimes Z_F)/T_\partial(E, F) \xrightarrow{\cong} H_{E \otimes F},$$

where $T_\partial(E, F) = B_E \otimes Z_F + Z_E \otimes B_F$. Hence,

$$\bar{\sigma} \circ \rho = \sigma \tag{2.31}$$

where ρ denotes the canonical projection

$$\rho : Z_E \otimes Z_F \to (Z_E \otimes Z_F)/T_\partial(E, F).$$

Consider the bilinear mapping

$$\psi : H_E \times H_F \to (Z_E \otimes Z_F)/T_\partial(E, F)$$

defined by

$$\psi(\pi_E z_1, \pi_F z_2) = \rho(z_1 \otimes z_2) \qquad z_1 \in Z_E, z_2 \in Z_F. \tag{2.32}$$

Then Formulas (2.32), (2.31), and (2.23) yield

$$\begin{aligned} \bar{\sigma}\psi(\pi_E z_1, \pi_F z_2) &= \bar{\sigma}\rho(z_1 \otimes z_2) \\ &= \sigma(z_1 \otimes z_2) \\ &= \pi_{E \otimes F}(z_1 \otimes z_2) \\ &= \varphi(\pi_E z_1, \pi_F z_2) \end{aligned}$$

and so it follows that $\bar{\sigma}\psi = \varphi$. Hence we have the commutative diagram

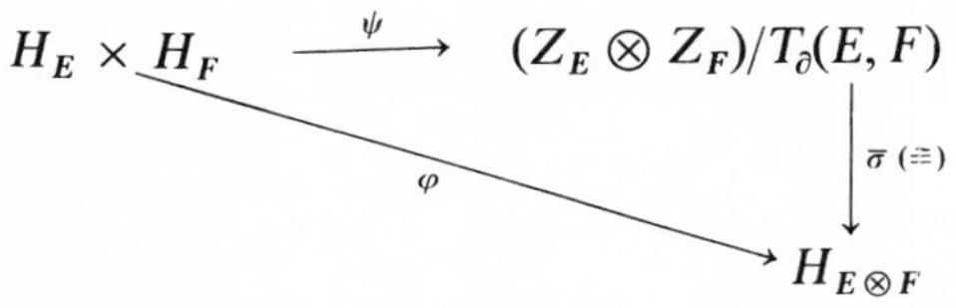

Since the pair $(Z_E \otimes Z_F/T_\partial(E, F), \psi)$ is the tensor product for H_E and H_F (see (1.13)) and $\bar{\sigma}$ is a linear isomorphism, it follows that the pair $(H_{E \otimes F}, \varphi)$ is also the tensor product of H_E and H_F. The result is restated in the

Künneth theorem: *Let (E, ∂_E) and (F, ∂_F) be two differential spaces and $(E \otimes F, \partial_{E \otimes F})$ be their tensor product. Then the pair $(H_{E \otimes F}, \varphi)$ is the tensor product of H_E and H_F, where $\varphi : H_E \times H_F \to H_{E \otimes F}$ denotes the bilinear mapping induced by the bilinear mapping $E \times F \to E \otimes F$.*

In view of the above theorem we may denote the mapping φ by $\otimes$. Then we have the relations

$$\pi_{E \otimes F}(z_1 \otimes z_2) = \pi_E z_1 \otimes \pi_F z_2$$

and

$$H_{E \otimes F} = H_E \otimes H_F.$$

2.11. Tensor Products of Dual Differential Spaces

Suppose that (E, ∂_E), (E^*, ∂_E^*) and (F, ∂_F), (F^*, ∂_F^*) are two pairs of differential spaces dual with respect to scalar products $\langle , \rangle$ and suppose further that ω, ω^* is a pair of dual involutions in E and E^* respectively. Then the induced differential operators D, D^* in $E \otimes F$ and $E^* \otimes F^*$ are given by

$$D = \partial_E \otimes \iota + \omega \otimes \partial_F$$

and

$$D^* = \partial_E^* \otimes \iota + \omega^* \otimes \partial_F^*.$$

It follows from Section 1.23 that D and D^* are again dual, and hence $(E \otimes F, D)$ and $(E^* \otimes F^*, D^*)$ are dual differential spaces.

As an immediate consequence we have that there is induced a bilinear form Φ in $H_{E^* \otimes F^*} \times H_{E \otimes F}$ such that

$$\Phi(\rho z^*, \pi z) = \langle z^*, z \rangle \qquad z^* \in Z_{E^* \otimes F^*}, z \in Z_{E \otimes F}, \tag{2.33}$$

where

$$\rho : Z_{E^* \otimes F^*} \to H_{E^* \otimes F^*} \quad \text{and} \quad \pi : Z_{E \otimes F} \to H_{E \otimes F}$$

are the canonical projections. On the other hand, consider the bilinear functions Φ_1 and Φ_2 in $H_{E^*} \times H_E$ and $H_{F^*} \times H_F$ defined by

$$\Phi_1(\rho_1 u^*, \pi_1 u) = \langle u^*, u \rangle \qquad u^* \in Z_{E^*}, u \in Z_E$$

and

$$\Phi_2(\rho_2 v^*, \pi_2 v) = \langle v^*, v) \qquad v^* \in Z_{F^*}, v \in Z_F.$$

For the tensor product of the bilinear functions Φ_1 and Φ_2 (see Section 1.22) we obtain

$$\begin{aligned}(\Phi_1 \otimes \Phi_2)(\rho_1 u^* \otimes \rho_2 v^*, \pi_1 u \otimes \pi_2 v) &= \Phi_1(\rho_1 u^*, \pi_1 u)\Phi_2(\rho_2 v^*, \pi_2 v) \\ &= \langle u^*, u \rangle \langle v^*, v \rangle \\ &= (u^* \otimes v^*, u \otimes v \rangle. \end{aligned} \tag{2.34}$$

On the other hand, Formula (2.33) shows that

$$\begin{aligned}\Phi(\rho_1 u^* \otimes \rho_2 v^*, \pi_1 u \otimes \pi_2 v) &= \Phi(\rho(u^* \otimes v^*), \pi(u \otimes v)) \qquad (2.35) \\ &= \langle u^* \otimes v^*, u \otimes v \rangle. \end{aligned}$$

Comparing (2.34) and (2.35) we find that

$$\Phi = \Phi_1 \otimes \Phi_2.$$

In particular, since Φ_1 and Φ_2 are nondegenerate (see Section 1.22) so is Φ. In view of Section 6.9 of *Linear Algebra* the nondegeneracy of Φ is also derivable from the duality of D and D^*.

2.12. Graded Differential Spaces

Consider two graded differential spaces $E = \sum_i E_i$ and $F = \sum_j F_j$. We shall assume that the operators ∂_E and ∂_F are homogeneous of odd degree k. Then the canonical involution, ω, defined by

$$\omega x = (-1)^p x \qquad x \in E_p$$

satisfies Condition (2.19). In fact,

$$(\partial_E \omega + \omega\, \partial_E)x = (-1)^p\, \partial_E x + (-1)^{p+k}\, \partial_E x = 0.$$

The induced differential operator D in $E \otimes F$ is given by

$$D(x \otimes y) = \partial_E x \otimes y + (-1)^p x \otimes \partial_F y \quad x \in E_p,\, y \in F. \tag{2.36}$$

Clearly D is again homogeneous of degree k. In general, the induced differential operator in the tensor product of graded differential spaces will mean the operator defined with respect to the canonical involution, ω, in E.

The gradations of E and F induce gradations in $H(E)$ and $H(F)$ respectively defined by

$$H(E) = \sum_i H_i(E) \qquad H_i(E) = \pi_1 Z_i(E) \tag{2.37}$$

and

$$H(F) = \sum_j H_j(F) \qquad H_j(F) = \pi_2 Z_j(F). \tag{2.38}$$

Similarly, we have (in the induced simple gradation)

$$H(E \otimes F) = \sum_k H_k(E \otimes F) \qquad H_k(E \otimes F) = \pi Z_k(E \otimes F). \tag{2.39}$$

Formulas (2.39), (2.37), and (2.38) yield in view of the Künneth theorem

$$\begin{aligned}\sum_k H_k(E \otimes F) &= \left(\sum_i H_i(E)\right) \otimes \left(\sum_j H_j(F)\right)\\ &= \sum_k \sum_{i+j=k} H_i(E) \otimes H_j(F).\end{aligned} \tag{2.40}$$

Since

$$H_i(E) \otimes H_j(F) \subset H_{i+j}(E \otimes F)$$

we obtain from (2.40) the Künneth formula for *graded differential spaces*

$$H_k(E \otimes F) = \sum_{i+j=k} H_i(E) \otimes H_j(F). \tag{2.41}$$

In particular, the gradation in $H(E \otimes F)$ determined by the gradation in $E \otimes F$ coincides with the gradation obtained by identifying $H(E \otimes F)$ with the tensor product of the graded spaces $H(E)$ and $H(F)$.

Now assume that E and F are almost finite positively graded spaces. Then so are $H(E)$, $H(F)$ and $H(E \otimes F)$. Moreover, if $P_{H(E)}$, $P_{H(F)}$ and $P_{H(E\otimes F)}$ are the corresponding Poincaré polynomials, then

$$P_{H(E\otimes F)} = P_{H(E)}P_{H(F)}.$$

2.13. Dual Graded Differential Spaces

Suppose now that $E^* = (\sum_i E_i, \partial_E^*)$ and $F^* = (\sum_j F_j, \partial_F^*)$ are two graded differential spaces which are dual to the graded differential spaces E and F respectively. Then ∂_F^*, ∂_E^* will be of degree $-k$ and the canonical involution ω^* of E^* is dual to the canonical involution of E, ($k = \deg \partial_E = \deg \partial_F$).

It follows that the differential operator

$$D^* = \partial_E^* \otimes \iota + \omega^* \otimes \partial_F^*$$

coincides with the differential operator in the graded space $E^* \otimes F^*$ defined in Section 2.11. Hence, $(E \otimes F, D)$ and $(E^* \otimes F^*, D^*)$ are dual differential spaces and graded spaces as well. Moreover, we have from Section 2.7 that $E \otimes F$ and $E^* \otimes F^*$ are dual graded spaces (i.e., the scalar product respects the gradation). Thus these spaces are dual graded differential spaces.

Problem

Let ∂ be a differential operator in a finite-dimensional vector space E. Define differential operators ∂_1 and ∂_2 in the space $L(E; E)$ by

$$\partial_1\varphi = \partial \circ \varphi \quad \text{and} \quad \partial_2\varphi = \varphi \circ \partial \qquad \varphi \in L(E; E)$$

and let H_1, H_2 be the corresponding homology spaces. Prove that

$$H_1 \cong E^* \otimes H(E) \quad \text{and} \quad H_2 \cong H(E^*) \otimes E$$

($H(E)$ and $H(E^*)$ are the homology spaces of E and E^*).

Tensor Products of Differential Algebras

2.14. The Structure Map of the Homology Algebra

Suppose (A, ∂_A) is a differential algebra with respect to an involution ω_A (see Section 6.12 of *Linear Algebra*). Then the differential operator ∂_A is an antiderivation with respect to ω_A; i.e.,

$$\partial_A(xy) = \partial_A x \cdot y + \omega_A x \cdot \partial_A y \qquad x, y \in A. \tag{2.42}$$

Introducing the differential-space $(A \otimes A, \partial_{A \otimes A})$ we can rewrite (2.42) in the form

$$\mu_A \partial_{A \otimes A} = \partial_A \mu_A,$$

where μ_A denotes the structure map of A. Now consider the homology algebra $H(A)$ (see Section 2.10). The equation

$$\pi_A z_1 \cdot \pi_A z_2 = \pi_A(z_1 z_2) \qquad z_1, z_2 \in Z(A)$$

shows that the structure map of $H(A)$ is given by

$$\mu_{H(A)} \circ (\pi_A \otimes \pi_A) = \pi_A \circ \mu_{Z(A)}, \tag{2.43}$$

where $\pi_A\colon Z(A) \to H(A)$ denotes the canonical projection and $\mu_{Z(A)}$ is the structure map of the algebra $Z(A)$. Since $Z(A)$ is a subalgebra of A it is clear that $\mu_{Z(A)}$ is the restriction of the structure map μ_A to the subspace $Z(A) \otimes Z(A)$ of $A \otimes A$.

2.15. Tensor Products of Differential Algebras

Suppose now that (A, ∂_A) and (B, ∂_B) are two differential algebras with respect to involutions ω_A and ω_B. Then the induced differential operator in the space $(A \otimes B, \partial_{A \otimes B})$, is given by

$$\partial_{A \otimes B} = \partial_A \otimes \iota + \omega_A \otimes \partial_B.$$

Now consider the canonical tensor product $A \otimes B$. Recall that $\omega_{A \otimes B}$ is an involution in $A \otimes B$ and that $\partial_{A \otimes B}$ is an antiderivation with respect to $\omega_{A \otimes B}$. Hence $(A \otimes B, \partial_{A \otimes B})$ is a differential algebra, and so an algebra structure is induced in $H(A \otimes B)$. The structure map, $\mu_{H(A \otimes B)}$, of $H(A \otimes B)$ is given by

$$\mu_{H(A \otimes B)} \circ (\pi_{A \otimes B} \otimes \pi_{A \otimes B}) = \pi_{A \otimes B} \circ \mu_{Z(A \otimes B)}, \tag{2.44}$$

where $\pi_{A \otimes B}\colon Z(A \otimes B) \to H(A \otimes B)$ denotes the canonical projection, and $\mu_{Z(A \otimes B)}$ is the structure map for the algebra $Z(A \otimes B)$ (cf. (2.43)).

2.16. The Algebra $H(A) \otimes H(B)$

It follows from the Künneth formula that the vector space $H(A \otimes B)$ may be considered to be the tensor product of the spaces $H(A)$ and $H(B)$,

$$H(A \otimes B) = H(A) \otimes H(B).$$

In this section it will be shown that $H(A \otimes B)$ as an algebra is the canonical tensor product of the algebras $H(A)$ and $H(B)$.

The structure map of the algebra $H(A) \otimes H(B)$ is given by

$$\mu_{H(A) \otimes H(B)} = (\mu_{H(A)} \otimes \mu_{H(B)})S_H, \tag{2.45}$$

where S_H denotes the flip operator for the pair $H(A)$, $H(B)$ (see Section 2.2). It has to be shown that

$$\mu_{H(A \otimes B)} = \mu_{H(A) \otimes H(B)}. \tag{2.46}$$

To simplify notation we set

$$\mu_{H(A)} = \sigma, \qquad \mu_{H(B)} = \tau, \qquad \mu_{H(A \otimes B)} = \rho.$$

Then we obtain from (2.43) that

$$\begin{aligned}(\sigma \otimes \tau)(\pi_A \otimes \pi_A \otimes \pi_B \otimes \pi_B) &= \sigma(\pi_A \otimes \pi_A) \otimes \tau(\pi_B \otimes \pi_B) \\ &= \pi_A \mu_{Z(A)} \otimes \pi_B \mu_{Z(B)} \\ &= (\pi_A \otimes \pi_B)(\mu_{Z(A)} \otimes \mu_{Z(B)}).\end{aligned}$$

Next we observe that

$$S_H(\pi_A \otimes \pi_B \otimes \pi_A \otimes \pi_B) = (\pi_A \otimes \pi_A \otimes \pi_B \otimes \pi_B)S_Z, \tag{2.47}$$

where S_Z is the flip operator for $Z(A)$ and $Z(B)$. The preceding two equations yield

$$(\sigma \otimes \tau)S_H(\pi_A \otimes \pi_B \otimes \pi_A \otimes \pi_B) = (\pi_A \otimes \pi_B)(\mu_{Z(A)} \otimes \mu_{Z(B)})S_Z. \tag{2.48}$$

But

$$(\mu_{Z(A)} \otimes \mu_{Z(B)})S_Z = \mu_{Z(A) \otimes Z(B)}$$

and so (2.48) implies that

$$(\sigma \otimes \tau)S_H(\pi_A \otimes \pi_B \otimes \pi_A \otimes \pi_B) = (\pi_A \otimes \pi_B)\mu_{Z(A) \otimes Z(B)}. \tag{2.49}$$

On the other hand it follows from (2.44) that

$$\rho(\pi_{A \otimes B} \otimes \pi_{A \otimes B}) = \pi_{A \otimes B}\mu_{Z(A \otimes B)}.$$

Restricting this relation to the subspace $Z(A) \otimes Z(B) \otimes Z(A) \otimes Z(B)$, and observing that the restriction of $\pi_{A \otimes B}$ to $Z(A) \otimes Z(B)$ is $\pi_A \otimes \pi_B$ (see (2.23)), we obtain

$$\rho(\pi_A \otimes \pi_B \otimes \pi_A \otimes \pi_B) = (\pi_A \otimes \pi_B)\mu_{Z(A) \otimes Z(B)}. \tag{2.50}$$

Combining (2.49) and (2.50), we find that

$$((\sigma \otimes \tau)S_H - \rho)(\pi_A \otimes \pi_B \otimes \pi_A \otimes \pi_B) = 0. \tag{2.51}$$

But

$$\begin{aligned}\pi_A \otimes \pi_B \otimes \pi_A \otimes \pi_B : Z(A) \otimes Z(B) \otimes Z(A) \otimes Z(B)& \\ \to H(A) \otimes H(B) \otimes H(A) \otimes H(B)&\end{aligned}$$

is a surjective mapping, and so it follows from (2.51) that

$$(\sigma \otimes \tau)S_H = \rho,$$

i.e.,

$$(\mu_{H(A)} \otimes \mu_{H(B)})S_H = \mu_{H(A \otimes B)}. \tag{2.52}$$

Relations (2.45) and (2.52) yield (2.46).

2.17. Graded Differential Algebras

Let $A = (\sum_p A_p, \partial_A)$ and $B = (\sum_q B_q, \partial_B)$ be two graded differential algebras and assume that ∂_A and ∂_B are both antiderivations of odd degree k. Consider the anticommutative tensor product $A \hat{\otimes} B$. Then the structure map of the algebra $A \hat{\otimes} B$ is given by

$$\mu_{A \hat{\otimes} B} = (\mu_A \otimes \mu_B) \circ Q,$$

where Q denotes the anticommutative flip operator for A and B. It follows from Section 2.8 that $A \hat{\otimes} B$ is a graded anticommutative algebra and, since k is odd, $\partial_{A \hat{\otimes} B}$ is an antiderivation of degree k. Hence $(A \hat{\otimes} B, \partial_{A \hat{\otimes} B})$ is a graded differential algebra. It will be shown that $H(A \hat{\otimes} B)$ is the anticommutative tensor product of the graded algebras $H(A)$ and $H(B)$.

Let Q_H and Q_Z be the anticommutative flip operators for the pairs $H(A)$, $H(B)$ and $Z(A)$, $Z(B)$. Then we have that

$$Q_H \circ (\pi_A \otimes \pi_B \otimes \pi_A \otimes \pi_B) = (\pi_A \otimes \pi_A \otimes \pi_B \otimes \pi_B) \circ Q_Z,$$

where $\pi_A : Z(A) \to H(A)$ and $\pi_B : Z(B) \to H(B)$ denote the canonical projections. With this formula the proof coincides with the proof for the analogous result in Section 2.16.

3 Tensor Algebra

In this chapter except where noted otherwise all vector spaces will be defined over an arbitrary field Γ.

Tensors

3.1.

Definition. Let E be a vector space and consider for each $p \geq 2$ the pair $(\otimes^p E, \otimes^p)$ where

$$\otimes^p E = \underbrace{E \otimes \cdots \otimes E}_{p}.$$

We extend the definition of $\otimes^p E$ to the case $p = 1$ and $p = 0$ by setting $\otimes^1 E = E$ and $\otimes^0 E = \Gamma$. The pair $(\otimes^p E, \otimes^p)$ is called the pth *tensorial power of* E. The space $\otimes^p E$ is also called the pth tensorial power of E and its elements are called *tensors of degree* p.

A tensor of the form $x_1 \otimes \cdots \otimes x_p$, $p \geqq 1$, and tensors of degree zero are called *decomposable*.

For every pair (p, q) there is a unique bilinear mapping

$$\beta : \otimes^p E \times \otimes^q E \to \otimes^{p+q} E$$

such that

$$\beta(x_1 \otimes \cdots \otimes x_p, x_{p+1} \otimes \cdots \otimes x_{p+q}) = x_1 \otimes \cdots \otimes x_{p+q} \qquad x_i \in E. \quad (3.1)$$

Moreover, the pair $(\otimes^{p+q} E, \beta)$ is the tensor product of $\otimes^p E$ and $\otimes^q E$ (see Section 1.20). Hence we may write $u \otimes v$ instead of $\beta(u, v)$ for $u \in \otimes^p E$. Then (3.1) reads

$$(x_1 \otimes \cdots \otimes x_p) \otimes (x_{p+1} \otimes \cdots \otimes x_{p+q}) = x_1 \otimes \cdots \otimes x_{p+q}. \quad (3.2)$$

The tensor $u \otimes v$ is called the product of the tensors u and v. The product (3.2) is associative, as follows from the definitions.

However, it is not commutative except for the case $\dim E = 1$. (In fact, if $x \in E$ and $y \in E$ are linearly independent vectors, then the products $x \otimes y$ and $y \otimes x$ are also linearly independent and hence $x \otimes y \neq y \otimes x$). Finally, we notice that the product $\lambda \otimes z$ ($\lambda \in \otimes^0 E = \Gamma$, $z \in \otimes^p E$) is the vector in $\otimes^p E$ obtained by multiplying the vector z by the scalar λ. In particular, the scalar 1 acts as identity.

Let $\{e_\nu\}_{\nu \in I}$ be a basis of E. Then the products $e_{\nu_1} \otimes \cdots \otimes e_{\nu_p}$, ($\nu_i \in I$) form a basis of $\otimes^p E$. (see Section 1.20). In particular, if E has finite dimension and e_ν ($\nu = 1, \ldots, n$) is a basis of E then the products $e_{\nu_1} \otimes \cdots \otimes e_{\nu_p}$ ($\nu_i = 1, \ldots, n$) form a basis of $\otimes^p E$ and $\dim \otimes^p E = n^p$ ($n = \dim E$). Every tensor $z \in \otimes^p E$ can be uniquely written as a sum

$$z = \sum_{(\nu)} \zeta^{\nu_1, \ldots, \nu_p} e_{\nu_1} \otimes \cdots \otimes e_{\nu_p}.$$

The coefficients $\zeta^{\nu_1, \ldots, \nu_p}$ are called *components* of z with respect to the basis e_ν.

3.2. Tensor Algebra

Suppose that $(\tilde{\otimes}^p E, \tilde{\otimes}^p)$ is a pth tensorial power of E ($p = 0, 1, \ldots$) and consider the direct sum

$$\otimes E = \bigoplus_{p=0}^{\infty} \tilde{\otimes}^p E.$$

The elements of $\otimes E$ are the sequences

$$(z_0, z_1, \ldots) \qquad z_p \in \tilde{\otimes}^p E \qquad (p = 0, 1, \ldots)$$

such that only finitely many z_p are different from zero in each sequence. If $i_p\colon \tilde{\otimes}^p E \to \otimes E$ denotes the canonical injection we can write

$$\otimes E = \sum_{p=0}^{\infty} i_p \tilde{\otimes}^p E.$$

Since the pair $(i_p \tilde{\otimes}^p E, i_p \tilde{\otimes}^p)$ is again a pth tensorial power of E we denote the p-linear mapping $i_p \tilde{\otimes}^p$ by $\otimes^p$ and the vector space $i_p \tilde{\otimes}^p E$ by $\otimes^p E$. Then the above equation reads

$$\otimes E = \sum_{p=0}^{\infty} \otimes^p E.$$

By assigning the degree p to the elements of $\otimes^p E$ we obtain a positive gradation in the space $\otimes E$.

We now define a bilinear mapping

$$(u, v) \to uv \qquad u, v, uv \in \otimes E$$

by

$$uv = \sum_{p,q} u_p \otimes v_q \qquad u = \sum_p u_p, v = \sum_q v_q.$$

This multiplication makes $\otimes E$ into an associative (but noncommutative, if $\dim E \geqq 2$) algebra in which the sequence $(1, 0, \ldots)$ acts as a unit element. It is clear from the definition that $\otimes E$ is a positively graded algebra. $\otimes E$ is called a *tensor algebra* over the vector space E. From now on we shall identify $\otimes^0 E$ with Γ and $\otimes^1 E$ with E. Then Γ and E are subspaces of $\otimes E$, and the elements of E, together with the scalar 1 generate (in the algebraic sense) the algebra $\otimes E$.

If E has finite dimension the gradation of $\otimes E$ is almost finite and the Poincaré series of the graded space $\otimes E$ is given by

$$P(t) = \sum_{p=0}^{\infty} n^p t^p = \frac{1}{1 - nt}, \qquad n = \dim E.$$

Remark: The reader should observe that if $E \neq 0$ the bilinear mapping $\beta:(\otimes E) \times (\otimes E) \to \otimes E$ which is defined by the multiplication is *not* a tensor product. In fact, if β_{pq} denotes the restriction of β to $(\otimes^p E) \times (\otimes^q E)$, we have

$$\operatorname{Im} \beta_{pq} = \operatorname{Im} \beta_{qp} = \otimes^{p+q} E.$$

Set $E_1 = \otimes^p E$ and $F_1 = \otimes^q E$, $p \neq q$. Then

$$E_1 \cap F_1 = 0,$$

while $\beta(E_1 \times F_1) = \beta(F_1 \times E_1)$. Hence if β were a tensor product it would follow that $E_1 = 0$ or $F_1 = 0$, whence $E = 0$.

3.3. The Universal Property of $\otimes E$

Consider an arbitrary associative algebra A, with unit element e, and a linear map $\eta: E \to A$. Then there exists precisely one homomorphism $h: \otimes E \to A$ such that $h(1) = e$ and $h \circ i = \eta$; i.e., such that the diagram

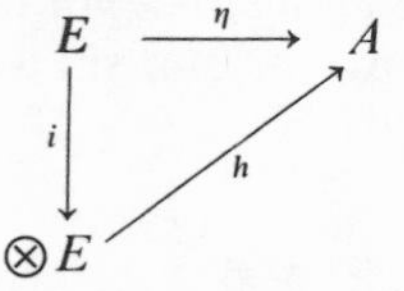

is commutative, where i denotes the injection of E into $\otimes E$. To define h consider the p-linear mapping

$$E \times \cdots \times E \to A$$

given by

$$(x_1, \ldots, x_p) \to \eta x_1 \cdots \eta x_p.$$

In view of $\otimes_2$ there exists a linear map $h_p : \otimes^p E \to A$ such that

$$h_p(x_1 \otimes \cdots \otimes x_p) = \eta x_1 \cdots \eta x_p.$$

We extend the definition of h_p to the cases $p = 1$ and $p = 0$ by setting

$$h_1 = \eta \quad \text{and} \quad h_0(\lambda) = \lambda e \quad \text{for all } \lambda \in \Gamma.$$

Then a homomorphism $h : \otimes E \to A$ is given by

$$hu = \sum_p h_p u_p \qquad u_p \in \otimes^p E, u = \sum_p u_p.$$

In fact, if $u, v \in \otimes E$ are decomposable, then it is clear that

$$h(uv) = hu \cdot hv.$$

Since every element in $\otimes E$ is the sum of decomposable tensors, and h is linear, it follows that h preserves products.

To show that h is uniquely determined by η, we notice that the conditions $h \circ i = \eta$ and $h(1) = e$ determine h in $\otimes^1 E = E$ and in $\otimes^0 E = \Gamma$. But $\otimes E$ is generated by 1 and the vectors of E; consequently, h is uniquely determined in $\otimes E$.

3.4. Universal Pairs

Now let U be an associative algebra with unit element 1 and $\varepsilon : E \to U$ be a linear map. We shall say that the pair (ε, U) has the *universal tensor algebra property with respect to E* if the following conditions are satisfied:

T₁: The space $\operatorname{Im} \varepsilon$ together with the unit element 1 generates U (in the algebraic sense),

T₂: If η is a linear map of E into an associative algebra A with unit element e then there is a homomorphism $h : U \to A$ such that $h(1) = e$ and the diagram

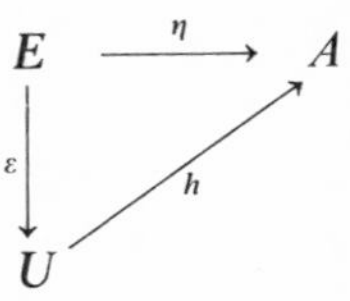

(3.3)

is commutative.

The properties $\mathbf{T}_1$ and $\mathbf{T}_2$ are equivalent to the property:

T: If η is a linear mapping of E into an associative algebra A with unit element, then there exists a *unique* homomorphism $h : U \to A$ such that $h(1) = e$ and Diagram (3.3) is commutative.

It is clear that $\mathbf{T}_1$ and $\mathbf{T}_2$ imply $\mathbf{T}$. Conversely, assume that the pair (ε, U) satisfies the condition $\mathbf{T}$. Then $\mathbf{T}_2$ follows immediately. To prove $\mathbf{T}_1$ consider the subalgebra V of U generated by Im ε and the unit element. Then ε can be considered as a linear map of E into V (which, to avoid confusion, we denote by ε_V) and hence there is a homomorphism $h: U \to V$ such that $h(1) = 1$ and $h \circ \varepsilon = \varepsilon_V$. If $j: V \to U$ denotes the inclusion map we have $\varepsilon = j \circ \varepsilon_V$ and hence it follows that $\varepsilon = (j \circ h) \circ \varepsilon$. In diagram form we have

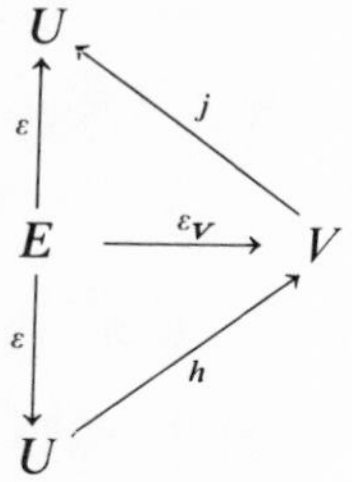

and hence the diagram

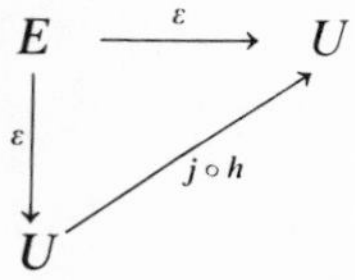

is commutative. On the other hand, we have the commutative diagram

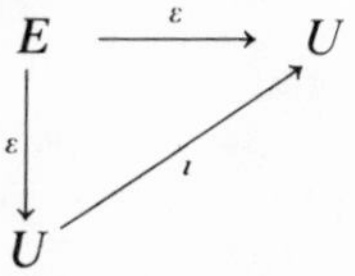

where ι is the identity map of U. Since $j \circ h$ is an endomorphism of the algebra U, the uniqueness part of $\mathbf{T}$ implies that

$$j \circ h = \iota.$$

Consequently, j is an onto map, and hence $U = V$. This proves $\mathbf{T}_1$.

We shall now prove the following

Uniqueness theorem. *Let (ε, U) and (ε', U') be two universal pairs for E. Then there exists precisely one isomorphism $f: U \to U'$ such that*

$$f \circ \varepsilon = \varepsilon'.$$

Proof. In view of $\mathbf{T}$ there exist unique homomorphisms $f: U \to U'$ and $g: U' \to U$ such that

$$f \circ \varepsilon = \varepsilon' \quad \text{and} \quad g \circ \varepsilon' = \varepsilon.$$

Hence, $g \circ f$ is an endomorphism of U which reduces to the identity in Im ε.

Since the space Im ε generates U, it follows that $g \circ f = \iota$. In the same way it is shown that $f \circ g = \iota'$, the identity map of U'. Hence, f is an isomorphism of U onto U' and $g = f^{-1}$. The uniqueness theorem is thereby proved. □

Since $(i, \otimes E)$ is a universal pair for E it follows from the above uniqueness theorem that *for every pair* (ε, U) *there is precisely one isomorphism* $f: \otimes E \to U$ *such that* $f \circ i = \varepsilon$.

Since $\otimes E$ is a graded algebra, a gradation is induced in U by the isomorphism f. The algebra U furnished with this gradation is a graded algebra and $f: \otimes E \to U$ is a homogeneous isomorphism of degree zero.

In view of the uniqueness theorem, the universal algebra U is usually called the tensor algebra over E and is denoted by $\otimes E$.

3.5. The Induced Homomorphism

Let $\varphi: E \to F$ be a linear map from the vector space E to a second vector space F. Then φ extends in a unique way to a homomorphism

$$\varphi_\otimes : \otimes E \to \otimes F$$

such that $\varphi_\otimes(1) = 1$. In fact, consider the linear map $\eta: E \to \otimes F$ given by $\eta = j \circ \varphi$ (where j denotes the inclusion map) and apply the result of Section 3.3.

Clearly, the homomorphism $\varphi_\otimes$ is homogeneous of degree zero. It follows from the definition of $\varphi_\otimes$ that

$$\varphi_\otimes(x_1 \otimes \cdots \otimes x_p) = \varphi x_1 \otimes \cdots \otimes \varphi x_p \qquad x_i \in E.$$

Now let G be a third vector space, $\otimes G$ be a tensor algebra over G and $\psi: F \to G$ be a linear map. Then it is clear from the definitions that

$$(\psi \circ \varphi)_\otimes = \psi_\otimes \circ \varphi_\otimes. \tag{3.4}$$

If $F = E$ and ι is the identity map then $\iota_\otimes$ is the identity map of $\otimes E$,

$$\iota_\otimes = \iota. \tag{3.5}$$

It follows from (3.4) and (3.5) that $\varphi_\otimes$ is injective (surjective) whenever φ is injective (surjective). In fact, if φ is injective there exists a linear map $\psi: E \leftarrow F$ such that $\psi \circ \varphi = \iota$. Formulas (3.4) and (3.5) now imply that

$$\psi_\otimes \circ \varphi_\otimes = \iota_\otimes = \iota$$

and hence $\varphi_\otimes$ is injective.

It is easy to see that

$$\operatorname{Im} \varphi_\otimes = \otimes(\operatorname{Im} \varphi).$$

Hence $\varphi_\otimes$ is surjective whenever φ is.

3.6. The Derivation Induced by a Linear Transformation

Let φ be a linear transformation of E. Then φ can be extended in a unique way to a derivation $\theta_{\otimes}(\varphi)$ in the algebra $\otimes E$. To construct $\theta_{\otimes}(\varphi)$, we notice that for each $p \geq 2$ a p-linear mapping

$$\underbrace{E \times \cdots \times E}_{p} \to \otimes^p E$$

is defined by

$$(x_1, \ldots, x_p) \to \sum_{i=1}^{p} x_1 \otimes \cdots \otimes \varphi x_i \otimes \cdots \otimes x_p.$$

This p-linear mapping induces a linear map $\theta_{\textcircled{p}}(\varphi)\colon \otimes^p E \to \otimes^p E$ such that

$$\theta_{\textcircled{p}}(\varphi)(x_1 \otimes \cdots \otimes x_p) = \sum_{i=1}^{p} x_1 \otimes \cdots \otimes \varphi x_i \otimes \cdots \otimes x_p.$$

We extend this definition to the case $p = 1$ and $p = 0$ by setting

$$\theta_{\textcircled{1}}(\varphi) = \varphi \quad \text{and} \quad \theta_{\textcircled{0}}(\varphi) = 0$$

and define $\theta_{\otimes}(\varphi)$ to be the linear transformation of $\otimes E$ into itself which extends the $\theta_{\textcircled{p}}(\varphi)$.

It remains to be shown that $\theta_{\otimes}(\varphi)$ is a derivation, i.e., that

$$\theta_{\otimes}(\varphi)(u \cdot v) = \theta_{\otimes}(\varphi)u \cdot v + u \cdot \theta_{\otimes}(\varphi)v. \tag{3.6}$$

For the proof, we may assume (as before) that u and v are decomposable so that

$$u = x_1 \otimes \cdots \otimes x_p \quad \text{and} \quad v = x_{p+1} \otimes \cdots \otimes x_{p+q} \qquad x_i \in E.$$

Let us first assume that $p \geq 1$ and $q \geq 1$. We then obtain

$$\begin{aligned} \theta_{\otimes}(\varphi)(u \cdot v) &= \theta_{\otimes}(\varphi)(x_1 \otimes \cdots \otimes x_{p+q}) \\ &= \sum_{i=1}^{p+q} x_1 \otimes \cdots \otimes \varphi x_i \otimes \cdots \otimes x_{p+q} \\ &= \sum_{i=1}^{p} x_1 \otimes \cdots \otimes \varphi x_i \otimes \cdots \otimes x_{p+q} \\ &\quad + \sum_{i=p+1}^{p+q} x_1 \otimes \cdots \otimes x_p \otimes \cdots \otimes \varphi x_i \otimes \cdots \otimes x_{p+q} \\ &= \theta_{\otimes}(\varphi)u \cdot v + u \cdot \theta_{\otimes}(\varphi)v. \end{aligned}$$

Formula (3.6) remains true if $p = 0$ or $q = 0$. In the case $p = 0$, for instance, we have

$$\theta_{\otimes}(\varphi)(\lambda v) = \lambda \theta_{\otimes}(\varphi)v = \theta_{\otimes}(\varphi)\lambda \cdot v + \lambda \cdot \theta_{\otimes}(\varphi)v.$$

Since the algebra $\otimes E$ is generated by the elements of E and the unit element, it follows that the extension of φ into a derivation in the algebra $\otimes E$ is unique (see Section 5.6 of *Linear Algebra*).

Let ψ be a second linear transformation of E. Then

$$\theta_{\otimes}(\lambda\varphi + \mu\psi) = \lambda\theta_{\otimes}(\varphi) + \mu\theta_{\otimes}(\psi) \qquad \lambda, \mu \in \Gamma \tag{3.7}$$

and

$$\theta_{\otimes}(\varphi \circ \psi - \psi \circ \varphi) = \theta_{\otimes}(\varphi) \circ \theta_{\otimes}(\psi) - \theta_{\otimes}(\psi) \circ \theta_{\otimes}(\varphi). \tag{3.8}$$

Formula (3.7) follows immediately from the definition of $\theta_{\otimes}$. To prove (3.8) we remark first that the operator on each side is a derivation (see Section 5.6 of *Linear Algebra*). Hence it is sufficient to show that

$$\theta_{①}(\varphi \circ \psi - \psi \circ \varphi) = \theta_{①}(\varphi) \circ \theta_{①}(\psi) - \theta_{①}(\psi) \circ \theta_{①}(\varphi).$$

But this is immediately clear since

$$\theta_{①}(\varphi) = \varphi \quad \text{and} \quad \theta_{①}(\psi) = \psi.$$

3.7. Tensor Algebra Over a G-Graded Vector Space

Suppose $E = \sum_\alpha E_\alpha$ is a G-graded vector space and let $\otimes E = \sum_p \otimes^p E$ be a tensor algebra over E. Then a G-gradation is induced in each $\otimes^p E$ $(p \geq 1)$ by

$$\otimes^p E = \sum_\alpha (\otimes^p E)_\alpha \quad \text{where } (\otimes^p E)_\alpha = \sum_{\alpha_1 + \cdots + \alpha_p = \alpha} E_{\alpha_1} \otimes \cdots \otimes E_{\alpha_p}.$$

It will be convenient to extend this gradation to $\otimes^0 E = \Gamma$ by assigning the degree 0 to the elements of Γ. The induced G-gradation in the direct sum $\otimes E$ is given by $\otimes E = \sum_\alpha (\otimes E)_\alpha$ where $(\otimes E)_\alpha = \sum_p (\otimes^p E)_\alpha$. Then $\otimes E$ becomes a G-graded algebra. In fact, let

$$u = x_{\alpha_1} \otimes \cdots \otimes x_{\alpha_p} \in (\otimes E)_\alpha \quad \text{and} \quad v = y_{\beta_1} \otimes \cdots \otimes y_{\beta_q} \in (\otimes E)_\beta$$

be two decomposable elements. Then

$$uv = x_{\alpha_1} \otimes \cdots \otimes x_{\alpha_p} \otimes y_{\beta_1} \otimes \cdots \otimes y_{\beta_q} \in (\otimes E)_{\alpha+\beta}$$

and thus we have

$$\deg(uv) = \deg u + \deg v.$$

The algebra $\otimes E$ together with this gradation is called the *G-graded tensor algebra over E*. If $G = \mathbb{Z}$ and all the vectors of E are of degree one then the induced gradation of $\otimes E$ coincides with the gradation defined in Section 3.2.

Problems

1. Let $u_1 = a_1 \otimes b_1$ and $u_2 = a_2 \otimes b_2$ be two decomposable tensors and assume that $u_1 \neq 0$. Prove that $u_1 + u_2$ is decomposable if and only if $a_2 = \lambda a_1$ or $b_2 = \mu b_1$ ($\lambda, \mu \in \Gamma$).

2. Assume that $a_1 \otimes \cdots \otimes a_p \neq 0$. Prove that

$$a_1 \otimes \cdots \otimes a_p = b_1 \otimes \cdots \otimes b_p$$

if and only if

$$b_i = \lambda_i a_i \qquad i = 1, \ldots, p, \lambda_1, \ldots, \lambda_p = 1 \ (\lambda_i \in \Gamma).$$

3. Use Formula (3.8) to prove that

$$\operatorname{tr}(\varphi \otimes \psi) = \operatorname{tr} \varphi \cdot \operatorname{tr} \psi.$$

Tensors Over a Pair of Dual Spaces

3.8.

Definition. Suppose that E and E^* are vector spaces, dual with respect to a scalar product $\langle , \rangle$, and let $\otimes E = \sum \otimes^p E$ and $\otimes E^* = \sum \otimes^p E^*$ be tensor algebras over E and E^*, respectively. According to Section 1.22 there is induced between $\otimes^p E$ and $\otimes^p E^*$ for each $p \geq 1$ a unique scalar product such that

$$\langle x^{*1} \otimes \cdots \otimes x^{*p}, x_1 \otimes \cdots \otimes x_p \rangle = \langle x^{*1}, x_1 \rangle \cdots \langle x^{*p}, x_p \rangle. \tag{3.9}$$

We extend the definition of $\langle , \rangle$ to the case $p = 0$ by setting

$$\langle \lambda, \mu \rangle = \lambda\mu \qquad \lambda, \mu \in \otimes^0 E = \Gamma.$$

The scalar products between $\otimes^p E$ and $\otimes^p E^*$ can be extended in a unique way to a gradation-respecting scalar product $\langle , \rangle$ between the spaces $\otimes E$ and $\otimes E^*$ (see Section 6.5 of *Linear Algebra*) and this scalar product is given by

$$\langle u^*, v \rangle = \sum_p \langle u^{*p}, v_p \rangle, \qquad u^* = \sum_p u^{*p}, v = \sum_p v_p.$$

Now assume that E has finite dimension and let e_ν, $e^{*\nu}$ be a pair of dual bases for E and E^*. Then the scalar product between the induced basis vectors $e_{\nu_1} \otimes \cdots \otimes e_{\nu_p}$ and $e^{*\mu_1} \otimes \cdots \otimes e^{*\mu_p}$ is given by

$$\langle e^{*\mu_1} \otimes \cdots \otimes e^{*\mu_p}, e_{\nu_1} \otimes \cdots \otimes e_{\nu_p} \rangle = \delta^{\mu_1}_{\nu_1} \cdots \delta^{\mu_p}_{\nu_p}. \tag{3.10}$$

This formula shows that the bases $e_{\nu_1} \otimes \cdots \otimes e_{\nu_p}$ and $e^{*\mu_1} \otimes \cdots \otimes e^{*\mu_p}$ are again dual. It follows from (3.10) that the scalar product of two tensors

$$u = \sum_{(\nu)} \xi^{\nu_1, \ldots, \nu_p} e_{\nu_1} \otimes \cdots \otimes e_{\nu_p} \quad \text{and} \quad u^* = \sum_{(\mu)} \eta_{\mu_1, \ldots, \mu_p} e^{*\mu_1} \otimes \cdots \otimes e^{*\mu_p}$$

is given by

$$\langle u^*, u\rangle = \sum_{(\nu)} \xi^{\nu_1, \dots, \nu_p}\eta_{\nu_1, \dots, \nu_p}.$$

3.9. The Induced Homomorphism

Suppose that F, F^* is a second pair of spaces dual with respect to a scalar product (again denoted by $\langle , \rangle$). Let $\varphi: E \to F$, $\varphi^*: E^* \leftarrow F^*$ be a dual pair of linear maps. The homomorphism $(\varphi^*)_\otimes : \otimes E^* \leftarrow \otimes F^*$ induced by φ^* will be denoted by $\varphi^\otimes$. Then we have

$$\varphi^\otimes(y^{*1} \otimes \cdots \otimes y^{*p}) = \varphi^* y^{*1} \otimes \cdots \otimes \varphi^* y^{*p} \qquad y^{*i} \in F^*. \tag{3.11}$$

Now it will be shown that the homomorphisms

$$\varphi_\otimes : \otimes E \to \otimes F \quad \text{and} \quad \varphi^\otimes : \otimes E^* \leftarrow \otimes F^*$$

form a dual pair,

$$(\varphi^*)_\otimes = (\varphi_\otimes)^*. \tag{3.12}$$

Let $u \in \otimes E$ and $v^* \in \otimes F^*$ be arbitrary elements. It has to be shown that

$$\langle v^*, \varphi_\otimes u\rangle = \langle \varphi^\otimes v^*, u\rangle.$$

Since $\varphi_\otimes$ and $\varphi^\otimes$ are both homogeneous of degree zero we may assume that u and v^* are both homogeneous of the same degree p. Moreover, we may assume that u and v^* are decomposable, $u = x_1 \otimes \cdots \otimes x_p$ and $v^* = y^{*1} \otimes \cdots \otimes y^{*p}$. Then Formulas (3.11), (3.8) and (3.9) yield

$$\begin{aligned}
\langle \varphi^\otimes v^*, u\rangle &= \langle \varphi^\otimes(y^{*1} \otimes \cdots \otimes y^{*p}), x_1 \otimes \cdots \otimes x_p\rangle \\
&= \langle \varphi^* y^{*1} \otimes \cdots \otimes \varphi^* y^{*p}, x_1 \otimes \cdots \otimes x_p\rangle \\
&= \langle \varphi^* y^{*1}, x_1\rangle \cdots \langle \varphi^* y^{*p}, x_p\rangle = \langle y^{*1}, \varphi x_1\rangle \cdots \langle y^{*p}, \varphi x_p\rangle \\
&= \langle y^{*1} \otimes \cdots \otimes y^{*p}, \varphi x_1 \otimes \cdots \otimes \varphi x_p\rangle = \langle v^*, \varphi_\otimes u\rangle
\end{aligned}$$

whence

$$\langle v^*, \varphi_\otimes u\rangle = \langle \varphi^\otimes v^*, u\rangle.$$

If G, G^* is a third pair of dual spaces and

$$\psi: F \to G, \qquad \psi^*: F^* \leftarrow G^*$$

is a second pair of dual mappings, we have in view of (3.12) and (3.4) that

$$(\psi \circ \varphi)^\otimes = [(\psi \circ \varphi)^*]_\otimes = (\varphi^* \circ \psi^*)_\otimes = (\varphi^*)_\otimes \circ (\psi^*)_\otimes = \varphi^\otimes \circ \psi^\otimes$$

whence

$$(\psi \circ \varphi)^\otimes = \varphi^\otimes \circ \psi^\otimes.$$

3.10. The Induced Derivation

Consider again a pair of dual mappings, $\varphi:E \to E$, $\varphi^*:E^* \leftarrow E^*$. The derivation in the algebra $\otimes E^*$ which is induced by φ^* will be denoted by $\theta^{\otimes}(\varphi)$, $\theta^{\otimes}(\varphi) = \theta_{\otimes}(\varphi^*)$. Then

$$\theta^{\otimes}(\varphi)(x^{*1} \otimes \cdots \otimes x^{*p}) = \sum_{i=1}^{p} x^{*1} \otimes \cdots \varphi^* x^{*i} \cdots \otimes x^{*p}. \tag{3.13}$$

The derivations

$$\theta_{\otimes}(\varphi): \otimes E \to \otimes E$$

and

$$\theta^{\otimes}(\varphi): \otimes E^* \leftarrow \otimes E^*$$

again form a dual pair,

$$\theta_{\otimes}(\varphi^*) = [\theta_{\otimes}(\varphi)]^*. \tag{3.14}$$

As in Section 3.9 we have to prove that

$$\langle \theta^{\otimes}(\varphi)v^*, u \rangle = \langle v^*, \theta_{\otimes}(\varphi)u \rangle \qquad v^* \in \otimes E^*, u \in \otimes E$$

and we may assume that v^* and u are of the form

$$v^* = x^{*1} \otimes \cdots \otimes x^{*p} \qquad u = x_1 \otimes \cdots \otimes x_p.$$

Then we obtain from (3.13) that

$$\begin{aligned}
\langle \theta^{\otimes}(\varphi)v^*, u \rangle &= \left\langle \sum_{i=1}^{p} x^{*1} \otimes \cdots \varphi^* x^{*i} \cdots \otimes x^{*p}, x_1 \otimes \cdots \otimes x_p \right\rangle \\
&= \sum_{i=1}^{p} \langle x^{*1}, x_1 \rangle \cdots \langle \varphi^* x^{*i}, x_i \rangle \cdots \langle x^{*p}, x_p \rangle \\
&= \sum_{i=1}^{p} \langle x^{*1}, x_1 \rangle \cdots \langle x^{*i}, \varphi x_i \rangle \cdots \langle x^{*p}, x_p \rangle \\
&= \left\langle x^{*1} \otimes \cdots \otimes x^{*p}, \sum_{i=1}^{p} x_1 \otimes \cdots \varphi x_i \cdots \otimes x_p \right\rangle \\
&= \langle v^*, \theta_{\otimes}(\varphi)u \rangle
\end{aligned}$$

whence

$$\langle \theta^{\otimes}(\varphi)v^*, u \rangle = \langle v^*, \theta_{\otimes}(\varphi)u \rangle.$$

If $\psi:E \to E$, $\psi^*:E^* \leftarrow E^*$ is a second dual pair of linear maps, then it follows from (3.14) and (3.8) that

$$\begin{aligned}
\theta^{\otimes}(\varphi\psi - \psi\varphi) &= \theta_{\otimes}(\psi^*\varphi^* - \varphi^*\psi^*) = \theta_{\otimes}(\psi^*)\theta_{\otimes}(\varphi^*) - \theta_{\otimes}(\varphi^*)\theta_{\otimes}(\psi^*) \\
&= \theta^{\otimes}(\psi)\theta^{\otimes}(\varphi) - \theta^{\otimes}(\varphi)\theta^{\otimes}(\psi)
\end{aligned}$$

whence

$$\theta^{\otimes}(\varphi\psi - \psi\varphi) = \theta^{\otimes}(\psi)\theta^{\otimes}(\varphi) - \theta^{\otimes}(\varphi)\theta^{\otimes}(\psi).$$

PROBLEM

Let E, E^* be a pair of dual n-dimensional vector spaces, and consider a tensor $u \in \otimes^p E$. Define a subspace $E_u \subset E^*$ as follows: a vector $x^* \in E^*$ is to be contained in E_u if and only if

$$\langle u, x^* \otimes v^* \rangle = 0 \quad \text{for every} \quad v^* \in \otimes^{p-1} E^*.$$

Show that u is decomposable if and only if $\dim E_u = p - 1$.

Mixed Tensors

3.11.

Definition. Let E^*, E be a pair of dual vector spaces and consider, for every pair (p, q), $p \geq 1$, $q \geq 1$, the space

$$\otimes^p_q(E^*, E) = (\otimes^p E^*) \otimes (\otimes^q E).$$

Extend the definition of $\otimes^p_q(E^*, E)$ to the cases $q = 0$ and $p = 0$ by setting

$$\otimes^p_0(E^*, E) = \otimes^p E^* \quad \text{and} \quad \otimes^0_q(E^*, E) = \otimes^q E.$$

The elements of $\otimes^p_q(E^*, E)$ are called *mixed tensors* over the pair (E^*, E) and are said to be *homogeneous of bidegree* (p, q). The number $p + q$ is called the *total degree.*

A tensor of the form

$$\omega = x_1^* \otimes \cdots \otimes x_p^* \otimes x_1 \otimes \cdots \otimes x_q \qquad x_i^* \in E^*, x_j \in E,$$

is called *decomposable.*

The scalar product between E^* and E induces a scalar product between $\otimes^p_q(E^*, E)$ and $\otimes^q_p(E^*, E)$ determined by

$$\langle u^* \otimes v, v^* \otimes u \rangle = \langle u^*, u \rangle \langle v^*, v \rangle \tag{3.15}$$

(see Section 1.22). Thus any two spaces $\otimes^p_q(E^*, E)$ and $\otimes^q_p(E^*, E)$ are dual. In particular, every space $\otimes^p_p(E^*, E)$ is self-dual. Finally note that

$$\langle z_1, z_2 \rangle = \langle z_2, z_1 \rangle \qquad z_1 \in \otimes^p_q(E^*, E), z_2 \in \otimes^q_p(E^*, E)$$

as follows from the definition.

3.12. The Mixed Tensor Algebra

The *mixed tensor algebra* over the pair E^*, E is defined to be the canonical tensor product of the algebras $\otimes E^*$ and $\otimes E$ (see Section 2.2). It will be denoted by $\otimes(E^*, E)$,

$$\otimes(E^*, E) = (\otimes E^*) \otimes (\otimes E).$$

Thus $\otimes(E^*, E)$ is an associative (noncommutative) algebra with $1 \otimes 1$ as unit element. It is (algebraically) generated by the elements $1 \otimes 1$, $x^* \otimes 1$, and $1 \otimes x$ with $x^* \in E^*$ and $x \in E$.

Now let

$$i^p : \otimes^p E^* \to \otimes E^*, \qquad i_q : \otimes^q E \to \otimes E$$

and

$$i_q^p : (\otimes^p E^*) \otimes (\otimes^q E) \to \otimes(E^*, E)$$

be the inclusion maps and identify the spaces $\otimes^p E^*$, $\otimes^q E$, and $\otimes_q^p(E^*, E)$ with their images under these maps. Then we have the direct decomposition

$$\otimes(E^*, E) = \sum_{p,q} (\otimes^p E^*) \otimes (\otimes^q E).$$

3.13. Contraction

Assume that $p \geq 1$ and $q \geq 1$. Fix a pair (i, j) with $1 \leq i \leq p$ and $1 \leq j \leq q$ and consider the $(p + q)$-linear mapping

$$\Phi_j^i : \underbrace{E^* \times \cdots \times E^*}_{p} \times \underbrace{E \times \cdots \times E}_{q} \to \otimes_{q-1}^{p-1}(E^*, E)$$

given by

$$\Phi_j^i(x_1^*, \ldots, x_p^*, x_1 \cdots x_q) = \langle x_i^*, x_j \rangle x_1^* \otimes \cdots \otimes \widehat{x_i^*} \otimes \cdots \otimes x_p^* \otimes x_1 \otimes \cdots \otimes \widehat{x_j} \otimes \cdots \otimes x_q.$$

In view of the universal property, Φ_j^i determines a linear map

$$C_j^i : \otimes_q^p(E^*, E) \to \otimes_{q-1}^{p-1}(E^*, E).$$

C_j^i is called the *contraction operator* with respect to the pair (i, j) and the tensor $\Phi_j^i(\omega)$ is called the *contraction* of ω with respect to (i, j). In particular,

$$C_1^1(x^* \otimes x) = \langle x^*, x \rangle \qquad x^* \in E^*, x \in E.$$

Now assume that E has finite dimension and let $\{e^{*\nu}\}$, $\{e_\nu\}$ be a pair of dual bases of E^* and E. Then the products

$$e_{\mu_1, \ldots, \mu_q}^{\nu_1, \ldots, \nu_p} = e^{*\nu_1} \otimes \cdots \otimes e^{*\nu_p} \otimes e_{\mu_1} \otimes \cdots \otimes e_{\mu_q}$$

form a basis of $\otimes_q^p(E^*, E)$. Thus every tensor $\omega \in \otimes_q^p(E^*, E)$ can be written in the form

$$\omega = \sum_{(\nu)(\mu)} \zeta_{\nu_1, \ldots, \nu_p}^{\mu_1, \ldots, \mu_q} e_{\mu_1, \ldots, \mu_q}^{\nu_1, \ldots, \nu_p} \qquad \zeta_{\nu_1, \ldots, \nu_p}^{\mu_1, \ldots, \mu_q} \in \Gamma.$$

The scalars $\zeta_{\nu_1, \ldots, \nu_p}^{\mu_1, \ldots, \mu_q}$ are called the *components* of ω with respect to the basis $\{e_\nu\}$. Since

$$C_j^i e_{\mu_1, \ldots, \mu_q}^{\nu_1, \ldots, \nu_p} = \delta_{\mu_j}^{\nu_i} e_{\mu_1, \ldots, \hat{\mu}_j, \ldots, \mu_q}^{\nu_1, \ldots, \hat{\nu}_i, \ldots, \nu_p},$$

it follows that the components of the contraction $C_j^i \omega$ are given by

$$\eta_{\nu_1, \ldots, \nu_{p-1}}^{\mu_1, \ldots, \mu_{q-1}} = \sum_{\lambda=1}^{n} \zeta_{\nu_1, \ldots, \nu_{j-1} \lambda \nu_j, \ldots, \nu_{p-1}}^{\mu_1, \ldots, \mu_{i-1} \lambda \mu_i, \ldots, \mu_{q-1}}.$$

3.14. Tensorial Maps

Let E be an n-dimensional vector space. Then every linear automorphism α of E determines for each pair (p, q) a linear automorphism T_α of $\otimes_q^p(E^*, E)$ given by

$$T_\alpha(u^* \otimes u) = (\alpha^\otimes)^{-1} u^* \otimes \alpha_\otimes u.$$

It is easily checked that

$$T_{\beta \circ \alpha} = T_\beta \circ T_\alpha$$

and

$$T_\iota = \iota.$$

A linear map

$$\Phi: \otimes_q^p(E^*, E) \to \otimes_s^r(E^*, E)$$

is called *tensorial*, if it satisfies

$$\Phi \circ T_\alpha = T_\alpha \circ \Phi$$

for all linear automorphisms α of E.

As an example consider the contraction operator C_j^i. For the sake of simplicity we let $i = j = 1$ and write $C_1^1 = C$. Then we have for $z = x_1^* \otimes \cdots \otimes x_p^* \otimes x_1 \otimes \cdots \otimes x_q$

$$T_\alpha(z) = \alpha^{*-1} x_1^* \otimes \cdots \otimes \alpha^{*-1} x_p^* \otimes \alpha x_1 \otimes \cdots \otimes \alpha x_q$$

whence

$$\begin{aligned} (CT_\alpha)(z) &= \langle \alpha^{*-1} x_1^*, \alpha x_1 \rangle \alpha^{*-1} x_2^* \otimes \cdots \otimes \alpha^{*-1} x_p^* \otimes \alpha x_2 \otimes \cdots \otimes \alpha x_q \\ &= \langle x_1^*, x_1 \rangle \alpha^{*-1} x_2^* \otimes \cdots \otimes \alpha^{*-1} x_p^* \otimes \alpha x_2 \otimes \cdots \otimes \alpha x_q \\ &= (T_\alpha C)(z). \end{aligned}$$

Thus C is a tensorial map.

PROBLEMS

1. Let $\{e_\nu\}$, $\{\bar{e}_\nu\}$ $(\nu = 1, \ldots, n)$ be two bases of E and consider a tensor $w \in \otimes^p_q(E, E^*)$. Assume that the bases $\{e_\nu\}$, $\{\bar{e}_\nu\}$ and the dual bases $\{e^{*\nu}\}$, $\{\bar{e}^{*\nu}\}$ are connected by the relations

$$\bar{e}_\nu = \sum_\lambda \alpha^\lambda_\nu e_\lambda \qquad \bar{e}^{*\nu} = \sum_\lambda \beta^\nu_\lambda e^{*\lambda}.$$

Prove the following transformation formulas for the components of w:

$$\bar{\xi}^{\nu_1, \ldots \nu_p}_{\mu_1, \ldots \mu_q} = \sum_{(\lambda), (\kappa)} \beta^{\nu_1}_{\lambda_1} \cdots \beta^{\nu_p}_{\lambda_p} \alpha^{\kappa_1}_{\mu_1} \cdots \alpha^{\kappa_q}_{\mu_q} \xi^{\lambda_1, \ldots, \lambda_p}_{\kappa_1, \ldots, \kappa_q}.$$

2. Using the formula in problem 1 show explicitly that the components of a contracted tensor satisfy the same transformation formula.

3. Show that $\langle u^*, u \rangle = (C^1_1)^p(u \otimes u^*)$ for all $u \in \otimes^p E$, $u^* \in \otimes^p E^*$.

4. Assume that

$$\Phi: \otimes^p_q(E, E^*) \to \otimes^r_s(E, E^*) \qquad \Phi^*: \otimes^q_p(E, E^*) \leftarrow \otimes^s_r(E, E^*)$$

is a pair of dual mappings and that Φ is tensorial. Prove that Φ^* is also tensorial.

5. Show that sum, composition, and tensor product of tensorial mappings is again tensorial.

6. Let

$$\Phi: \otimes^p_q(E, E^*) \to \otimes^r_s(E, E^*)$$

be a nonzero tensorial mapping. Prove that

$$r - p = s - q.$$

7. Consider the linear map

$$\mu(a): z \to a \otimes z,$$

where $a \in E \otimes E^*$ is a fixed tensor. Show that $\mu(a)$ is tensorial if and only if $a = \lambda \cdot t$, where t is the unit tensor.

8. Prove that every tensorial mapping $\Phi: E \otimes E^* \to \Gamma$ is of the form

$$\Phi = \lambda \cdot C,$$

where C is the contraction operator.

9. Verify the relations

$$C^i_j \circ C^k_l = \begin{cases} C^{k-1}_{l-1} \circ C^i_j & \text{if} \quad i < k, j < l, \\ C^{k-1}_l \circ C^i_{j-1} & \text{if} \quad i < k, j \geq l, \\ C^k_{l-1} \circ C^{i+1}_j & \text{if} \quad i \geq k, j < l. \end{cases}$$

10. Consider the bilinear mapping

$$\otimes_q^p(E, E^*) \times \otimes_q^p(F, F^*) \to \otimes_q^p(E \otimes F, E^* \otimes F^*)$$

defined by

$$(x_1 \otimes \cdots \otimes x_p \otimes x^{*1} \otimes \cdots \otimes x^{*q}, y_1 \otimes \cdots \otimes y_p \otimes y^{*1} \otimes \cdots \otimes y^{*q}) \to (x_1 \otimes y_1) \otimes \cdots \otimes (x_p \otimes y_p) \otimes (x^{*1} \otimes y^{*1}) \otimes \cdots \otimes (x^{*q} \otimes y^{*q}).$$

(a) Show that this mapping is a tensor product
(b) Prove that

$$C_j^i(u \otimes v) = C_j^i(u) \otimes C_j^i(v) \qquad u \in \otimes_q^p(E, E^*), v \in \otimes_q^p(F, F^*)$$

11. A tensor $u \in \otimes_q^p(E, E^*)$ is called *invariant* if $T_\alpha(u) = u$ for every linear automorphism (see Section 3.14).
(a) Show that if $u \neq 0$ is an invariant tensor then $p = q$.
(b) If E has finite dimension show that a tensor $u \in E \otimes E^*$ is invariant if and only if $u = \lambda t$ where t is the unit tensor.
(c) If E has infinite dimension show that every invariant tensor $u \in E \otimes E^*$ must be zero.
(d) Assume that E has finite dimension. Show that a tensor u is invariant if and only if the components of u are the same with respect to all pairs of dual bases.

Tensor Algebra Over an Inner Product Space

3.15. The Induced Inner Product

Let E be an inner product space and consider the pth tensorial power $\otimes^p E$. The inner product in E induces an inner product in $\otimes^p E$ such that

$$(x_1 \otimes \cdots \otimes x_p, y_1 \otimes \cdots \otimes y_p) = (x_1, y_1) \cdots (x_p, y_p)$$

(see Section 1.25). In particular, if E is an n-dimensional Euclidean space and if e_ν $(\nu = 1, \ldots, n)$ is an orthonormal basis of E, then the products $e_{\nu_1} \otimes \cdots \otimes e_{\nu_p}$ form an orthonormal basis of $\otimes^p E$ and so $\otimes^p E$ becomes a Euclidean space as well.

The inner products in the spaces $\otimes^p E$ determine an inner product in $\otimes E$ given by

$$(u, v) = \sum_p (u_p, v_p) \qquad u, v \in \otimes E,$$

where

$$u = \sum_p u_p \quad \text{and} \quad v = \sum_p v_p \qquad u_p, v_p \in \otimes^p E.$$

3.16. The Isomorphism $\tau_{\otimes}$

Let E be an n-dimensional inner product space with dual space E^*. Then the inner product in E determines a linear isomorphism $\tau: E \xrightarrow{\cong} E^*$ given by

$$\langle \tau x, y \rangle = (x, y) \qquad x, y \in E.$$

This isomorphism τ in turn induces a linear isomorphism

$$\tau_{\otimes}: \otimes E \xrightarrow{\cong} \otimes E^*.$$

This isomorphism satisfies the relation

$$\langle \tau_{\otimes} u, v \rangle = (u, v) \qquad u, v \in \otimes E. \tag{3.16}$$

In fact, let

$$u = x_1 \otimes \cdots \otimes x_p \quad \text{and} \quad v = y_1 \otimes \cdots \otimes y_p$$

be decomposable tensors. Then we have

$$\begin{aligned} \langle \tau_{\otimes} u, v \rangle &= \langle \tau x_1 \otimes \cdots \otimes \tau x_p, y_1 \otimes \cdots \otimes y_p \rangle \\ &= \langle \tau x_1, y_1 \rangle \cdots \langle \tau x_p, y_p \rangle = (x_1, y_1) \cdots (x_p, y_p) \\ &= (x_1 \otimes \cdots \otimes x_p, y_1 \otimes \cdots \otimes y_p) = (u, v). \end{aligned}$$

Relation (3.16) shows that the restriction of $\tau_{\otimes}$ to $\otimes^p E$ coincides with the isomorphism $\otimes^p E \xrightarrow{\cong} \otimes^p E^*$ induced by the inner product in $\otimes^p E$. Finally note that

$$\tau^{\otimes} = \tau_{\otimes}$$

since $\tau^* = \tau$.

3.17. The Metric Tensors

Let E be an n-dimensional Euclidean space. Choose an orthonormal basis $\{e_\nu\}$ $(\nu = 1, \ldots, n)$ and set

$$g = \sum_\nu e_\nu \otimes e_\nu.$$

Since an orthonormal basis is self-dual (with respect to the inner product in E), it follows that the tensor g is independent of the choice of the orthonormal basis $\{e_\nu\}$. It is called the *contravariant metric tensor of E*. Similarly, the *covariant metric tensor* is defined by

$$g^* = \sum_\nu e^{*\nu} \otimes e^{*\nu},$$

where $\{e^{*\nu}\}$ $(\nu = 1, \ldots, n)$ is an orthonormal basis of E^* with respect to the induced inner product.

The inner product of two vectors x and y can be expressed in the form

$$(x, y) = \langle g^*, x \otimes y \rangle.$$

In fact, write

$$x = \sum_\nu \xi^\nu e_\nu \quad \text{and} \quad y = \sum_\nu \eta^\nu e_\nu.$$

Then we have

$$\langle g^*, x \otimes y \rangle = \sum_\nu \langle e^{*\nu} \otimes e^{*\nu}, x \otimes y \rangle = \sum_\nu \langle e^{*\nu}, x \rangle \langle e^{*\nu}, y \rangle$$
$$= \sum_\nu \xi^\nu \eta^\nu = (x, y).$$

The same argument shows that

$$(x^*, y^*) = \langle x^* \otimes y^*, g \rangle \qquad x^*, y^* \in E^*.$$

PROBLEMS

1. Given a Euclidean space E prove that

$$(u_1 \otimes v_1, u_2 \otimes v_2) = (u_1, u_2)(v_1, v_2) \qquad u_1, u_2 \in \otimes^p E, v_1, v_2 \in \otimes^q E.$$

2. Let E be a Euclidean space and E^* be a dual space. Consider the metric tensors $g \in E \otimes E$ and $g^* \in E^* \otimes E^*$.
   (a) Show that

$$C_2^1(g \otimes g^*) = t,$$

   where t is the unit tensor.
   (b) Prove that the metric tensor of $\otimes^p E$ is given by $\underbrace{g \otimes \cdots \otimes g}_{p}$.

3. Verify the formula

$$(\mu(t)u, \mu(t)v) = n(u, v) \qquad u, v \in \otimes E,$$

   where t is the unit tensor of E, $n = \dim E$ (see Section 1.26).

4. Let E be a Euclidean space and E^* be a dual space. Consider the space $L(E; E) = E^* \otimes E$ (see Section 1.27).
   (a) Show that the adjoint of a linear transformation $\varphi = a^* \otimes b$ is given by $\tilde{\varphi} = \tau b \otimes \tau^{-1} a^*$.
   (b) Show that the induced inner product in $L(E; E)$ is given by

$$(\varphi, \psi) = \operatorname{tr}(\varphi \circ \tilde{\psi}).$$

5. Given a Euclidean space E and an arbitrary basis $\{e_\nu\}$ $(\nu = 1, \ldots, n)$ show that

$$(u, v) = \sum_{(\nu)(\mu)} g_{\nu_1\mu_1} \cdots g_{\nu_p\mu_p} \xi^{\nu_1, \ldots, \nu_p} \eta^{\mu_1, \ldots, \mu_p},$$

   where

$$u = \sum_{(\nu)} \xi^{\nu_1, \ldots, \nu_p} e_{\nu_1} \otimes \cdots \otimes e_{\nu_p} \quad \text{and} \quad v = \sum_{(\mu)} \eta^{\mu_1, \ldots, \mu_p} e_{\mu_1} \otimes \cdots \otimes e_{\mu_p}.$$

6. Show that the components of the metric tensor g^* with respect to an arbitrary basis x_ν ($\nu = 1, \ldots, n$) of E are the inner products (x_ν, x_μ).

7. Let E be a Euclidean space and E^* be a dual space. Given a pair of dual bases $\{x_\nu\}$, $\{x^{*\nu}\}$ ($\nu = 1, \ldots, n$) show that

$$g^* = \sum_{\nu, \mu} (x_\nu, x_\mu) x^{*\nu} \otimes x^{*\mu}$$

and

$$g = \sum_{\nu, \mu} (x^{*\nu}, x^{*\mu}) x_\nu \otimes x_\mu .$$

8. Let E, E^* be a pair of dual spaces and assume that E is a Euclidean space. Consider the linear isomorphism $\tau: E \to E^*$ defined by

$$\langle \tau x, y \rangle = (x, y) \qquad x, y \in E.$$

Prove that the isomorphism $\tau_\otimes : \otimes^p E \to \otimes^p E^*$ coincides with the linear isomorphism $\otimes^p E \to \otimes^p E^*$ which is induced by the inner product in $\otimes^p E$. Show that $\tau^\otimes = \tau_\otimes$ and that $(\tau^\otimes)^2 = \iota$.

The Algebra of Multilinear Functions

In Sections 3.18–3.23 E denotes a finite-dimensional vector space.

3.18. The Algebra $T^\cdot(E)$

Consider for each $p \geq 1$ the space $T^p(E)$ of p-linear functions

$$\Phi: \underbrace{E \times \cdots \times E}_{p} \to \Gamma .$$

In particular $T^1(E) = E^*$. It will be convenient to extend the definition of $T^p(E)$ to $p = 0$ by setting $T^0(E) = \Gamma$. The *product* of a p-linear function Φ and a q-linear function Ψ is the $(p + q)$-linear function $\Phi \cdot \Psi$ given by

$$(\Phi \cdot \Psi)(x_1, \ldots, x_{p+q}) = \Phi(x_1, \ldots, x_p) \cdot \Psi(x_{p+1}, \ldots, x_{p+q}). \tag{3.17}$$

In the cases $p = 0$ or $q = 0$ we define the multiplication to be the ordinary multiplication by scalars. This multiplication makes the direct sum

$$T^\cdot(E) = \sum_{p=0}^{\infty} T^p(E)$$

into an associative (noncommutative) algebra with the scalar 1 as unit element.

A linear map $\varphi: E \to F$ induces a homomorphism $\varphi^*: T^\cdot(E) \leftarrow T^\cdot(F)$ given by

$$(\varphi^* \Psi)(x_1, \ldots, x_p) = \Psi(\varphi x_1, \ldots, \varphi x_p) \qquad \Psi \in T^p(F)$$

as follows directly from the definitions.

Moreover, a linear transformation φ of E determines a derivation $\theta^T(\varphi)$ in the algebra $T^{\bullet}(E)$ given by

$$(\theta^T(\varphi)\Phi)(x_1, \ldots, x_p) = \sum_{\nu=1}^{p} \Phi(x_1, \ldots, \varphi x_\nu, \ldots, x_p).$$

3.19. The Substitution Operators

Fix a vector $h \in E$ and consider the operators $i_\nu(h): T^{\bullet}(E) \to T^{\bullet}(E)$ given by

$$(i_\nu(h)\Phi)(x_1, \ldots, x_{p-1}) = \Phi(\underbrace{x_1, \ldots, h}_{\nu}, \ldots, x_{p-1}).$$

$i_\nu(h)$ is called the νth *substitution operator* in $T^{\bullet}(E)$ corresponding to the vector h. Clearly,

$$i_\nu(h)(\Phi + \Psi) = i_\nu(h)\Phi + i_\nu(h)\Psi \qquad \Phi, \Psi \in T^{\bullet}(E).$$

Moreover, it follows from the definition that for $\Phi \in T^p(E)$

$$i_\nu(h)(\Phi \cdot \Psi) = \begin{cases} i_\nu(h)\Phi \cdot \Psi & \nu \leq p, \\ \Phi \cdot i_{\nu-p}(h)\Psi & \nu \geq p + 1. \end{cases} \tag{3.18}$$

This relation implies that for $f_\nu \in E^*$

$$i_\nu(h)(f_1 \cdot \ldots \cdot f_p) = f_\nu(h) f_1 \cdot \ldots \cdot \hat{f}_\nu \cdot \ldots \cdot f_p.$$

Now define operators $i_A(h)$ and $i_S(h)$ on $T^{\bullet}(E)$ by

$$i_A(h) = \sum_{\nu=1}^{p} (-1)^{\nu-1} i_\nu(h)$$

and

$$i_S(h) = \sum_{\nu=1}^{p} i_\nu(h) \qquad (p = 1, 2, \ldots).$$

Then we obtain from (3.18) the formulas

$$i_A(h)(\Phi \cdot \Psi) = i_A(h)\Phi \cdot \Psi + (-1)^p \Phi \cdot i_A(h)\Psi$$

and

$$i_S(h)(\Phi \cdot \Psi) = i_S(h)\Phi \cdot \Psi + \Phi \cdot i_S(h)\Psi$$

for all $\Phi \in T^p(E)$, $\Psi \in T^{\bullet}(E)$.

3.20. The Isomorphism $\otimes^p E^* \xrightarrow{\cong} T^p(E)$

In this section we shall establish an isomorphism between the space $T^p(E)$ and the pth tensorial power over E^*. Consider the p-linear mapping

$$\varphi: \underbrace{E^* \times \cdots \times E^*}_{p} \to T^p(E)$$

given by

$$\varphi(f_1, \ldots, f_p) = f_1 \cdots f_p \qquad f_\nu \in E^*.$$

It has to be shown that this map has the universal property. Fix a basis $\{e_1, \ldots, e_n\}$ in E and let $\{e^{*1}, \ldots, e^{*n}\}$ be the dual basis.

We show that the products $e^{*\nu_1} \cdots e^{*\nu_p}$ ($\nu_i = 1, \ldots, n$) form a basis of $T^p(E)$. In fact, every vector $x \in E$ can be written in the form

$$x = \sum_\nu e^{*\nu}(x)e_\nu \qquad e^{*\nu} \in E^{*\nu}.$$

Now let $\Phi \in T^p(E)$. Then we have

$$\begin{aligned}\Phi(x_1, \ldots, x_p) &= \Phi\left(\sum_\nu e^{*\nu}(x_1)e_\nu, \ldots, \sum_\nu e^{*\nu}(x_p)e_\nu\right) \\ &= \sum_{(\nu)} e^{*\nu_1}(x_1) \cdots e^{*\nu_p}(x_p)\Phi(e_{\nu_1}, \ldots, e_{\nu_p}).\end{aligned}$$

This equation can be written in the form

$$\Phi = \sum_{(\nu)} \Phi(e_{\nu_1}, \ldots, e_{\nu_p})e^{*\nu_1} \cdots e^{*\nu_p}$$

and so it shows that the products $e^{*\nu_1} \cdots e^{*\nu_p}$ generate $T^p(E)$.

On the other hand, assume a relation

$$\sum_{(\nu)} \lambda_{\nu_1, \ldots, \nu_p} e^{*\nu_1} \cdots e^{*\nu_p} = 0.$$

Then

$$\sum_{(\nu)} \lambda_{\nu_1, \ldots, \nu_p} \langle e^{*\nu_1}, x_1 \rangle \cdots \langle e^{*\nu_p}, x_p \rangle = 0 \qquad x_i \in E.$$

Now fix a p-tuple $(\alpha_1, \ldots, \alpha_p)$ and set $x_i = e_{\alpha_i}$ ($i = 1, \ldots, p$). Then the relation above implies that $\lambda_{\alpha_1, \ldots, \alpha_p} = 0$. Thus the products $e^{*\nu_1} \cdots e^{*\nu_p}$ form a basis of $T^p(E)$. In particular,

$$\dim T^p(E) = n^p \qquad n = \dim E.$$

Now consider the linear map $\alpha: \otimes^p E^* \to T^p(E)$ induced by φ. Then we have the commutative diagram

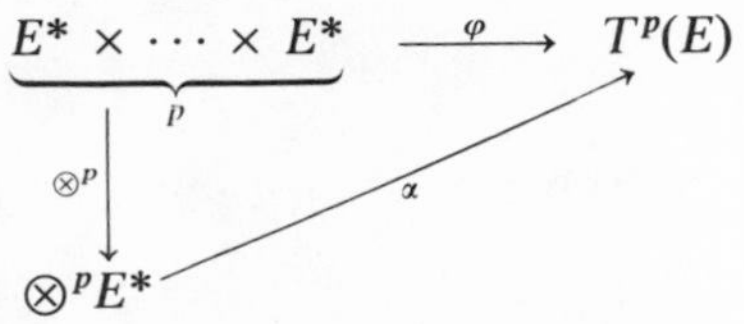

In particular

$$\alpha(e^{*\nu_1} \otimes \cdots \otimes e^{*\nu_p}) = e^{*\nu_1} \cdots e^{*\nu_p}.$$

Since the products $e^{*\nu_1} \cdots e^{*\nu_p}$ form a basis of $T^p(E)$ it follows that α is an isomorphism. Finally observe that, since $\alpha(u^* \otimes v^*) = \alpha(u^*) \cdot \alpha(v^*)$ α is an algebra isomorphism

$$\alpha : \otimes E^* \xrightarrow{\cong} T^{\cdot}(E).$$

3.21. The Algebra $T_{\cdot}(E)$

Let $T_p(E)(p \geq 1)$ denote the space of p-linear functions in the dual space E^* and set $T_0(E) = \Gamma$. Observe that the space $T_1(E)$ is canonically isomorphic to E under the correspondence $a \mapsto f_a$ given by

$$f_a(x^*) = \langle x^*, a \rangle \qquad a \in E.$$

Applying the results of Section 3.18 with E replaced by E^* we obtain a multiplication between the spaces $T_p(E)(p \geq 0)$ which makes the direct sum

$$T_{\cdot}(E) = \sum_{p=0}^{\infty} T_p(E)$$

into an associative algebra.

A linear map $\varphi: E \to F$ induces a homomorphism $\varphi_*: T_{\cdot}(E) \to T_{\cdot}(F)$ given by

$$(\varphi_* \Phi)(y_1^*, \ldots, y_p^*) = \Phi(\varphi^* y_1^*, \ldots, \varphi^* y_p^*) \qquad \Phi \in T_p(E)$$

and a linear transformation φ of E determines a derivation in $T_{\cdot}(E)$ given by

$$(\theta_T(\varphi)\Phi)(x_1^*, \ldots, x_p^*) = \sum_{\nu=1}^{p} \Phi(x_1^*, \ldots, \varphi^* x_\nu^*, \ldots, x_p^*).$$

Finally note that $T_{\cdot}(E)$ is isomorphic to the tensor algebra over E (see Section 3.20).

3.22. The Duality Between $T^p(E)$ and $T_p(E)$

Fix a pair of dual bases $\{e^{*\nu}\}$, $\{e_\nu\}$ in E and E^* and consider the bilinear function $\langle , \rangle: T^p(E) \times T_p(E) \to \Gamma$ given by

$$\langle \Phi, \Psi \rangle = \sum_{(\nu)} \Phi(e_{\nu_1}, \ldots, e_{\nu_p}) \Psi(e^{*\nu_1}, \ldots, e^{*\nu_p}).$$

Then we have in particular

$$\begin{aligned}\langle \Phi, x_1 \cdot \ldots \cdot x_p \rangle &= \sum_{(\nu)} \Phi(e_{\nu_1}, \ldots, e_{\nu_p}) \langle e^{*\nu_1}, x_1 \rangle \cdots \langle e^{*\nu_p}, x_p \rangle \\ &= \Phi\left(\sum_{\nu_1} \langle e^{*\nu_1}, x_1 \rangle e_{\nu_1}, \ldots, \sum_{\nu_p} \langle e^{*\nu_p}, x_p \rangle e_{\nu_p}\right).\end{aligned}$$

Since

$$\sum_{\nu} \langle e^{*\nu}, x \rangle e_\nu = x \qquad x \in E,$$

it follows that

$$\langle \Phi, x_1 \cdot \ldots \cdot x_p \rangle = \Phi(x_1, \ldots, x_p) \qquad \Phi \in T^p(E),\ x_\nu \in E.$$

Similarly

$$\langle f_1 \cdot \ldots \cdot f_p \Psi \rangle = \Psi(f_1, \ldots, f_p) \qquad \Psi \in T_p(E),\ f_\nu \in E^*.$$

These relations imply that the bilinear function $\langle , \rangle$ does not depend on the choice of the dual bases $\{e_\nu\}$, $\{e^{*\nu}\}$ $(\nu = 1, \ldots, n)$ and that it is non-degenerate. Thus it defines a scalar product in the spaces $T^p(E)$ and $T_p(E)$.

On the other hand, we have a scalar product between the spaces $\otimes^p E^*$ and $\otimes^p E$ (see Section 3.8). A simple computation shows that the isomorphisms $\otimes^p E^* \xrightarrow{\cong} T^p(E)$ and $\otimes^p E \xrightarrow{\cong} T_p(E)$ preserve the scalar products.

3.23. The Algebra $T(E)$

Denote by $T_q^p(E)$ the space of $(p + q)$-linear functions

$$\Phi: \underbrace{E \times \cdots \times E}_{p} \times \underbrace{E^* \times \cdots \times E^*}_{q} \to \Gamma.$$

In particular,

$$T_0^p(E) = T^p(E) \quad \text{and} \quad T_q^0(E) = T_q(E)$$

(see Section 3.18).

The *product* of two multilinear functions $\Phi \in T_q^p(E)$ and $\Psi \in T_s^r(E)$ is defined by

$$\begin{aligned}(\Phi . \Psi)(x_1, \ldots, x_{p+r}, x_1^*, \ldots, x_{q+s}^*) \\ = \Phi(x_1, \ldots, x_p, x_1^*, \ldots, x_q^*)\Psi(x_1, \ldots, x_r, x_1^*, \ldots, x_s^*).\end{aligned}$$

Form the direct sum $T(E)$ of the spaces $T_q^p(E)$ and identify each $T_q^p(E)$ with its image under the inclusion map. Then the multiplication above makes $T(E)$ into an associative algebra.

Now consider the bilinear mapping $T^p(E) \times T_q(E) \to T_q^p(E)$ defined by this multiplication. It follows from Section 1.20 that this bilinear mapping has the universal property. Thus,

$$T_q^p(E) \cong T^p(E) \otimes T_q(E)$$

and

$$T(E) \cong T^{\boldsymbol{\cdot}}(E) \otimes T_{\boldsymbol{\cdot}}(E).$$

Moreover, since for $\Phi_1, \Phi_2 \in T^{\cdot}(E)$ and $\Psi_1, \Psi_2 \in T_{\cdot}(E)$

$$(\Phi_1.\Psi_1).(\Phi_2.\Psi_2) = (\Phi_1.\Phi_2).(\Psi_1.\Psi_2),$$

$T(E)$ is the *canonical tensor product* of the algebras $T^{\cdot}(E)$ and $T_{\cdot}(E)$.

Finally, let $\alpha: E \xrightarrow{\cong} F$ be a linear isomorphism. Then an induced isomorphism $T_\alpha: T(E) \xrightarrow{\cong} T(F)$ is explicitly given by

$$(T_\alpha \Phi)(y_1, \ldots, y_p, y_1^*, \ldots, y_q^*) = \Phi(\alpha^{-1}y_1, \ldots, \alpha^{-1}y_p, \alpha^* y_1^*, \ldots, \alpha^* y_q^*)$$

for all $\Phi \in T_q^p(E)$, $y_i \in F$, $y^{*j} \in F^*$.

4 Skew-Symmetry and Symmetry in the Tensor Algebra

All vector spaces in this chapter are defined over a field of characteristic zero.

Skew-Symmetric Tensors

4.1. The Space $N^p(E)$

Given a vector space E consider the pth tensorial power $\otimes^p E$ and let S_p denote the permutation group on p letters. Then every permutation $\sigma \in S_p$ determines a linear automorphism of $\otimes^p E$ (also denoted by σ) defined by

$$\sigma(x_1 \otimes \cdots \otimes x_p) = x_{\sigma^{-1}(1)} \otimes \cdots \otimes x_{\sigma^{-1}(p)} \qquad x_\nu \in E.$$

As an immediate consequence of the definition we have the formulas

$$(\tau\sigma)u = \tau(\sigma u) \qquad \sigma, \tau \in S_p,\ u \in \otimes^p E$$

and

$$\iota u = u$$

(ι being the identity permutation).

Now consider the subspace $N^p(E)$ of $\otimes^p E$ generated by all products $x_1 \otimes \cdots \otimes x_p$ such that $x_i = x_j$ for at least one pair $i \neq j$. Clearly $N^p(E)$ is stable under σ for each $\sigma \in S_p$.

It will be shown that for each $u \in \otimes^p E$ and each $\sigma \in S_p$

$$\boxed{u - \varepsilon_\sigma \sigma u \in N^p(E)} \tag{4.1}$$

For the proof we may assume that u is decomposable, $u = x_1 \otimes \cdots \otimes x_p$.

Consider first the case of a transposition $\tau: i \rightleftarrows j$. Then we have

$$
\begin{aligned}
u - \varepsilon_\tau \tau u &= x_1 \otimes \cdots \otimes x_i \otimes \cdots \otimes x_j \otimes \cdots \otimes x_p \\
&\quad + x_1 \otimes \cdots \otimes x_j \otimes \cdots \otimes x_i \otimes \cdots \otimes x_p \\
&= x_1 \otimes \cdots \otimes (x_i + x_j) \otimes \cdots \otimes (x_i + x_j) \otimes \cdots \otimes x_p \\
&\quad - x_1 \otimes \cdots \otimes x_i \otimes \cdots \otimes x_i \otimes \cdots \otimes x_p \\
&\quad - x_1 \otimes \cdots \otimes x_j \otimes \cdots \otimes x_j \otimes \cdots \otimes x_p \in N^p(E).
\end{aligned}
$$

Now assume that $u - \varepsilon_\sigma \sigma u \in N^p(E)$ for all permutations σ which are products of m transpositions and consider the permutation $\rho = \tau\sigma$ where τ is a transposition and σ is a product of m transpositions. By hypothesis we have

$$u - \varepsilon_\sigma \sigma u \in N^p(E).$$

Since $N^p(E)$ is stable under τ it follows that

$$\tau u - \varepsilon_\sigma \tau\sigma u \in N^p(E),$$

whence

$$\varepsilon_\tau \tau u - \varepsilon_{\tau\sigma} \tau\sigma u \in N^p(E).$$

On the other hand we have

$$u - \varepsilon_\tau \tau u \in N^p(E).$$

Adding these relations we obtain

$$u - \varepsilon_{\tau\sigma} \tau\sigma u \in N^p(E).$$

Now (4.1) follows by induction.

4.2. The Alternator

A tensor $u \in \otimes^p E$ is called *skew-symmetric* if

$$\sigma u = \varepsilon_\sigma u \quad \text{for every } \sigma \in S_p.$$

The skew-symmetric tensors of degree p form a subspace $X^p(E)$ of $\otimes^p E$.

The *alternator* in $\otimes^p E$ is the linear map $\pi_A: \otimes^p E \to \otimes^p E$ given by

$$\pi_A = \frac{1}{p!} \sum_\sigma \varepsilon_\sigma \sigma.$$

It follows from the definition that for $\tau \in S_p$

$$
\begin{aligned}
\pi_A \tau &= \frac{1}{p!} \sum_\sigma \varepsilon_\sigma \sigma\tau = \frac{1}{p!} \varepsilon_\tau \sum_\sigma \varepsilon_{\sigma\tau} \sigma\tau \\
&= \frac{1}{p!} \varepsilon_\tau \sum_\rho \varepsilon_\rho \rho = \varepsilon_\tau \pi_A.
\end{aligned}
$$

Thus we have

$$\pi_A \circ \tau = \varepsilon_\tau \pi_A \qquad \tau \in S_p. \tag{4.2}$$

Similarly,

$$\tau \circ \pi_A = \varepsilon_\tau \pi_A \qquad \tau \in S_p. \tag{4.3}$$

Next we establish the relations

$$\ker \pi_A = N^p(E) \tag{4.4}$$

and

$$\operatorname{Im} \pi_A = X^p(E). \tag{4.5}$$

In fact, let $u = x_1 \otimes \cdots \otimes x_p$ be a generator of $N^p(E)$. Then, for some transposition τ, $\tau u = u$ and so Formula (4.2) implies that $\pi_A u = -\pi_A u$ whence $\pi_A(u) = 0$. This shows that $N^p(E) \subset \ker \pi_A$.

On the other hand, it follows from the definition of π_A that for $u \in \otimes^p E$

$$\pi_A u - u = \frac{1}{p!} \sum_\sigma (\varepsilon_\sigma \sigma u - u) \in N^p(E).$$

Hence, if $\pi_A u = 0$, then $u \in N^p(E)$. Thus (4.4) is established.

To prove Formula (4.5) observe that, in view of (4.3), $\operatorname{Im} \pi_A \subset X^p(E)$. On the other hand, if $u \in X^p(E)$ then $\pi_A u = u$ and so $X^p(E) \subset \operatorname{Im} \pi_A$.

Next note that Formula (4.3) implies that

$$\pi_A^2 = \pi_A$$

and so π_A is a projection operator. Hence we obtain from Relations (4.4) and (4.5) the direct decomposition

$$\otimes^p E = X^p(E) \oplus N^p(E). \tag{4.6}$$

Thus every tensor $u \in \otimes^p E$ can be uniquely decomposed in the form

$$u = v + w \qquad v \in X^p(E),\ w \in N^p(E).$$

The tensor $v = \pi_A u$ is called the *skew part of* u.

4.3. Dual Spaces

Suppose now that E, E^* is a pair of dual spaces, and let π^A be the alternator for $\otimes^p E^*, p \geq 2$. If $x^{*1} \otimes \cdots \otimes x^{*p}$ and $x_1 \otimes \cdots \otimes x_p$ are two decomposable tensors in $\otimes^p E^*$ and $\otimes^p E$ respectively, we have, for any $\sigma \in S_p$,

$$\begin{aligned}\langle x^{*1} \otimes \cdots \otimes x^{*p}, \sigma(x_1 \otimes \cdots \otimes x_p)\rangle &= \langle x^{*1}, x_{\sigma^{-1}(1)}\rangle \cdots \langle x^{*p}, x_{\sigma^{-1}(p)}\rangle \\ &= \langle x^{*\sigma(1)}, x_1\rangle \cdots \langle x^{*\sigma(p)}, x_p\rangle \\ &= \langle \sigma^{-1}(x^{*1} \otimes \cdots \otimes x^{*p}), x_1 \otimes \cdots \otimes x_p\rangle\end{aligned}$$

and hence we obtain the relation

$$\langle u^*, \sigma u\rangle = \langle \sigma^{-1}u^*, u\rangle \qquad u^* \in \otimes^p E^*, u \in \otimes^p E,$$

which shows that σ and σ^{-1} are dual operators.

Since

$$\pi_A = \frac{1}{p!}\sum_\sigma \varepsilon_\sigma \sigma \quad \text{and} \quad \pi^A = \frac{1}{p!}\sum_\sigma \varepsilon_\sigma \sigma = \frac{1}{p!}\sum_\sigma \varepsilon_\sigma \sigma^{-1},$$

it follows that π_A and π^A are dual operators as well, i.e.,

$$\langle u^*, \pi_A u\rangle = \langle \pi^A u^*, u\rangle \qquad u^* \in \otimes^p E^*, u \in \otimes^p E. \tag{4.7}$$

The duality of π_A and π^A implies that the restriction of the scalar product to the subspaces $\operatorname{Im} \pi_A = X^p(E)$ and $\operatorname{Im} \pi^A = X^p(E^*)$ is again nondegenerate and hence a duality is induced between $X^p(E)$ and $X^p(E^*)$.

Suppose now that

$$u = x^{*1} \otimes \cdots \otimes x^{*p} \quad \text{and} \quad u = x_1 \otimes \cdots \otimes x_p$$

are decomposable tensors in $\otimes^p E^*$ and $\otimes^p E$ respectively. Then we obtain from (4.7) the formula

$$\begin{aligned}
&\langle \pi^A(x^{*1} \otimes \cdots \otimes x^{*p}), \pi_A(x_1 \otimes \cdots \otimes x_p)\rangle \\
&\quad = \langle x^{*1} \otimes \cdots \otimes x^{*p}, \pi_A^2(x_1 \otimes \cdots \otimes x_p)\rangle \\
&\quad = \langle x^{*1} \otimes \cdots \otimes x^{*p}, \pi_A(x_1 \otimes \cdots \otimes x_p)\rangle \\
&\quad = \frac{1}{p!}\sum_\sigma \varepsilon_\sigma \langle x^{*1}, x_{\sigma^{-1}(1)}\rangle \cdots \langle x^{*p}, x_{\sigma^{-1}(p)}\rangle,
\end{aligned}$$

whence

$$\boxed{\langle \pi^A(x^{*1} \otimes \cdots \otimes x^{*p}), \pi_A(x_1 \otimes \cdots \otimes x_p)\rangle = \frac{1}{p!}\det(\langle x^{*i}, x_j\rangle).} \tag{4.8}$$

4.4. The Skew-Symmetric Part of a Product

Let $\otimes E = \sum_{p=0}^{\infty} \otimes^p E$ be a tensor algebra over E and consider the subspaces $N^p(E) \subset \otimes^p E$, $p \geq 2$. It will be convenient to extend the definition to the cases $p = 1$ and $p = 0$ by setting $N^1(E) = N^0(E) = 0$. Accordingly we define π_A to be the identity on $\otimes^1 E$ and $\otimes^0 E$ and then the previously established formulas continue to hold in the cases $p = 1$ and $p = 0$.

It follows from the definition of $N^p(E)$ that

$$\begin{aligned}
N^p(E) \otimes \otimes^q E &\subset N^{p+q}(E) \\
\otimes^p E \otimes N^q(E) &\subset N^{p+q}(E)
\end{aligned} \qquad p \geq 0, q \geq 0. \tag{4.9}$$

Now let $u \in \otimes^p E$ and $v \in \otimes^q E$ be two arbitrary tensors. Then we can write

$$u = \pi_A u + u_1 \qquad u_1 \in N^p(E)$$

and

$$v = \pi_A v + v_1 \qquad v_1 \in N^q(E),$$

whence

$$u \otimes v = \pi_A u \otimes \pi_A v + \pi_A u \otimes v_1 + u_1 \otimes \pi_A v + u_1 \otimes v_1.$$

Applying the projection π_A to this equation and observing Relations (4.9) and (4.4), we obtain the formula

$$\pi_A(u \otimes v) = \pi_A(\pi_A u \otimes \pi_A v). \tag{4.10}$$

Since π_A is a projection operator, it follows that

$$\pi_A(\pi_A u \otimes v) = \pi_A(u \otimes v) = \pi_A(u \otimes \pi_A v). \tag{4.11}$$

Problem

Show that the mapping $\sigma: \otimes^p E \to \otimes^p E$ is tensorial (see Section 3.14), where $\otimes^p E$ is considered as a subspace of $\otimes(E, E^*)$. If the dimension of E is finite, prove that σ is generated by the operators $\mu(t)$ and C_j^i, t being the unit tensor for E and E^*.

The Factor Algebra $\otimes E/N(E)$

4.5. The Ideal $N(E)$

Consider the direct sum

$$N(E) = \sum_p N^p(E).$$

Formulas (4.9) imply that $N(E)$ is a graded ideal in the graded algebra $\otimes E$.

Suppose now that $u \in \otimes^p E$ and $v \in \otimes^q E$ are two arbitrary tensors. Then we have

$$u \otimes v - (-1)^{pq} v \otimes u \in N^{p+q}(E). \tag{4.12}$$

In fact, if σ is the permutation given by

$$(1, \ldots, p, p+1, \ldots, p+q) \to (q+1, \ldots, q+p, 1, \ldots, q)$$

it follows that

$$\sigma(u \otimes v) = v \otimes u \quad \text{and} \quad \varepsilon_\sigma = (-1)^{pq}$$

and thus Formula (4.1) yields

$$u \otimes v - (-1)^{pq} v \otimes u = u \otimes v - \varepsilon_\sigma \sigma(u \otimes v) \in N^{p+q}(E).$$

Applying the operator π_A to (4.12) we obtain the formula

$$\pi_A(u \otimes v) = (-1)^{pq} \pi_A(v \otimes u) \qquad u \in \otimes^p E,\ v \in \otimes^q E. \tag{4.13}$$

4.6. The Algebra $\otimes E/N(E)$

Consider the canonical projection

$$\pi : \otimes E \to \otimes E/N(E). \tag{4.14}$$

Since $N(E)$ is an ideal in $\otimes E$, a multiplication is induced in $\otimes E/N(E)$ by

$$\pi a \cdot \pi b = \pi(a \otimes b) \qquad a, b \in \otimes E. \tag{4.15}$$

It follows from (4.15) that this multiplication is associative and that $\pi(1)$ is a unit element. Since the ideal $N(E)$ is graded in the graded algebra $\otimes E$, a gradation is induced in the factor algebra $\otimes E/N(E)$ by

$$\otimes E/N(E) = \sum_p \pi(\otimes^p E)$$

and so $\otimes E/N(E)$ becomes a graded algebra. Since $N^1(E) = N^0(E) = 0$, we have in particular that $\pi(\otimes^1 E)$ and $\pi(\otimes^0 E)$ are isomorphic to $\otimes^1 E = E$ and $\otimes^0 E = \Gamma$ respectively. Consequently, we shall identify $\pi(\otimes^1 E)$ and $\pi(\otimes^0 E)$ with E and Γ respectively.

From (4.13) we obtain the commutation relation

$$uv = (-1)^{pq} vu, \tag{4.16}$$

for every two homogeneous elements of degree p and q in the algebra $\otimes E/N(E)$.

4.7. Skew-Symmetric Tensors

Define the subspace $X(E) \subset \otimes E$ by

$$X(E) = \sum_p X^p(E)$$

and extend the projection operators $\pi_A : \otimes^p E \to \otimes^p E$ ($\pi_A = \iota$ in $\otimes^1 E$ and $\otimes^0 E$) to a linear map $\pi_A : \otimes E \to \otimes E$. Then we have

$$\ker \pi_A = N(E)$$

and

$$\operatorname{Im} \pi_A = X(E).$$

Moreover, π_A is a projection operator and

$$\otimes E = N(E) \oplus X(E).$$

If ρ denotes the restriction of the projection π to the subspace $X(E)$ then

$$\rho: X(E) \to \otimes E/N(E)$$

is a homogeneous linear isomorphism of degree zero. Let $\pi_X: \otimes E \to X(E)$ be the restriction of π_X to $\otimes E, X(E)$. Then we have the following commutative diagram:

$$\begin{array}{ccc} \otimes E & \xrightarrow{\pi_X} & X(E) \\ {\scriptstyle \pi}\downarrow & \overset{\cong}{\swarrow}{\scriptstyle \rho} & \\ \otimes E/N(E) & & \end{array} \tag{4.17}$$

4.8. The Induced Scalar Product

Let E, E^* be a pair of dual vector spaces. Then a scalar product is induced in $\otimes E$, $\otimes E^*$ (see Section 3.8). It follows from (4.7) that the restriction of this scalar product to the subspaces $X(E)$ and $X(E^*)$ is again nondegenerate. Since $\rho: X(E) \xrightarrow{\cong} \otimes E/N(E)$ is a linear isomorphism, a scalar product $\langle , \rangle$ in the pair $\otimes E/N(E)$, $\otimes E^*/N(E^*)$ is induced by

$$\langle \rho u^*, \rho u \rangle = p! \langle u^*, u \rangle \qquad u^* \in X^p(E^*),\ u \in X^p(E). \tag{4.18}$$

Clearly the scalar product (4.18) respects the gradation. Moreover it follows from (4.17) and (4.18) that

$$\begin{aligned} \langle \pi u^*, \pi u \rangle &= \langle \rho \pi^X u^*, \rho \pi_X u \rangle \\ &= \langle \rho \pi^A u^*, \rho \pi_A u \rangle = p! \langle \pi^A u^*, \pi_A u \rangle \qquad u^* \in \otimes^p E^*,\ u \in \otimes^p E. \end{aligned} \tag{4.19}$$

Now assume that u and u^* are decomposable,

$$u = x_1 \otimes \cdots \otimes x_p, \qquad u^* = x^{*1} \otimes \cdots \otimes x^{*p}.$$

Then formulas (4.19) and (4.8) yield

$$\boxed{\langle \pi(x^{*1} \otimes \cdots \otimes x^{*p}), \pi(x_1 \otimes \cdots \otimes x_p) \rangle = \det(\langle x^{*i}, x_j \rangle).} \tag{4.20}$$

Suppose now that E is an inner product space. Then E is dual to itself with respect to the inner product and hence we may set $E^* = E$. It follows from Section 3.15 that the induced scalar product in $\otimes E$ is again non-degenerate and, hence, so is its restriction to the subspace $X(E)$. Hence an inner product is determined in the factor space $\otimes E/N(E)$ such that

$$(\pi u, \pi v) = p!(u, v) \qquad u, v \in X(E)$$

(cf. Formula 4.19). Formula (4.20) yields the relation

$$(\pi(x_1 \otimes \cdots \otimes x_p), \pi(y_1 \otimes \cdots \otimes y_p)) = \det(x_i, y_j) \qquad x_i \in E, y_j \in E.$$

PROBLEM

Define a multiplication in $X(E)$ such that the linear map $\rho: X(E) \to \otimes E/N(E)$ becomes an isomorphism. (This multiplication is necessarily uniquely determined.) Prove that

$$\pi_A u \cdot \pi_A v = \pi_A(u \otimes v) \qquad u, v \in \otimes E.$$

Symmetric Tensors

4.9. The Space $M^p(E)$

Consider the subspace $M^p(E)$ of $\otimes^p E$ generated (linearly) by the tensors $u - \tau u$ where $u \in \otimes^p E$ and τ is a transposition. The space $M^p(E)$ is stable under every transposition. In fact, if $v = u - \tau u$ is a generator of $M^p(E)$ and τ' is a transposition we have

$$\tau' v = (\tau' u - u) - (\tau u - u) + (\tau u - \tau'\tau u) \in M^p(E).$$

The same argument as in Section 4.1 shows that

$$\boxed{u - \sigma u \in M^p(E)} \tag{4.21}$$

for every $u \in \otimes^p E$ and every permutation σ.

4.10. The Symmetrizer

A tensor $u \in \otimes^p E$ is called *symmetric*, if

$$\sigma u = u \qquad \sigma \in S_p.$$

The symmetric tensors form a subspace $Y^p(E)$ of $\otimes^p E$.

Next consider the linear map $\pi_S: \otimes^p E \to \otimes^p E$ given by

$$\pi_S = \frac{1}{p!} \sum_\sigma \sigma. \tag{4.22}$$

An argument similar to the one given in Section 4.2 shows that

$$\ker \pi_S = M^p(E) \tag{4.23}$$

and

$$\operatorname{Im} \pi_S = Y^p(E). \tag{4.24}$$

Moreover, π_S is a projection operator,

$$\pi_S^2 = \pi_S. \tag{4.25}$$

Thus we have the direct decomposition

$$\otimes^p E = Y^p(E) \oplus M^p(E). \tag{4.26}$$

The operator π_S is called the *symmetrizer* in $\otimes^p E$ and $\pi_S u$ is called the *symmetric part of u*.

4.11. Dual Spaces

Suppose now that E, E^* is a pair of dual spaces and let π^S be the symmetrizer for $\otimes^p E^*$ ($p \geq 2$). The same argument as that used in Section 4.3 shows that π_S and π^S are dual operators,

$$\langle u^*, \pi_S u\rangle = \langle \pi^S u^*, u\rangle \qquad u^* \in \otimes^p E^*, u \in \otimes^p E. \tag{4.27}$$

It follows from (4.27) that the restriction of the scalar product $\langle , \rangle$ to the subspaces $Y^p(E^*)$, $Y^p(E)$ is again nondegenerate.

Now let

$$u^* = x^{*1} \otimes \cdots \otimes x^{*p} \quad \text{and} \quad u = x_1 \otimes \cdots \otimes x_p$$

be decomposable tensors. Then we obtain from (4.27) and (4.25) that

$$\begin{aligned}
&\langle \pi^S(x^{*1} \otimes \cdots \otimes x^{*p}), \pi_S(x_1 \otimes \cdots \otimes x_p)\rangle \\
&= \langle x^{*1} \otimes \cdots \otimes x^{*p}, \pi_S^2(x_1 \otimes \cdots \otimes x_p)\rangle \\
&= \langle x^{*1} \otimes \cdots \otimes x^{*p}, \pi_S(x_1 \otimes \cdots \otimes x_p)\rangle \\
&= \frac{1}{p!}\sum_\sigma \langle x^{*1}, x_{\sigma(1)}\rangle \cdots \langle x^{*p}, x_{\sigma(p)}\rangle.
\end{aligned} \tag{4.28}$$

Introducing the *permanent* of a $p \times p$-matrix α_i^j by

$$\operatorname{perm}(\alpha_i^j) = \sum_\sigma \alpha_{\sigma(1)}^1 \cdots \alpha_{\sigma(p)}^p,$$

we can rewrite (4.28) in the form

$$\langle \pi^S(x^{*1} \otimes \cdots \otimes x^{*p}), \pi_S(x_1 \otimes \cdots \otimes x_p)\rangle = \frac{1}{p!}\operatorname{perm}(\langle x^{*i}, x_j\rangle). \tag{4.29}$$

4.12. The Symmetric Part of a Product

Let $\otimes E$ be a tensor algebra over E, and consider the subspaces $M^p(E) \subset \otimes^p E$. As before we set $M^0(E) = M^1(E) = 0$, and define π_S to be the identity on $\otimes^0 E$ and $\otimes^1 E$. Then the formulas developed above continue to hold.

Now let $v = u - \tau u$ be any generator of $M^p(E)$, $p \geq 2$, and let $w \in \otimes^q E$ be an arbitrary tensor. Then we have

$$v \otimes w = u \otimes w - \tau u \otimes w = u \otimes w - \tau'(u \otimes w),$$

where $\tau' \in S_{p+q}$ denotes the transposition given by

$$\tau'(\nu) = \begin{cases} \tau(\nu) & 1 \leq \nu \leq p, \\ \nu & p + 1 \leq \nu \leq p + q. \end{cases}$$

It follows that

$$M^p(E) \otimes \otimes^q E \subset M^{p+q}(E). \tag{4.30a}$$

Similarly we obtain

$$\otimes^p E \otimes M^q(E) \subset M^{p+q}(E). \tag{4.30b}$$

Now by the same argument as in Section 4.4 it is shown that

$$\begin{aligned} \pi_S(u \otimes v) &= \pi_S(\pi_S u \otimes \pi_S v) \\ &= \pi_S(u \otimes \pi_r v) = \pi_S(\pi_S u \otimes v) \qquad u \in \otimes^p E,\ v \in \otimes^q E. \end{aligned} \tag{4.31}$$

PROBLEMS

1. Show that a bilinear mapping $\varphi: E \times E \to G$ can be uniquely written in the form

$$\varphi = \varphi_1 + \varphi_2,$$

where φ_1 is symmetric and φ_2 is skew-symmetric.

2. For each $p \geq 2$ prove that

$$X^p(E) \cap Y^p(E) = 0.$$

3. Let $\varphi: E \times \cdots \times E \to F$ be a symmetric p-linear mapping such that

$$\varphi(x, \ldots, x) = 0 \qquad x \in E.$$

Show that $\varphi = 0$.

The Factor Algebra $\otimes E/M(E)$

4.13. The Ideal $M(E)$

Consider the direct sum

$$M(E) = \sum_p M^p(E).$$

Formulas (4.30a) and (4.30b) imply that $M(E)$ is a graded ideal in the graded algebra $\otimes E$.

Suppose now that $u \in \otimes E$ and $v \in \otimes E$ are two arbitrary tensors. A calculation similar to that in Section 4.5 shows that

$$u \otimes v - v \otimes u \in M(E). \tag{4.32}$$

4.14. The Algebra $\otimes E/M(E)$

Consider the canonical projection

$$\pi : \otimes E \to \otimes E/M(E).$$

Since $M(E)$ is an ideal in $\otimes E$, a multiplication is induced in $\otimes E/M(E)$ such that

$$\pi(a \otimes b) = \pi a \cdot \pi b \qquad a, b \in \otimes E. \tag{4.33}$$

It follows from (4.33) that this multiplication is associative, and that $\pi(1)$ is a unit element. (4.32) implies that the multiplication is commutative as well.

Since $M(E)$ is graded, a gradation is induced in the factor algebra by

$$\otimes E/M(E) = \sum_p \pi(\otimes^p E)$$

and so $\otimes E/M(E)$ becomes a graded algebra. Since $M^1(E) = M^0(E) = 0$, the restriction of π to $\otimes^1 E = E$ and $\otimes^0 E = \Gamma$ is an isomorphism. Consequently we identify $\pi(\otimes^1 E)$ with E and $\pi(\otimes^0 E)$ with Γ.

4.15. Symmetric Tensors

Let $Y(E) \subset \otimes E$ be the space defined by

$$Y(E) = \sum_p Y^p(E).$$

Extend the projection operators $\pi_S : \otimes^p E \to \otimes^p E$ ($\pi_S = \iota$ in $\otimes^1 E$ and $\otimes^0 E$) to a linear map $\pi_S : \otimes E \to \otimes E$. Then

$$\ker \pi_S = M(E)$$

$$\operatorname{Im} \pi_S = Y(E)$$

and

$$\otimes E = M(E) \oplus Y(E).$$

The restriction σ of π to the subspace $Y(E)$ is a homogeneous linear isomorphism

$$\sigma : Y(E) \to \otimes E/M(E)$$

of degree zero. If $\pi_Y : \otimes E \to Y(E)$ denotes the restriction of π_S to $\otimes E$, $Y(E)$ we have the following commutative diagram:

$$\begin{array}{ccc} \otimes E & \xrightarrow{\pi_Y} & Y(E) \\ {\scriptstyle\pi}\downarrow & \swarrow {\scriptstyle\cong\,\sigma} & \\ \otimes E/M(E) & & \end{array} \tag{4.34}$$

4.16. The Induced Scalar Product

Let E and E^* be a pair of dual vector spaces and consider the induced scalar product in $\otimes E$ and $\otimes E^*$. According to Section 4.11 the restriction of this scalar product to the subspaces $Y(E)$ and $Y(E^*)$ is again nondegenerate. Consequently, a scalar product is induced in the factor spaces $\otimes E/M(E)$, $\otimes E^*/M(E^*)$ such that

$$\langle \sigma u^*, \sigma u \rangle = p!\langle u^*, u \rangle \qquad u^* \in Y^p(E^*), u \in Y^p(E). \tag{4.35}$$

Clearly, the scalar product (4.35) respects the gradations. Moreover, it follows from (4.35) and (4.34) that

$$\langle \pi u^*, \pi u \rangle = p!\langle \pi^S u^*, \pi_S u \rangle \qquad u^* \in \otimes E^*, u \in \otimes E. \tag{4.36}$$

Now let u^* and u be decomposable,

$$u = x_1 \otimes \cdots \otimes x_p, \qquad u^* = x^{*1} \otimes \cdots \otimes x^{*p}.$$

Then formulas (4.36) and (4.29) yield

$$\langle \pi(x^{*1} \otimes \cdots \otimes x^{*p}), \pi(x_1 \otimes \cdots \otimes x_p) \rangle = \operatorname{perm}(\langle x^{*i}, x_j \rangle).$$

5 Exterior Algebra

For this chapter E denotes a vector space over a field of characteristic zero.

Skew-Symmetric Mappings

5.1. Skew-Symmetric Mappings

Let E and F be two vector spaces and let

$$\varphi : \underbrace{E \times \cdots \times E}_{p} \to F$$

be a p-linear mapping. Then every permutation $\sigma \in S_p$ determines another p-linear mapping $\sigma\varphi$ given by

$$\sigma\varphi(x_1, \ldots, x_p) = \varphi(x_{\sigma(1)}, \ldots, x_{\sigma(p)}).$$

It follows immediately that

$$(\tau\sigma)\varphi = \tau(\sigma\varphi),$$

for every two permutations, and that

$$\iota\varphi = \varphi,$$

where ι is the identity permutation. A p-linear mapping φ is called *skew-symmetric* if

$$\sigma\varphi = \varepsilon_\sigma \varphi$$

for every permutation σ where $\varepsilon_\sigma = +1$ or -1 if the permutation is, respectively, even or odd.

A p-linear mapping φ is skew-symmetric if and only if

$$\tau\varphi = -\varphi \tag{5.1}$$

for every transposition τ. In fact, since a transposition is an odd permutation it follows that every skew symmetric mapping satisfies (5.1). Conversely, assume that φ is a p-linear mapping which satisfies (5.1) and let σ be an arbitrary permutation. Now σ is a product of m transpositions where m is even (odd) if σ is an even (odd) permutation. It follows that φ satisfies $\sigma\varphi = \varepsilon_\sigma \varphi$, and hence φ is skew-symmetric.

This result implies that a p-linear mapping, φ, is skew-symmetric if and only if

$$\varphi(x_1, \ldots, x_p) = 0, \tag{5.2}$$

whenever $x_i = x_j$ for at least one pair $i \neq j$. In fact, suppose that φ is skew-symmetric and assume that $x_i = x_j$ $(i \neq j)$. Let τ be the transposition $i \rightleftarrows j$. Then

$$\varphi(x_1, \ldots, x_p) = -\tau\varphi(x_1, \ldots, x_p) = -\varphi(x_1, \ldots, x_p),$$

whence $\varphi(x_1, \ldots, x_p) = 0$. Conversely, assume that φ satisfies (5.2). Then if $\tau : i \rightleftarrows j$ is any transposition, it follows that

$$\begin{aligned}\varphi(x_1, \ldots, x_i, \ldots, x_j, \ldots, x_p) &+ \varphi(x_1, \ldots, x_j, \ldots, x_i, \ldots, x_p)\\ &= \varphi(x_1, \ldots, x_i + x_j, \ldots, x_i + x_j, \ldots, x_p)\\ &\quad -\varphi(x_1, \ldots, x_i, \ldots, x_i, \ldots, x_p) - \varphi(x_1, \ldots, x_j, \ldots, x_j, \ldots, x_p) = 0,\end{aligned}$$

i.e.,

$$\varphi + \tau\varphi = 0.$$

Hence φ is skew-symmetric.

Since every transposition τ is a product of an odd number of transpositions of the form $i \rightleftarrows i + 1$, it follows that a p-linear mapping φ is skew-symmetric if and only if

$$\varphi(x_1, \ldots, x_p) = 0, \quad \text{whenever } x_i = x_{i+1} \qquad 1 \le i \le p - 1.$$

Formula (5.2) implies that a p-linear mapping, φ, is skew-symmetric if and only if

$$\varphi(x_1, \ldots, x_p) = 0, \tag{5.3}$$

whenever the x_i are linearly dependent. In fact, it is clear that this condition implies that φ is skew-symmetric. Conversely, let φ be a skew-symmetric map. Then if $x_1, \ldots, x_p$ are any linearly dependent vectors, we have

$$x_j = \sum_{i \neq j} \lambda^i x_i \qquad \text{(for some } j\text{)}.$$

Without loss of generality, we may assume $j = p$. It follows that

$$\varphi(x_1, \ldots, x_p) = \sum_{i=1}^{p-1} \lambda^i \varphi(x_1, \ldots, x_{p-1}, x_i) = 0,$$

which proves (5.3).

From every p-linear mapping φ we can obtain a skew-symmetric p-linear mapping A_φ by setting

$$A_\varphi = \frac{1}{p!}\sum_\sigma \varepsilon_\sigma \sigma\varphi.$$

To show that A_φ is indeed skew-symmetric let ρ be an arbitrary permutation. Then it follows that

$$\begin{aligned} p!\,\rho(A_\varphi) &= \sum_\sigma \varepsilon_\sigma \rho(\sigma\varphi) = \sum_\sigma \varepsilon_\sigma(\rho\sigma)\varphi \\ &= \varepsilon_\rho \sum_\sigma \varepsilon_{\sigma\rho}(\rho\sigma)\varphi = \varepsilon_\rho \sum_\tau \varepsilon_\tau \tau\varphi = \varepsilon_\rho p!\,A_\varphi, \end{aligned}$$

whence $\rho(A_\varphi) = \varepsilon_\rho A_\varphi$.

The mapping A_φ is called the *skew-symmetric part* of φ, and the operator $A:\varphi \to A_\varphi$ is called the *antisymmetry operator*. If φ is skew-symmetric itself we have $\sigma\varphi = \varepsilon_\sigma \varphi$ for every σ and hence it follows that

$$A_\varphi = \frac{1}{p!}\sum_\sigma \varphi = \varphi.$$

This equation shows that a skew-symmetric mapping coincides with its skew-symmetric part. Since A_φ is skew-symmetric it follows that $A^2 = A$.

Proposition 5.1.1. Let

$$\varphi:\underbrace{E \times \cdots \times E}_{p} \to F$$

be a p-linear mapping , where F is an arbitrary vector space, and let $f:\otimes^p E \to F$ be the linear map induced by φ. Then φ is skew-symmetric if and only if $N^p(E) \subset \ker f$.

PROOF. φ is skew-symmetric if and only if $\varphi(x_1, \ldots, x_p) = 0$, whenever $x_i = x_j$ for some pair (i, j), $i \neq j$. But

$$\varphi(x_1, \ldots, x_p) = f(x_1 \otimes \cdots \otimes x_p)$$

and so φ is skew-symmetric if and only if f is zero on the generators of $N^p(E)$; i.e., if and only if $N^p(E) \subset \ker f$. □

As an example of a p-linear skew-symmetric mapping, consider the p-linear mapping

$$\psi_A:\underbrace{E \times \cdots \times E}_{p} \to \otimes^p E$$

defined by

$$\psi_A = \pi_A \circ \otimes^p,$$

where π_A is the alternator (see Section 4.2). Since $\ker \pi_A = N^p(E)$, it follows from the proposition that ψ_A is skew-symmetric.

PROBLEMS

1. Denote by $L_p(E; F)$ the space of all p-linear mappings $\varphi: E \times \cdots \times E \to F$ and by $A_p(E; F)$ the subspace of skew-symmetric mappings. Assume that

$$T: L_p(E; F) \to L_p(E; F)$$

is a linear map such that

$$T\varphi = \varphi \qquad \varphi \in A_p(E; F)$$

and

$$T(\sigma\varphi) = \varepsilon_\sigma T(\varphi) \qquad \sigma \in S_p,\ \varphi \in L_p(E; F).$$

Show that T is the antisymmetry operator.

2. Let E be an n-dimensional vector space and suppose that $\Delta \neq 0$ is a determinant function in E.
   (a) Given an n-linear skew-symmetric mapping $\varphi: E \times \cdots \times E \to F$ show that there is a unique vector $b \in F$ such that

$$\varphi(x_1, \ldots, x_p) = \Delta(x_1, \ldots, x_n)b.$$

   (b) Show that every $(n-1)$-linear skew-symmetric function Φ can be written in the form

$$\Phi(x_1, \ldots, x_{n-1}) = \Delta(x_1, \ldots, x_{n-1}, a_\Phi),$$

where $a_\Phi \in E$ is a fixed vector. *Hint*: Consider the n-linear mapping φ defined by

$$\varphi(x_1, \ldots, x_n) = \sum_{i=1}^{n} (-1)^{n-1}\Phi(x_1, \ldots, \hat{x}_i, \ldots, x_n)x_i.$$

Exterior Algebra

5.2. The Universal Property

Let

$$\wedge^p: \underbrace{E \times \cdots \times E}_{p} \to A$$

be a skew-symmetric p-linear map from E to a vector space A. We shall say that $\wedge^p$ has the *universal property* (for skew-symmetric maps) if it satisfies the following conditions:

$\wedge_1$: The vectors $\wedge^p(x_1, \ldots, x_p)$ $(x_i \in E)$ generate A, or equivalently, $\operatorname{Im} \wedge^p = A$.

$\wedge_2$: If φ is a skew-symmetric p-linear mapping from E into any vector space H, then there exists a linear map $f:A \to H$ such that the diagram

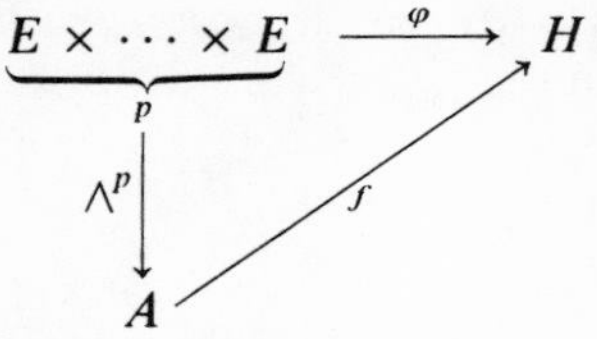

commutes.

Conditions $\wedge_1$ and $\wedge_2$ are equivalent to the following condition (the proof being the same as in Section 1.4)

$\wedge$: If

$$\varphi:\underbrace{E \times \cdots \times E}_{p} \to H$$

is a skew-symmetric p-linear mapping, there is a *unique* linear map $f:A \to H$ such that the diagram above commutes.

The skew-symmetry of $\wedge^p$ implies that

$$\wedge^p(x_1, \ldots, x_p) = 0$$

whenever the vectors $x_1, \ldots, x_p$ are linearly dependent. On the other hand, if the vectors $x_1, \ldots, x_p$ are linearly independent, then $\wedge^p(x_1, \ldots, x_p) \neq 0$. In fact, assume that $\wedge^p(x_1, \ldots, x_p) = 0$. Then, by $\wedge_2$, $\psi(x_1, \ldots, x_p) = 0$ for every skew-symmetric p-linear mapping ψ. In particular,

$$\pi_A(x_1 \otimes \cdots \otimes x_p) = 0.$$

Thus, by Section 4.2, the vectors x_ν are linearly dependent.

5.3. Uniqueness and Existence

Suppose that

$$\wedge^p:\underbrace{E \times \cdots \times E}_{p} \to A \quad \text{and} \quad \tilde{\wedge}^p:\underbrace{E \times \cdots \times E}_{p} \to \tilde{A}$$

are two p-linear mappings with the universal property. Then there are linear maps

$$f:A \to \tilde{A} \quad \text{and} \quad g:\tilde{A} \to A$$

such that

$$f \circ \wedge^p = \tilde{\wedge}^p \quad \text{and} \quad g \circ \tilde{\wedge}^p = \wedge^p.$$

Now condition $\wedge_1$ implies that

$$g \circ f = \iota \quad \text{and} \quad f \circ g = \iota.$$

Thus f and g are inverse isomorphisms.

To prove existence, set

$$\wedge^p E = \otimes^p E / N^p(E)$$

(see Section 4.1) and let $\wedge^p$ denote the p-linear mapping

$$\underbrace{E \times \cdots \times E}_{p} \to \wedge^p E$$

defined by

$$\wedge^p(x_1, \ldots, x_p) = \pi(x_1 \otimes \cdots \otimes x_p), \tag{5.4}$$

where π denotes the projection (see Section 4.6). In view of Proposition 5.1.1, $\wedge^p$ is skew-symmetric. Property $\wedge_1$ follows directly from the definition. To verify $\wedge_2$, let

$$\varphi : \underbrace{E \times \cdots \times E}_{p} \to H$$

be any p-linear skew-symmetric mapping. Then φ determines a linear map $h : \otimes^p E \to H$ such that

$$h(x_1 \otimes \cdots \otimes x_p) = \varphi(x_1, \ldots, x_p) \qquad x_\nu \in E \tag{5.5}$$

(see Section 1.20). Since φ is skew-symmetric, it reduces to zero in the subspace $N^p(E)$ and so it induces a linear map $f : \wedge^p E \to H$ such that

$$f \circ \pi = h.$$

Combining this relation with (5.5) we obtain

$$\varphi(x_1, \ldots, x_p) = f\pi(x_1 \otimes \cdots \otimes x_p) = f \wedge^p(x_1, \ldots, x_p)$$

whence $\varphi = f \circ \wedge^p$.

Definition. The pth *exterior power of* E is a pair $(A, \wedge^p)$, where

$$\wedge^p : \underbrace{E \times \cdots \times E}_{p} \to A$$

is a skew-symmetric p-linear mapping with the universal property. The space A (which is uniquely determined up to an isomorphism) will also be called the pth exterior algebra of E and is denoted by $\wedge^p E$.

The elements of $\wedge^p E$ are called *p-vectors*.

Next we shall give a description of the pth exterior power of E in terms of the subspace $X^p(E) \subset \otimes^p E$ of skew-symmetric tensors. Consider the skew-symmetric p-linear mapping $\varphi\colon E \times \cdots \times E \to \otimes^p E$ given by

$$\varphi(x_1, \ldots, x_p) = \pi_A(x_1 \otimes \cdots \otimes x_p) \qquad x_\nu \in E,$$

where π_A denotes the alternator (see Section 4.2). By the universal property it induces a linear map $\eta\colon \wedge^p E \to \otimes^p E$. We show that the diagram

$$\begin{array}{ccc} & \otimes^p E & \\ {\scriptstyle \pi} \swarrow & & \searrow {\scriptstyle \pi_A} \\ \wedge^p E & \xrightarrow[\eta]{} & \otimes^p E \end{array}$$

commutes. In fact,

$$\begin{aligned} \eta\pi(x_1 \otimes \cdots \otimes x_p) &= \eta \wedge^p(x_1, \ldots, x_p) = \varphi(x_1, \ldots, x_p) \\ &= \pi_A(x_1 \otimes \cdots \otimes x_p) \end{aligned}$$

and so

$$\eta \circ \pi = \pi_A. \tag{5.6}$$

This relation implies that $\operatorname{Im} \eta = \operatorname{Im} \pi_A$ (since π is surjective). But $\operatorname{Im} \pi_A = X^p(E)$ and so we have

$$\operatorname{Im} \eta = X^p(E).$$

On the other hand, it is easy to check the relation

$$\pi \circ \eta = \iota.$$

This implies that η is injective. Thus η determines a linear isomorphism from $\wedge^p E$ onto $X^p(E)$.

$$\eta\colon \wedge^p E \xrightarrow{\cong} X^p(E).$$

Remark. Since η is injective, the formula $\eta \circ \pi = \pi_A$ implies that $\ker \pi_A = \ker \pi$. Thus we have the relation $\ker \pi_A = N^p(E)$ which was proved in Section 4.2 in a different way.

5.4. Exterior Algebra

Extend the definition of $\wedge^p E$ to the cases $p = 1$ and $p = 0$ by setting $\wedge^1 E = E$ and $\wedge^0 E = \Gamma$. Consider the direct sum, $\wedge E$, of the spaces $\wedge^p E$ $(p = 0, 1, \ldots)$ and identify each $\wedge^p E$ with its image under the inclusion map. Then we can write

$$\wedge E = \sum_{p=0}^{\infty} \wedge^p E.$$

The projections $\pi: \otimes^p E \to \wedge^p E$ (with kernel $N^p(E)$) determine a projection

$$\pi: \otimes E \to \wedge E$$

with kernel $N(E) = \sum_p N^p(E)$. Thus we have a linear isomorphism

$$f: \otimes E/N(E) \xrightarrow{\cong} \wedge E.$$

Now recall from Section 4.6 that $\otimes E/N(E)$ is an associative algebra. Hence there is a unique multiplication in $\wedge E$, denoted by $\wedge$, such that f becomes an algebra isomorphism. Thus we have

$$u \wedge v = \pi(\tilde{u} \otimes \tilde{v}) \qquad u \in \wedge E, v \in \wedge E,$$

where $\tilde{u} \in \otimes E$, $\tilde{v} \in \otimes E$ are elements such that $\pi\tilde{u} = u$ and $\pi\tilde{v} = v$. This multiplication makes $\wedge E$ into an associative algebra with the scalar 1 as unit element. It is called the *exterior algebra over E*. It is generated (as an algebra) by the vectors $x \in E$ and the scalar 1.

Formula (5.4) can now be written in the form

$$\wedge^p(x_1, \ldots, x_p) = x_1 \wedge \cdots \wedge x_p \qquad x_v \in E.$$

From (4.16) we obtain the relation

$$u \wedge v = (-1)^{pq} v \wedge u \qquad u \in \wedge^p E, v \in \wedge^q E.$$

In particular,

$$u \wedge v = v \wedge u \quad \text{if } p \text{ or } q \text{ is even.} \tag{5.7}$$

and

$$u \wedge u = 0 \quad \text{if } p \text{ is odd.}$$

The kth *exterior power* of an element $u \in \wedge E$ is defined by

$$u^k = \frac{1}{k!}\underbrace{u \wedge \cdots \wedge u}_{k} \qquad k \geq 1,$$

$$u^0 = 1.$$

It follows that

$$u^k \wedge u^l = \binom{k+l}{k} u^{k+l} \qquad u \in \wedge E.$$

Now let $u \in \wedge^p E$ and $v \in \wedge^q E$ be arbitrary and assume that p *or* q is even. Then it follows from (5.7) that $u \wedge v = v \wedge u$. This yields the *binomial formula*

$$(u + v)^k = \sum_{i+j=k} u^i \wedge v^j$$

for $u \in \wedge^p E$, $v \in \wedge^q E$, p or q even.

Now we shall describe the exterior algebra $\wedge E$ in terms of the subspaces $X^p(E) \subset \otimes^p E$ (see Section 4.2). Consider the direct sum

$$X(E) = \sum_p X^p(E)$$

and define a multiplication in $X(E)$, denoted by $\cap$, by setting

$$a \cap b = \pi_A(a \otimes b) \qquad a, b \in X(E),$$

where π_A denotes the alternator. Recall from Section 5.3 the linear isomorphisms $\eta: \wedge^p E \xrightarrow{\cong} X^p(E)$. We shall show that η preserves products,

$$\eta(u \wedge v) = \eta(u) \cap \eta(v) \qquad u \in \wedge^p E, v \in \wedge^q E.$$

In fact, write

$$u = \pi\tilde{u}, \qquad v = \pi\tilde{v} \qquad \tilde{u}, \tilde{v} \in \otimes E.$$

Then we have, in view of the commutative triangle in Section 5.3 and Formula (4.10)

$$\begin{aligned} \eta(u \wedge v) &= \eta(\pi\tilde{u} \wedge \pi\tilde{v}) = \eta\pi(\tilde{u} \otimes \tilde{v}) \\ &= \pi_A(\tilde{u} \otimes \tilde{v}) = \pi_A(\pi_A \tilde{u} \otimes \pi_A \tilde{v}) \\ &= \pi_A \tilde{u} \cap \pi_A \tilde{v} = \eta\pi\tilde{u} \cap \eta\pi\tilde{v} = \eta u \cap \eta v. \end{aligned}$$

Thus η is an algebra isomorphism $\eta: \wedge E \xrightarrow{\cong} X(E)$.

5.5. The Universal Property of $\wedge E$

Let A be an associative algebra with unit element e and let

$$h: \wedge E \to A$$

be a homomorphism. Then a linear map $\varphi: E \to A$ is defined by

$$\varphi = h \circ i,$$

where i is the injection of E into $\wedge E$. It follows from (5.7) that

$$(\varphi x)^2 = 0 \qquad x \in E. \tag{5.8}$$

Conversely, assume that $\varphi: E \to A$ is a linear map satisfying (5.8).

Then there exists precisely one homomorphism $h: \wedge E \to A$ such that $h(1) = e$ and

$$\varphi = h \circ i.$$

For the proof we note first that (5.8) implies that

$$\varphi x \cdot \varphi y + \varphi y \cdot \varphi x = 0 \qquad x, y \in E. \tag{5.9}$$

In fact, if $x, y \in E$ are arbitrary elements, we have

$$\varphi x \cdot \varphi y + \varphi y \cdot \varphi x = \varphi(x + y) \cdot \varphi(x + y) - \varphi x \cdot \varphi x - \varphi y \cdot \varphi y = 0.$$

To define h consider, for every $p \geq 2$, the p-linear mapping

$$\alpha: \underbrace{E \times \cdots \times E}_{p} \to A$$

defined by

$$\alpha(x_1, \ldots, x_p) = \varphi x_1 \cdots \varphi x_p.$$

Then it follows from (5.9) that α is skew-symmetric and hence there exists a linear map $h^p: \wedge^p E \to A$ such that

$$h^p(x_1 \wedge \cdots \wedge x_p) = \varphi x_1 \cdots \varphi x_p \qquad p \geq 2.$$

Define h^1 and h^0 by $h^1 = \varphi$ and $h^0(1) = e$ and let $h: \wedge E \to A$ be the linear map whose restriction to $\wedge^p E$ is equal to h^p, $p \geq 0$.

To prove that h is a homomorphism, let

$$u = x_1 \wedge \cdots \wedge x_p \quad \text{and} \quad v = x_{p+1} \wedge \cdots \wedge x_{p+q}$$

be two decomposable elements. Then we have

$$\begin{aligned} h(u \wedge v) &= h(x_1 \wedge \cdots \wedge x_{p+q}) = \varphi x_1 \cdots \varphi x_{p+q} \\ &= (\varphi x_1 \cdots \varphi x_p) \cdot (\varphi x_{p+1} \cdots \varphi x_{p+q}) \\ &= h(x_1 \wedge \cdots \wedge x_p) \cdot h(x_{p+1} \wedge \cdots \wedge x_{p+q}) = hu \cdot hv. \end{aligned}$$

The uniqueness of h follows from the fact that the algebra $\wedge E$ is generated by the vectors of E and the scalar 1.

If A is a *positively graded* associative algebra, $A = \sum_p A^p$, and φ is a linear map of E into A^1 it follows that the extending homomorphism h is of degree zero.

Let U be an associative algebra with unit element 1 and let $\varepsilon: E \to U$ be a linear map with the following property: If $\varphi: E \to A$ is a linear map into any associative algebra A with unit element e such that $(\varphi x)^2 = 0$, $x \in E$, then there exists precisely one homomorphism $h: U \to A$ such that

$$h(1) = e \quad \text{and} \quad h \circ \varepsilon = \varphi.$$

Then we say that the pair (U, ε) has the *universal exterior algebra property*.

An argument completely analogous to that found in Section 3.4 shows that if the pairs (U, ε) and (U', ε') have the universal exterior algebra property then there exists a unique isomorphism $f: U \to U'$ such that $f \circ \varepsilon = \varepsilon'$.

It follows from the results of this section that the pair $(\wedge E, i)$ has the universal exterior algebra property, where $i: E \to \wedge E$ is the inclusion map. Now the above uniqueness theorem implies that for every universal pair (U, ε) there exists a unique isomorphism $f: \wedge E \to U$ such that $f \circ i = \varepsilon$.

5.6. Exterior Algebra Over Dual Spaces

Let E, E^* be a pair of dual vector spaces, and consider the exterior algebras over E and E^*. In view of the induced isomorphisms

$$f: \otimes E/N(E) \xrightarrow{\cong} \wedge E \quad \text{and} \quad g: \otimes E^*/N(E^*) \xrightarrow{\cong} E^*$$

it follows from Section 4.8 that a scalar product $\langle , \rangle$ may be defined between $\wedge E$ and $\wedge E^*$ such that

$$\begin{aligned} \langle x^{*1} \wedge \cdots \wedge x^{*p}, x_1 \wedge \cdots \wedge x_p \rangle &= \det(\langle x^{*i}, x_j \rangle) && p \geqq 1 \\ \langle \lambda, \mu \rangle &= \lambda\mu && \lambda, \mu \in \Gamma \\ \langle \wedge^p E^*, \wedge^q E \rangle &= 0 && \text{if } p \neq q. \end{aligned} \tag{5.10}$$

Condition $\wedge_1$ implies that the scalar product $\langle , \rangle$ is uniquely determined by (5.10). We also have that the restriction of $\langle , \rangle$ to the pair $\wedge^p E^*$, $\wedge^p E$ is nondegenerate for each p, and so induces a duality between these spaces. In particular, the restriction of $\langle , \rangle$ to $\wedge^1 E^* = E^*$ and $\wedge^1 E = E$ is just the original scalar product.

Now expanding the determinant in (5.10) by the ith row we obtain the formula

$$\langle x^{*1} \wedge \cdots \wedge x^{*p}, x_1 \wedge \cdots \wedge x_p \rangle = \sum_{j=1}^{p} (-1)^{i+j} \langle x^{*j}, x_i \rangle,$$
$$\langle x^{*1} \wedge \cdots \wedge \widehat{x^{*j}} \wedge \cdots \wedge x^{*p}, x_1 \wedge \cdots \wedge \hat{x}_i \wedge \cdots \wedge x_p \rangle \qquad p \geq 2. \tag{5.11}$$

5.7. Exterior Algebra Over a Vector Space of Finite Dimension

Suppose now that E is a vector space of dimension n and let $\{e_\nu\}$ $(\nu = 1, \ldots, n)$ be a basis of E. Then the products

$$e_{\nu_1} \wedge \cdots \wedge e_{\nu_p}, (\nu_1 < \cdots < \nu_p) \tag{5.12}$$

form a basis for $\wedge^p E$. In fact, it follows immediately from $\wedge_1$ that the products (5.12) generate $\wedge^p E$. To prove the linear independence, let E^* be a dual space of E. If $e^{*\nu}$ $(\nu = 1, \ldots, n)$ is the dual basis in E^* we have, in view of (5.10),

$$\langle e^{*\nu_1} \wedge \cdots \wedge e^{*\nu_p}, e_{\mu_1} \wedge \cdots \wedge e_{\mu_p} \rangle = \det(\langle e^{*\nu_i}, e_{\mu_j} \rangle) = \det(\delta^{\nu_i}_{\mu_j}). \tag{5.13}$$

This formula shows that the products $e_{\nu_1} \wedge \cdots \wedge e_{\nu_p}$ and $e^{*\nu_1} \wedge \cdots \wedge e^{*\nu_p}$ $(\nu_1 < \cdots < \nu_p)$ form dual bases of $\wedge^p E$ and $\wedge^p E^*$. Hence, the dimension of $\wedge^p E$ is given by

$$\dim \wedge^p E = \binom{n}{p} \qquad (0 \leq p \leq n). \tag{5.14}$$

For the dimension of the exterior algebra $\wedge E$ we obtain from (5.14)

$$\dim \wedge E = \sum_{p=0}^{n} \binom{n}{p} = 2^n,$$

while the Poincaré polynomial of the graded vector space $\wedge E$ is given by

$$P(t) = \sum_{p=0}^{n} \binom{n}{p} t^n = (1 + t)^n. \tag{5.15}$$

Every p-vector u can be uniquely represented in the form

$$u = \sum_{<} \xi^{\nu_1, \ldots, \nu_p} e_{\nu_1} \wedge \cdots \wedge e_{\nu_p},$$

where the symbol $<$ indicates that the indices $(\nu_1, \ldots, \nu_p)$ are subject to the condition $\nu_1 < \cdots < \nu_p$. The coefficients $\xi^{\nu_1, \ldots, \nu_p}$ are called the *components* of the p-vector u with respect to the basis $\{e_\nu\}$ of E. Formula (5.13) implies that the scalar product of two p-vectors

$$u = \sum_{<} \xi^{\nu_1, \ldots, \nu_p} e_{\nu_1} \wedge \cdots \wedge e_{\nu_p}$$

and

$$u^* = \sum_{<} \eta_{\nu_1, \ldots, \nu_p} e^{*\nu_1} \wedge \cdots \wedge e^{*\nu_p}$$

is given by

$$\langle u^*, u \rangle = \sum_{<} \xi^{\nu_1, \ldots, \nu_p} \eta_{\nu_1, \ldots, \nu_p}.$$

Inner product spaces. Suppose now that E is an inner product space and set $E^* = E$. Let $\wedge E$ be an exterior algebra over E. The isomorphism

$$f: \otimes E/N(E) \to \wedge E$$

(see Section 5.6) determines an inner product in $\wedge E$ such that

$$(x_1 \wedge \cdots \wedge x_p, y_1 \wedge \cdots \wedge y_p) = \det(x_i, y_j).$$

If E is Euclidean and $\{e_\nu\}$ $(\nu = 1, \ldots, n)$ is an orthonormal basis of E, it follows that the products

$$e_{\nu_1} \wedge \cdots \wedge e_{\nu_p} \qquad \nu_1 < \cdots < \nu_p$$

form an orthonormal basis of $\wedge^p E$.

Problems

1. Let E, E^* be a pair of n-dimensional dual vector spaces and consider the subspace $A \subset L(E^*; E)$ consisting of the transformations satisfying $\varphi^* = -\varphi$. Define a bilinear mapping $\varphi: E \times E \to A$ by

$$\varphi_{a,b} x^* = \langle x^*, a \rangle b - \langle x^*, b \rangle a.$$

Prove that the pair (A, φ) is a second exterior power of E.

2. Show that a 2-vector z is decomposable if and only if $z \wedge z = 0$.
3. Let E be a vector space of dimension n. Show that a 2-vector is decomposable if and only if the matrix of its components (with respect to a basis of E) has rank 1.
4. Establish the general *Lagrange identity*

$$\begin{vmatrix} \sum_{\nu=1}^{n} \xi_1^\nu \eta_\nu^1 & \cdots & \sum_{\nu=1}^{n} \xi_1^\nu \eta_\nu^p \\ \vdots & & \vdots \\ \sum_{\nu=1}^{n} \xi_p^\nu \eta_\nu^1 & \cdots & \sum_{\nu=1}^{n} \xi_p^\nu \eta_\nu^p \end{vmatrix} = \sum_{<} \det \begin{pmatrix} \xi_1^{\nu_1} & \cdots & \xi_1^{\nu_p} \\ \vdots & & \vdots \\ \xi_p^{\nu_1} & \cdots & \xi_p^{\nu_p} \end{pmatrix} \det \begin{pmatrix} \eta_1^{\nu_1} & \cdots & \eta_1^{\nu_p} \\ \vdots & & \vdots \\ \eta_p^{\nu_1} & \cdots & \eta_p^{\nu_p} \end{pmatrix}.$$

Hint: Employing a pair of dual bases $\{e^{*\nu}\}$, $\{e_\nu\}$ $(\nu = 1, \ldots, n)$, consider the vectors

$$x_i = \sum_\nu \xi_i^\nu e_\nu \quad \text{and} \quad y^{*i} = \sum_\nu \eta_\nu^i e^{*\nu} \qquad (i = 1, \ldots, p).$$

Evaluate the scalar product

$$\langle x_1 \wedge \cdots \wedge x_p, y^{*1} \wedge \cdots \wedge y^{*p} \rangle.$$

in two different ways.

5. Assume that E has finite dimension, and consider a differentiable mapping $\mathbb{R} \to \wedge^p E$, $t \mapsto u(t)$. Establish the formula

$$\frac{d}{dt} u^k = \frac{du}{dt} \wedge u^{k-1}.$$

6. Let E be any vector space and $\wedge E$ be an exterior algebra over E. Define a new multiplication in the space $\wedge E$ by

$$u \mathbin{\tilde{\wedge}} v = \frac{(p+q)!}{p!q!} u \wedge v \qquad u \in \wedge^p E, v \in \wedge^q E.$$

Prove that the resulting algebra $\tilde{\wedge} E$ is again an exterior algebra over E.

7. Consider the subspace $\sum_{k=0}^{\infty} \wedge^{2k} E$ of $\wedge E$. Show that this subspace is a commutative algebra which is algebraically generated by 1 together with a set of elements $\{w_\nu\}$ satisfying $w_\nu^2 = 0$. This algebra is called a *Nolting algebra* over E.

Homomorphisms, Derivations and Antiderivations

5.8. The Induced Homomorphism

Let $\wedge E$ and $\wedge F$ be exterior algebras over the vector spaces E and F and assume that a linear map $\varphi: E \to F$ is given. Then φ can be extended in a unique way to a homomorphism $\varphi_\wedge$, $\wedge E \to \wedge F$ such that $\varphi_\wedge(1) = 1$. To prove this consider the linear map $\eta: E \to \wedge F$ defined by $\eta = i \circ \varphi$ where i

denotes the injection of F into $\wedge F$. Then we have for every $x \in E$

$$\eta(x) \wedge \eta(x) = \varphi x \wedge \varphi x = 0 \qquad (\text{since} \quad \varphi x \in F).$$

Now it follows from the universal property of $\wedge E$ (see Section 5.5) that η (and hence φ) can be extended in a unique way to a homomorphism $\varphi_\wedge : \wedge E \to \wedge F$. Clearly $\varphi_\wedge$ is homogeneous of degree zero. Since $\varphi_\wedge$ preserves products we have

$$\varphi_\wedge(u \wedge v) = \varphi_\wedge u \wedge \varphi_\wedge v \qquad u, v \in \wedge E. \tag{5.16}$$

In terms of the multiplication operator (see Section 5.13) this can be written as

$$\varphi_\wedge \circ \mu(u) = \mu(\varphi_\wedge u) \circ \varphi_\wedge \qquad u \in \wedge E. \tag{5.17}$$

From (5.16) we obtain the formula

$$\varphi_\wedge(x_1 \wedge \cdots \wedge x_p) = \varphi x_1 \wedge \cdots \wedge \varphi x_p \qquad x_i \in E. \tag{5.18}$$

It follows immediately from (5.18) that $\operatorname{Im} \varphi_\wedge = \wedge(\operatorname{Im} \varphi)$. In particular, if φ is an onto map then so is $\varphi_\wedge$. The kernel of $\varphi_\wedge$ will be discussed in Section 5.24.

For the identity map $\iota : E \to E$ we have obviously

$$\iota_\wedge = \iota \tag{5.19}$$

while the homomorphism $(-\iota)_\wedge$ is given by $(-\iota)_\wedge u = (-1)^p u$, $u \in \wedge^p E$; hence $(-\iota)_\wedge$ is the canonical involution of the algebra $\wedge E$.

If ψ is a linear map of F into a third vector space G and $\wedge G$ is an exterior algebra over G, it is clear that

$$(\psi \circ \varphi)_\wedge = \psi_\wedge \circ \varphi_\wedge. \tag{5.20}$$

Formulas (5.20) and (5.19) imply that $\varphi_\wedge$ is injective whenever φ is injective. In fact, if φ is injective, there exists a linear map $\psi : E \leftarrow F$ such that $\psi \circ \varphi = \iota$. Now formulas (5.20) and (5.19) yield

$$\psi_\wedge \circ \varphi_\wedge = (\psi \circ \varphi)_\wedge = \iota_\wedge = \iota$$

and hence $\varphi_\wedge$ is injective.

In particular, if E_1 is a subspace of E, then the injection i: $E_1 \to E$ induces a monomorphism $i_\wedge : \wedge E_1 \to \wedge E$. Hence, $\wedge E_1$ can be identified with a subalgebra of $\wedge E$.

5.9. Dual Mappings

Suppose now that E^*, E and F^*, F are two pairs of dual spaces and that two dual mappings

$$\varphi : E \to F \quad \text{and} \quad \varphi^* : E^* \leftarrow F^*$$

are given. Consider the induced mappings

$$\varphi_\wedge : \wedge E \to \wedge F \quad \text{and} \quad (\varphi^*)_\wedge : \wedge E^* \leftarrow \wedge F^*.$$

The mapping $(\varphi^*)_\wedge$ will be denoted by $\varphi^\wedge$. Then $\varphi_\wedge$ and $\varphi^\wedge$ are again dual,

$$\varphi^\wedge = (\varphi_\wedge)^*. \tag{5.21}$$

In fact let $u \in \wedge E$ and $v^* \in \wedge E^*$ be two arbitrary elements. Since $\varphi_\wedge$ and $\varphi^\wedge$ are both homogeneous of degree zero we may assume that u and v^* are homogeneous of the same degree, say p. Furthermore, we may assume that the p-vectors u and v^* are both decomposable

$$u = x_1 \wedge \cdots \wedge x_p, \qquad v^* = y^{*1} \wedge \cdots \wedge y^{*p}.$$

Then relations (5.18) and (5.10) yield

$$\begin{aligned}\langle v^*, \varphi_\wedge u\rangle &= \langle y^{*1} \wedge \cdots \wedge y^{*p}, \varphi x_1 \wedge \cdots \wedge \varphi x_p\rangle \\ &= \det(\langle y^{*i}, \varphi x_j\rangle) = \det(\langle \varphi^* y^{*i}, x_j\rangle) \\ &= \langle \varphi^* y^{*1} \wedge \cdots \wedge \varphi^* y^{*p}, x_1 \wedge \cdots \wedge x_p\rangle = \langle \varphi^\wedge v^*, u\rangle\end{aligned}$$

whence

$$\langle v^*, \varphi_\wedge u\rangle = \langle \varphi^\wedge v^*, u\rangle \qquad v^* \in \wedge E^*, u \in \wedge E.$$

If G, G^* is a third pair of dual spaces with exterior algebras $\wedge G$ and $\wedge G^*$ and if $\psi: F \to G$, $\psi^*: F^* \leftarrow G^*$ is another pair of dual mappings, we have

$$(\psi \circ \varphi)^\wedge = ((\psi \circ \varphi)^*)_\wedge = (\varphi^* \circ \psi^*)_\wedge = (\varphi^*)_\wedge \circ (\psi^*)_\wedge = \varphi^\wedge \circ \psi^\wedge$$

whence

$$(\psi \circ \varphi)^\wedge = \varphi^\wedge \circ \psi^\wedge. \tag{5.22}$$

5.10. The Induced Derivation

Let $\varphi: E \to E$ be a linear transformation. Then φ can be extended in a unique way to a derivation, $\theta_\wedge(\varphi)$, in the algebra $\wedge E$. The uniqueness of $\theta_\wedge(\varphi)$ follows immediately from the fact that the algebra $\wedge E$ is generated by the vectors of E and the unit element 1.

To prove existence, consider the p-linear mapping

$$\psi_p : \underbrace{E \times \cdots \times E}_{p} \to \wedge^p E$$

defined by

$$\psi_p(x_1, \ldots, x_p) = \sum_{i=1}^{p} x_1 \wedge \cdots \wedge \varphi x_i \wedge \cdots \wedge x_p \qquad p \geq 2$$

$$\psi_1 = \varphi \quad \text{and} \quad \psi_0 = 0.$$

If for some $i < j$, $x_i = x_j = x$, then

$$\begin{aligned}\psi_p(x_1, \ldots, x, \ldots, x, \ldots, x_p) &= x_1 \wedge \cdots \wedge \varphi x \wedge \cdots \wedge x \wedge \cdots \wedge x_p \\ &\quad + x_1 \wedge \cdots \wedge x \wedge \cdots \wedge \varphi x \wedge \cdots \wedge x_p \\ &= (-1)^{j-i} x_1 \wedge \cdots \wedge x \wedge \varphi x \wedge \cdots \wedge x_p \\ &\quad + (-1)^{j-1-i} x_1 \wedge \cdots \wedge x \wedge \varphi x \wedge \cdots \wedge x_p \\ &= 0.\end{aligned}$$

Hence ψ_p is skew-symmetric. It follows that there is a linear map, $\theta_\wedge(\varphi)\colon \wedge E \to \wedge E$ such that

$$\theta_\wedge(\varphi)(x_1 \wedge \cdots \wedge x_p) = \sum_{i=1}^{p} x_1 \wedge \cdots \wedge \varphi x_i \wedge \cdots \wedge x_p \qquad p \geq 2 \tag{5.23}$$

and such that

$$\theta_\wedge(\varphi)x = \varphi x \qquad x \in E,$$

$$\theta_\wedge(\varphi)\lambda = 0 \qquad \lambda \in \Gamma.$$

Clearly $\theta_\wedge(\varphi)$ is a homogeneous (of degree 0) linear map extending φ. To prove that $\theta_\wedge(\varphi)$ is a derivation let $u = x_1 \wedge \cdots \wedge x_p$ and $v = y_1 \wedge \cdots \wedge y_q$ be arbitrary decomposable p- and q-vectors. Then

$$\begin{aligned}\theta_\wedge(\varphi)(u \wedge v) &= \theta_\wedge(\varphi)(x_1 \wedge \cdots \wedge x_p \wedge y_1 \wedge \cdots \wedge y_q) \\ &= \left(\sum_{i=1}^{p} x_1 \wedge \cdots \wedge \varphi x_i \wedge \cdots \wedge x_p\right) \wedge y_1 \wedge \cdots \wedge y_q \\ &\quad + \sum_{i=1}^{q} (x_1 \wedge \cdots \wedge x_p) \wedge (y_1 \wedge \cdots \wedge \varphi y_i \wedge \cdots \wedge y_q) \\ &= \theta_\wedge(\varphi)u \wedge v + u \wedge \theta_\wedge(\varphi)v.\end{aligned}$$

Now the linearity of $\theta_\wedge(\varphi)$ gives

$$\theta_\wedge(\varphi)(u \wedge v) = \theta_\wedge(\varphi)u \wedge v + u \wedge \theta_\wedge(\varphi)v \qquad u, v \in \wedge E$$

and hence $\theta_\wedge(\varphi)$ is a derivation. In terms of the multiplication operator this formula reads

$$\theta_\wedge(\varphi) \circ \mu(u) = \mu(\theta_\wedge(\varphi)u) + \mu(u) \circ \theta_\wedge(\varphi). \tag{5.24}$$

For the identity map we obtain that

$$\theta_\wedge(\iota)u = pu \qquad u \in \wedge^p E.$$

If $\psi\colon E \to E$ is a second linear transformation, then we have the relation

$$\theta_\wedge(\varphi\psi - \psi\varphi) = \theta_\wedge(\varphi)\theta_\wedge(\psi) - \theta_\wedge(\psi)\theta_\wedge(\varphi). \tag{5.25}$$

For the proof, we notice first that the operation on each side of (5.25) is a derivation in $\wedge E$ (see Section 5.6 of *Linear Algebra*) and consequently it is sufficient to consider the restriction of these operators to E. But in this case (5.25) is trivial.

Now suppose that $\varphi: E \to E$, $\varphi^*: E^* \leftarrow E^*$ is a dual pair of linear maps, and consider the induced derivations $\theta_\wedge(\varphi)$ and $\theta^\wedge(\varphi) = \theta_\wedge(\varphi^*)$. It will be shown that $\theta_\wedge(\varphi)$ and $\theta^\wedge(\varphi)$ again form a dual pair

$$\theta^\wedge(\varphi) = [\theta_\wedge(\varphi)]^*. \tag{5.26}$$

Let $u = x_1 \wedge \cdots \wedge x_p \in \wedge^p E$ and $u^* = x^{*1} \wedge \cdots \wedge x^{*p} \in \wedge^p E^*$. Then, in view of Formula (5.11), we have

$$\begin{aligned}
\langle u^*, \theta_\wedge(\varphi)u\rangle &= \sum_{i=1}^{p} \langle x^{*1} \wedge \cdots \wedge x^{*p}, x_1 \wedge \cdots \wedge \varphi x_i \wedge \cdots \wedge x_p\rangle \\
&= \sum_{i,j=1}^{p} (-1)^{i+j}\langle x^{*j}, \varphi x_i\rangle \langle x^{*1} \wedge \cdots \wedge \widehat{x^{*j}} \wedge \cdots \wedge x^{*p}, \\
&\qquad\qquad x_1 \wedge \cdots \wedge \widehat{\varphi x_i} \wedge \cdots \wedge x_p\rangle \\
&= \sum_{i,j=1}^{p} (-1)^{i+j}\langle \varphi^* x^{*j}, x_i\rangle \langle x^{*1} \wedge \cdots \wedge \widehat{x^{*j}} \wedge \cdots \wedge x^{*p}, \\
&\qquad\qquad x_1 \wedge \cdots \wedge \widehat{\varphi x_i} \wedge \cdots \wedge x_p\rangle \\
&= \sum_{j=1}^{p} \langle x^{*1} \wedge \cdots \wedge \varphi^* x^{*j} \wedge \cdots \wedge x^{*p}, x_1 \wedge \cdots \wedge x_p\rangle \\
&= \langle \theta^\wedge(\varphi)u^*, u\rangle,
\end{aligned}$$

whence

$$\langle u^*, \theta_\wedge(\varphi)u\rangle = \langle \theta^\wedge(\varphi)u^*, u\rangle \qquad u^* \in \wedge E^*, u \in \wedge E. \tag{5.27}$$

If $\psi: E \to E$ and $\psi^*: E^* \leftarrow E^*$ is a second pair of dual mappings, then formulas (5.25) and (5.26) yield

$$\begin{aligned}
\theta^\wedge(\varphi\psi - \psi\varphi) &= \theta_\wedge(\psi^*\varphi^* - \varphi^*\psi^*) = \theta_\wedge(\psi^*)\theta_\wedge(\varphi^*) - \theta_\wedge(\varphi^*)\theta_\wedge(\psi^*) \\
&= \theta^\wedge(\psi)\theta^\wedge(\varphi) - \theta^\wedge(\varphi)\theta^\wedge(\psi)
\end{aligned}$$

whence

$$\theta^\wedge(\varphi\psi - \psi\varphi) = \theta^\wedge(\psi)\theta^\wedge(\varphi) - \theta^\wedge(\varphi)\theta^\wedge(\psi). \tag{5.28}$$

5.11. Antiderivations

Let ω be a homogeneous involution of degree zero in $\wedge E$ and let Ω be a (not necessarily homogeneous) antiderivation with respect to ω. If φ denotes the restriction of Ω to E, we have

$$0 = \Omega(x \wedge x) = \varphi x \wedge x + \omega x \wedge \varphi x,$$

whence

$$\varphi x \wedge x + \omega x \wedge \varphi x = 0 \qquad x \in E. \tag{5.29}$$

Conversely, every linear map $\varphi: E \to \wedge E$ which satisfies (5.29) can be extended in a unique way into an antiderivation $\Omega(\varphi)$ (with respect to ω) of $\wedge E$. Since $\wedge E$ is generated by E and the scalar 1, the uniqueness follows at once. To prove the existence of $\Omega(\varphi)$ consider the p-linear mapping

$$\psi_p: \underbrace{E \times \cdots \times E}_{p} \to \wedge E$$

defined by

$$\psi_p(x_1, \ldots, x_p) = \sum_{\nu=1}^{p} \omega x_1 \wedge \cdots \wedge \omega x_{\nu-1} \wedge \varphi x_\nu \wedge x_{\nu+1} \wedge \cdots \wedge x_p \qquad p \geqq 2$$

$$\psi_1 = \varphi \quad \text{and} \quad \psi_0 = 0. \tag{5.30}$$

It will be shown that ψ_p is skew-symmetric. In view of Section 5.1 it is sufficient to verify that $\psi_p(x_1, \ldots, x_i, x_{i+1}, \ldots, x_p) = 0$ whenever $x_i = x_{i+1}$.

Since $\omega x \wedge \omega x = 0$ and $x \wedge x = 0$, we obtain from (5.30) that

$$\psi_p(x_1, \ldots, x, x, \ldots, x_p) = \omega x_1 \wedge \cdots \wedge \omega x_{i-1} \wedge (\varphi x \wedge x + \omega x \wedge \varphi x) \wedge x_{i+2} \wedge \cdots \wedge x_p$$

whence, in view of (5.29),

$$\psi_p(x_1, \ldots, x, x, \ldots, x_p) = 0.$$

The skew-symmetric mappings ψ_p induce a linear map $\Omega(\varphi): \wedge E \to \wedge E$ such that

$$\Omega(\varphi)(x_1 \wedge \cdots \wedge x_p) = \sum_{\nu=1}^{p} \omega x_1 \wedge \cdots \wedge \omega x_{\nu-1} \wedge \varphi x_\nu \wedge x_{\nu+1} \wedge \cdots \wedge x_p. \tag{5.31}$$

Now it will be shown that $\Omega(\varphi)$ is an antiderivation with respect to the involution ω. Let $u = x_1 \wedge \cdots \wedge x_p$ and $v = x_{p+1} \wedge \cdots \wedge x_{p+q}$ be two decomposable elements. Then

$$\begin{aligned}
\Omega(\varphi)(u \wedge v) &= \Omega(\varphi)(x_1 \wedge \cdots \wedge x_{p+q}) \\
&= \sum_{\nu=1}^{p} \omega x_1 \wedge \cdots \wedge \omega x_{\nu-1} \wedge \varphi x_\nu \wedge x_{\nu+1} \wedge \cdots \wedge x_{p+q} \\
&\quad + \sum_{\nu=p+1}^{q} \omega x_1 \wedge \cdots \wedge \omega x_{\nu-1} \wedge \varphi x_\nu \wedge x_{\nu+1} \wedge \cdots \wedge x_{p+q} \\
&= \Omega(\varphi)u \wedge v + \omega u \wedge \Omega(\varphi)v,
\end{aligned}$$

which completes the proof.

It is clear that $\Omega(\varphi)$ is homogeneous of degree k if and only if $\operatorname{Im} \varphi \subset \wedge^{k+1} E$. Assume now that $\Omega(\varphi)$ is in fact homogeneous of degree k.

The following two cases are of particular importance:

1. $\omega = \iota_\wedge$ (derivations): Condition (5.29) reduces to

$$\varphi x \wedge x + x \wedge \varphi x = 0$$

or, equivalently,

$$[1 + (-1)^{k+1}]\varphi x \wedge x = 0. \tag{5.32}$$

If k is even (5.32) always holds: *any linear map $\varphi: E \to \wedge^{2p+1} E$ can be extended in a unique way to a homogeneous derivation in $\wedge E$.*

Now assume that k is odd. Then equation (5.32) reads

$$\varphi x \wedge x = 0 \qquad x \in E,$$

whence

$$\varphi x \wedge y = -\varphi y \wedge x \qquad x, y \in E.$$

Since $k + 1$ is even we have

$$\varphi y \wedge x = x \wedge \varphi y.$$

It follows that

$$\varphi x \wedge y = -x \wedge \varphi y.$$

Now Formula (5.31) yields

$$\Omega(\varphi)(x_1 \wedge \cdots \wedge x_p) = \tfrac{1}{2}[1 + (-1)^{p+1}]\varphi x_1 \wedge x_2 \wedge \cdots \wedge x_p. \tag{5.33}$$

It follows in particular that the restriction of $\Omega(\varphi)$ to every subspace $\wedge^{2p} E$ is zero.

2. $\omega = (-\iota)_\wedge$ (antiderivations): Condition (5.29) becomes

$$\varphi x \wedge x = x \wedge \varphi x,$$

i.e.,

$$[1 + (-1)^k]\varphi x \wedge x = 0.$$

If k is odd this condition is always fulfilled: *any linear map $\varphi: E \to \wedge^{2p} E$ can be extended in a unique way to an antiderivation (with respect to the canonical involution) in $\wedge E$.*

Now assume that k is even. Then the above equation implies that

$$\varphi x \wedge x = 0,$$

whence

$$\varphi x \wedge y = -\varphi y \wedge x$$

or equivalently

$$\varphi x \wedge y = x \wedge \varphi y.$$

Formula (5.31) now yields

$$\Omega(\varphi)(x_1 \wedge \cdots \wedge x_p) = \tfrac{1}{2}[1 + (-1)^{p+1}]\varphi x_1 \wedge x_2 \wedge \cdots \wedge x_p. \qquad (5.34)$$

It follows again that the restriction of $\Omega(\varphi)$ to every subspace $\wedge^{2p}E$ is zero.

5.12. α-Antiderivations

Suppose now that F is a second vector space, ω_F is a homogeneous involution of degree zero in $\wedge F$ and that $\varphi: E \to \wedge E$, $\psi: F \to \wedge F$ are homogeneous mappings satisfying the conditions

$$\varphi x \wedge x + \omega_E x \wedge \varphi x = 0 \qquad x \in E,$$

$$\psi y \wedge y + \omega_F y \wedge \psi y = 0 \qquad y \in F.$$

Assume further that $\alpha: E \to F$ is a linear map such that

$$\omega_F \alpha_\wedge = \alpha_\wedge \omega_E \quad \text{and} \quad \psi\alpha = \alpha_\wedge \varphi. \qquad (5.35)$$

Then we have

$$\Omega_F(\psi) \circ \alpha_\wedge = \alpha_\wedge \circ \Omega_E(\varphi). \qquad (5.36)$$

In fact, it is easy to verify that the operators $\Omega_F(\psi) \circ \alpha_\wedge$ and $\alpha_\wedge \circ \Omega_E(\varphi)$ are α-antiderivations (see Section 5.8 of *Linear Algebra*). Relation (5.35) implies that the restrictions of these α-antiderivations to E coincide and so (5.36) follows.

Problems

1. Let E and F be two vector spaces, $\otimes E$, $\otimes F$ tensor algebras and $\wedge E$, $\wedge F$ exterior algebras over E and F respectively. Consider the projections $\pi_E: \otimes E \to \wedge E$ and $\pi_F: \otimes F \to \wedge F$ defined by

$$\pi_E(x_1 \otimes \cdots \otimes x_p) = x_1 \wedge \cdots \wedge x_p$$

and

$$\pi_F(y_1 \otimes \cdots \otimes y_q) = y_1 \wedge \cdots \wedge y_q.$$

(a) If $\varphi: E \to F$ is a linear map prove that

$$\pi_F \circ \varphi_\otimes = \varphi_\wedge \circ \pi_E.$$

(b) If $\varphi: E \to E$ is a linear map show that

$$\pi_F \circ \theta_\otimes(\varphi) = \theta_\wedge(\varphi) \circ \pi_E.$$

2. Let $a \in \wedge^k E$ be a fixed element, where k is odd. Define a linear map $\theta: \wedge E \to \wedge E$ by

$$\theta u = \begin{cases} a \wedge u & u \in \wedge^p E, \ p \text{ odd}, \\ 0 & u \in \wedge^p E, \ p \text{ even}. \end{cases}$$

Prove that θ is a derivation in the algebra $\wedge E$.

3. Suppose ω is an involution of degree zero in $\wedge E$ such that

$$\omega x \wedge x = 0 \qquad x \in E.$$

Prove that $\omega = \iota_\wedge$ or $\omega = (-\iota)_\wedge$.

4. Show that there does not exist an involution ω in $\wedge E$ such that (5.29) holds for every linear map

$$\varphi : E \to E.$$

Hint: First show that for such an ω the relation $\omega x \wedge x = 0$ must hold.

5. Let E and F be vector spaces of finite dimension and consider a linear map $\varphi : E \to F$ of rank r.
Prove that

$$r(\varphi_\wedge) = 2^r.$$

6. Let $\wedge E$ be the exterior algebra over E and consider an antiderivation Ω in $\wedge E$ of degree 1. Define a new multiplication $\tilde{\wedge}$ in $\wedge E$ by setting

$$u \mathbin{\tilde{\wedge}} v = \binom{p+q}{p} u \wedge v \qquad u \in \wedge^p E, v \in \wedge^q E.$$

Show that the operator $\tilde{\Omega}$ defined by $\tilde{\Omega} u = p \cdot \Omega u$, $u \in \wedge^p E$ is an antiderivation with respect to this multiplication.

7. Let $d : \otimes E \to \otimes E$ be a homogeneous linear map of degree 1 such that

$$d(u \otimes v) = du \otimes v + \sigma(u \otimes dv),$$

where σ is a fixed permutation such that $\varepsilon_\sigma = (-1)^{\deg u}$. Assume that

$$\pi_A d \pi_A = \pi_A d,$$

where π_A is the alternator (see Section 4.2). Define the operator δ by $\delta = \pi_A d$. Prove that δ is an antiderivation in the graded algebra $X(E)$ (cf. the problem in Section 4.8). Prove that δ is a differential operator if and only if $\pi_A d^2 = 0$.

8. Let $\psi : E \to E$ be a linear transformation of an n-dimensional space E. Prove that there exist uniquely determined transformations ψ_i $(i = 0, \ldots, n)$ of $\wedge E$ such that

$$(\psi - \lambda\iota)_\wedge = \sum_{i=0}^{n} \psi_i \lambda^{n-i} \qquad \lambda \in \Gamma.$$

If $\psi_i^{(n)}$ denotes the restriction of ψ_i to $\wedge^n E$, show that

$$\operatorname{tr} \psi_i^{(n)} = \alpha_i \qquad i = 0, \ldots, n,$$

where α_i is the ith characteristic coefficient of ψ (see Section 4.19 of *Linear Algebra*). Prove that

$$\psi_0 = (-1)_n \iota, \qquad \psi_1 = (-1)^{n-1} \theta_\wedge(\psi), \qquad \psi_n = \psi_\wedge.$$

The Operator $i(a)$

5.13.

Fix an element $a \in \wedge E$ and consider the linear transformation $\mu(a)$ of $\wedge E$ given by left multiplication with a,

$$\mu(a)u = a \wedge u \qquad u \in \wedge E.$$

Since the algebra $\wedge E$ is associative, we have the relation

$$\mu(a \wedge b) = \mu(a) \circ \mu(b) \qquad a, b \in \wedge E. \tag{5.37}$$

Now consider the dual map

$$i(a): \wedge E^* \leftarrow \wedge E^*.$$

It is determined by the equation

$$\langle i(a)u^*, v\rangle = \langle u^*, a \wedge v\rangle \qquad v \in \wedge E.$$

In particular,

$$i(\lambda)u^* = \lambda u^* \qquad \lambda \in \Gamma.$$

Now suppose that a is homogeneous of degree p. Then $i(a)$ restricts to linear maps

$$\wedge^r E^* \to \wedge^{r-p} E^*, \qquad r \geq p$$

and reduces to zero in $\wedge^r E^*$ if $r < p$. For $u^* \in \wedge^p E^*$ we have

$$i(a)u^* = \langle u^*, a\rangle.$$

Dualizing formula (5.37) we obtain

$$i(a \wedge b) = i(b) \circ i(a) \qquad a, b \in \wedge E. \tag{5.38}$$

In particular,

$$i(a \wedge b) = (-1)^{pq} i(b \wedge a) \qquad a \in \wedge^p E, b \in \wedge^q E. \tag{5.39}$$

Next, let F be a second vector space and let $\varphi: E \to F$, $\varphi^*: E^* \leftarrow F^*$ be a pair of dual maps. Since $\varphi_\wedge$ is a homomorphism,

$$\varphi_\wedge \circ \mu(a) = \mu(\varphi_\wedge a) \circ \varphi_\wedge \qquad a \in \wedge E.$$

Dualizing this relation yields

$$i(a) \circ \varphi^\wedge = \varphi^\wedge \circ i(\varphi_\wedge a) \qquad a \in \wedge E. \tag{5.40}$$

Finally, let φ and φ^* be a pair of dual linear transformations of E, and consider the induced derivations $\theta_\wedge(\varphi)$ and $\theta^\wedge(\varphi)$. Then Formula (5.24) implies that

$$i(a) \circ \theta^\wedge(\varphi) = i(\theta_\wedge(\varphi)a) + \theta^\wedge(\varphi) \circ i(a).$$

5.14. The Operator $i(h)$

In this section we shall consider the operator $i(h)$ for the special case $h \in E$. This operator is homogeneous of degree -1. Formula (5.39) implies that

$$i(h) \circ i(k) + i(k) \circ i(h) = 0 \qquad h, k \in E.$$

In particular,

$$i(h)^2 = 0.$$

Proposition 5.14.1. *The operator $i(h)$ is an antiderivation in the algebra $\wedge E^*$,*

$$i(h)(u^* \wedge v^*) = i(h)u^* \wedge v^* + (-1)^p u^* \wedge i(h)v^* \qquad u^* \in \wedge^p E^*, v^* \in \wedge E^*.$$

PROOF. Consider the linear map $\varphi_h: E^* \to \Gamma$ given by

$$\varphi_h x^* = \langle x^*, h \rangle, \qquad x^* \in E^*.$$

It follows from Section 5.11 that φ_h extends to an antiderivation Ω_h of degree -1 in $\wedge E^*$ (with respect to the canonical involution). We have to show that $\Omega(h) = i(h)$,

$$\langle \Omega_h u^*, v \rangle = \langle u^*, h \wedge v \rangle \qquad u^* \in \wedge E^*, v \in \wedge E. \tag{5.41}$$

We may assume that u^* and v are decomposable, $u^* = x_1^* \wedge \cdots \wedge x_p^*$ and $v = x_1 \wedge \cdots \wedge x_q$. If $p \neq q + 1$, both sides of (5.41) are zero and so only the case $p = q + 1$ has to be considered. Then we have, in view of (5.11),

$$\begin{aligned} \langle \Omega_h u^*, v \rangle &= \langle \Omega_h(x_1^* \wedge \cdots \wedge x_p^*), x_1 \wedge \cdots \wedge x_{p-1} \rangle \\ &= \sum_{\nu=1}^{p} (-1)^{\nu-1} \langle x_\nu^*, h \rangle \langle x_1^* \wedge \cdots \wedge \widehat{x_\nu^*} \wedge \cdots \wedge \\ &\qquad x_p^*, x_1 \wedge \cdots \wedge x_{p-1} \rangle \\ &= \langle x_1^* \wedge \cdots \wedge x_p^*, h \wedge x_1 \wedge \cdots \wedge x_{p-1} \rangle = \langle u^*, h \wedge v \rangle \end{aligned}$$

and so Formula (5.41) follows. □

Corollary I:

$$i(h) \circ \mu(h^*) + \mu(h^*) \circ i(h) = \langle h^*, h \rangle \iota \qquad h \in E, h^* \in E^*.$$

PROOF. Apply the proposition with $u^* = h^*$. □

Corollary II:

$$i(h)(x_1^* \wedge \cdots \wedge x_p^*) = \sum_{\nu=1}^{p} (-1)^{\nu-1} \langle x_\nu^*, h \rangle x_1^* \wedge \cdots \wedge \widehat{x_\nu^*} \wedge \cdots \wedge x_p^*.$$

Corollary III: *Let F be a subspace of E and let $F^\perp$ be its orthogonal complement. Identify $\wedge F$ and $\wedge F^\perp$ with subalgebras of $\wedge E$ and $\wedge E^*$ respectively. Then*

$$i(a)(u^* \wedge v^*) = (-1)^{pq} u^* \wedge i(a)v^* \qquad a \in \wedge^p F, u^* \in \wedge^q F^\perp, v^* \in \wedge E^*.$$

PROOF. We may assume that $a = y_1 \wedge \cdots \wedge y_p$, $y_\nu \in F$. Since $i(y)$ is an antiderivation, we have for $y \in F$

$$i(y)(u^* \wedge v^*) = i(y)u^* \wedge v^* + (-1)^q u^* \wedge i(y)v^*$$
$$= (-1)^q u^* \wedge i(y)v^*.$$

It follows that

$$i(a)(u^* \wedge v^*) = i(y_1 \wedge \cdots \wedge y_p)(u^* \wedge v^*)$$
$$= i(y_p) \cdots i(y_1)(u^* \wedge v^*)$$
$$= (-1)^{pq} u^* \wedge i(y_p) \cdots i(y_1)v^*$$
$$= (-1)^{pq} u^* \wedge i(a)v^*.$$

□

Proposition 5.14.2. *If an element $u^* \in \wedge^p E^*$ $(p \geq 1)$ satisfies the equation $i(h)u^* = 0$ for every $h \in E$, then $u^* = 0$.*

PROOF. Let $v \in \wedge E$ be arbitrary. Since $p \geq 1$, we can write

$$v = \sum_\nu h_\nu \wedge v_\nu \qquad h_\nu \in E, v_\nu \in \wedge^{p-1} E.$$

It follows that

$$\langle u^*, v \rangle = \sum_\nu \langle u^*, h_\nu \wedge v_\nu \rangle = \sum_\nu \langle i(h_\nu)u^*, v_\nu \rangle = 0$$

whence $u^* = 0$. □

PROBLEMS

1. Let $u^* \in \wedge^p E^*$, $p \geq 1$ be an element such that $i(a)u^* = 0$ for every $a \in \wedge^k E$, where $k \leq p$ is a fixed integer. Prove that $u^* = 0$.
Hint: Use the duality of the operators $i(a)$ and $\mu(a)$.

2. Let E, E^* be a pair of n-dimensional dual spaces and $\{e_\nu\}$, $\{e^{*\nu}\}$ be a pair of dual bases. Given a linear transformation $\varphi: E \to E$ show that

$$\theta_\wedge(\varphi) = \sum_\nu \mu(\varphi e_\nu) i(e^{*\nu})$$

$$\theta^\wedge(\varphi) = \sum_\nu \mu(e^{*\nu}) i(\varphi e_\nu).$$

3. Let E, E^* be a pair of n-dimensional dual spaces. Prove the following relations:

(a) $$\mu(x)i(x^*) + i(x^*)\mu(x) = \langle x^*, x \rangle \iota \qquad x^* \in E^*, x \in E$$

(b) $$\sum_\nu \mu(e_\nu) i(e^{*\nu}) u = pu \qquad u \in \wedge^p E$$

(c) $$\sum_\nu i(e^{*\nu}) \mu(e_\nu) u = (n - p)u \qquad u \in \wedge^p E.$$

4. Let E, E^* be a pair of dual spaces of dimension $n = 2m$. Prove the formula

$$\sum_{j=1}^{m} h^* \wedge i(h)u_j \wedge u_1 \wedge \cdots \wedge \widehat{u_j} \wedge \cdots \wedge u_m = \langle h^*, h\rangle u_1 \wedge \cdots \wedge u_m$$

$$h \in E, h^* \in E^*, u_j \in \wedge^2 E^* \qquad (j = 1, \ldots, m).$$

5. Show that the operator

$$i(a)\colon \wedge E^* \to \wedge E^* \qquad a \in \wedge^p E$$

is not an antiderivation unless $p = 1$.

6. Let $a \in \wedge^p E$ be arbitrary and assume that $p \le q$. Prove the formula

$$i(a)(x^{*1} \wedge \cdots \wedge x^{*q})$$
$$= \sum_{\nu_1 < \cdots < \nu_p} (-1)^{\sum_{i=1}^{p}(\nu_i - i)} \langle a, x^{*\nu_1} \wedge \cdots \wedge x^{*\nu_p}\rangle x^{*\nu_{p+1}} \wedge \cdots \wedge x^{*\nu_q},$$

where $(\nu_{p+1}, \ldots, \nu_q)$ denotes the complementary ordered $(q - p)$-tuple.

Exterior Algebra Over a Direct Sum

5.15.

Let E and F be vector spaces and consider the anticommutative tensor product of the graded algebras $\wedge E$ and $\wedge F$ (see Section 2.8.) On the other hand, we have the exterior algebra over the direct sum $E \oplus F$.

Theorem 5.15.1. *There is a canonical isomorphism between the graded algebras* $\wedge E \mathbin{\hat{\otimes}} \wedge F$ *and* $\wedge(E \oplus F)$.

PROOF. Let

$$i_1\colon E \to E \oplus F \quad \text{and} \quad i_2\colon F \to E \oplus F$$

be the inclusion maps. They extend to homomorphisms

$$(i_1)_\wedge\colon \wedge E \to \wedge(E \oplus F) \qquad \text{and} \qquad (i_2)_\wedge\colon \wedge F \to \wedge(E \oplus F).$$

Hence a linear map

$$f\colon \wedge E \mathbin{\hat{\otimes}} \wedge F \to \wedge(E \oplus F)$$

is given by

$$f(u \otimes v) = (i_1)_\wedge u \wedge (i_2)_\wedge v. \tag{5.42}$$

We show that f is an algebra homomorphism. In fact, let $u \in \wedge E$, $v \in \wedge^q F$, $u' \in \wedge^r E$, $v' \in \wedge F$. Then, if the multiplication in the algebra $\wedge E \,\hat{\otimes}\, \wedge F$ is also denoted by $\wedge$, we have

$$\begin{aligned} f[(u \otimes v) \wedge (u' \otimes v')] &= (-1)^{qr} f[(u \wedge u') \otimes (v \wedge v')] \\ &= (-1)^{qr}(i_1)_\wedge(u \wedge u') \wedge (i_2)_\wedge(v \wedge v') \\ &= (-1)^{qr}(i_1)_\wedge u \wedge (i_1)_\wedge u' \wedge (i_2)_\wedge v \wedge (i_2)_\wedge v' \\ &= (i_1)_\wedge u \wedge (i_2)_\wedge v \wedge (i_1)_\wedge u' \wedge (i_2)_\wedge v' \\ &= f(u \otimes v) \wedge f(u' \otimes v'). \end{aligned}$$

To show that f is an isomorphism we construct an inverse homomorphism. Consider the linear map $\eta: E \oplus F \to \wedge E \,\hat{\otimes}\, \wedge F$ given by

$$\eta(z) = \pi_1 z \otimes 1 + 1 \otimes \pi_2 z \qquad z \in E \oplus F, \tag{5.43}$$

where

$$\pi_1 : E \oplus F \to E \quad \text{and} \quad \pi_2 : E \oplus F \to F$$

are the canonical projections. Then we have

$$\begin{aligned} \eta(z) \wedge \eta(z) &= (\pi_1 z \otimes 1 + 1 \otimes \pi_2 z) \wedge (\pi_1 z \otimes 1 + 1 \otimes \pi_2 z) \\ &= (\pi_1 z \wedge \pi_1 z) \otimes 1 + \pi_1 z \otimes \pi_2 z - \pi_1 z \otimes \pi_2 z + 1 \otimes (\pi_2 z \wedge \pi_2 z) = 0. \end{aligned}$$

Hence η extends to a homomorphism

$$h: \wedge(E \oplus F) \to \wedge E \,\hat{\otimes}\, \wedge F.$$

Relations (5.42) and (5.43) imply that

$$hf(x \otimes 1) = h(i_1 x) = \eta(i_1 x) = x \qquad x \in E,$$

$$hf(1 \otimes y) = h(i_2 y) = \eta(i_2 y) = y \qquad y \in F,$$

$$fh(z) = f(\pi_1 z \otimes 1 + 1 \otimes \pi_2 z) = i_1 \pi_1 z + i_2 \pi_2 z = z \qquad z \in E \oplus F.$$

Since the vectors $x \otimes 1$ and $1 \otimes y$ together with the scalar 1 generate the algebra $\wedge E \,\hat{\otimes}\, \wedge F$ and the vectors $x \oplus y$ together with 1 generate the algebra $\wedge(E \oplus F)$ it follows that

$$h \circ f = \iota \quad \text{and} \quad f \circ h = \iota.$$

Thus f is an isomorphism and h is the inverse isomorphism. □

Corollary. *Let $E = E_1 \oplus E_2$ be a direct decomposition of E into two subspaces. Then $\wedge E \cong \wedge E_1 \,\hat{\otimes}\, \wedge E_2$.*

5.16. Direct Sums of Linear Maps

Suppose that E', F' is another pair of vector spaces and that linear maps $\varphi:E \to E'$, $\psi:F \to F'$ are given. Then the linear map $\varphi \oplus \psi:E \oplus F \to E' \oplus F'$ is given by

$$\varphi \oplus \psi = i'_1 \circ \varphi \circ \pi_1 + i'_2 \circ \psi \circ \pi_2, \tag{5.44}$$

where

$$i'_1:E' \to E' \oplus F' \quad \text{and} \quad i'_2:F' \to E' \oplus F'$$

denote the canonical injections. It will be shown that

$$(\varphi \oplus \psi)_\wedge = \varphi_\wedge \otimes \psi_\wedge. \tag{5.45}$$

From (5.44) we obtain

$$(\varphi \oplus \psi) \circ i_1 = i'_1 \circ \varphi \quad \text{and} \quad (\varphi \oplus \psi) \circ i_2 = i'_2 \circ \psi$$

whence

$$(\varphi \oplus \psi)_\wedge \circ (i_1)_\wedge = (i'_1)_\wedge \circ \varphi_\wedge \quad \text{and} \quad (\varphi \oplus \psi)_\wedge \circ (i_2)_\wedge = (i'_2)_\wedge \circ \psi_\wedge.$$

Now let $u \in \wedge E$ and $v \in \wedge F$ be two arbitrary elements. Then we have

$$\begin{aligned}(\varphi \oplus \psi)_\wedge(u \otimes v) &= (\varphi \oplus \psi)_\wedge((i_1)_\wedge u \wedge (i_2)_\wedge v)\\ &= [(\varphi \oplus \psi)_\wedge(i_1)_\wedge u] \wedge [(\varphi \oplus \psi)_\wedge(i_2)_\wedge v]\\ &= (i'_1)_\wedge \varphi_\wedge u \wedge (i'_2)_\wedge \psi_\wedge v\\ &= \varphi_\wedge u \otimes \psi_\wedge v\\ &= (\varphi_\wedge \otimes \psi_\wedge)(u \otimes v)\end{aligned}$$

whence (5.45).

Now consider a linear map $\varphi:E \to F$ and suppose that two direct decompositions

$$E = E_1 \oplus E_2 \quad \text{and} \quad F = F_1 \oplus F_2$$

are given such that $\varphi E_1 \subset F_1$ and $\varphi E_2 \subset F_2$. Then it follows from (5.45) that

$$\varphi_\wedge = (\varphi_1)_\wedge \otimes (\varphi_2)_\wedge, \tag{5.46}$$

where $\varphi_1:E_1 \to F_1$ and $\varphi_2:E_2 \to F_2$ are the restrictions of φ to E_1 and E_2. Formula (5.46) yields, in view of (1.12) and (1.11), that

$$\ker \varphi_\wedge = (\ker (\varphi_1)_\wedge) \otimes \wedge E_2 + \wedge E_1 \otimes (\ker (\varphi_2)_\wedge)$$

and

$$\operatorname{Im} \varphi_\wedge = \operatorname{Im} (\varphi_1)_\wedge \otimes \operatorname{Im} (\varphi_2)_\wedge.$$

5.17. Derivations

Suppose $\varphi:E \to E$ and $\psi:F \to F$ are two linear maps and consider the induced derivations $\theta_\wedge(\varphi): \wedge E \to \wedge E$ and $\theta_\wedge(\psi): \wedge F \to \wedge F$. Then

$$\boxed{\theta_\wedge(\varphi \oplus \psi) = \theta_\wedge(\varphi) \otimes \iota + \iota \otimes \theta_\wedge(\psi).} \tag{5.47}$$

In fact, Proposition 2.9.1 implies that the mapping on the right hand side of (5.47) is a derivation. Hence, it is sufficient to verify that the restrictions of the operators to $E \oplus F$ coincide.

Let $z \in E \oplus F$ be an arbitrary vector. Then

$$\begin{aligned}(\theta_\wedge(\varphi) \otimes \iota + \iota \otimes \theta_\wedge(\psi))z &= (\theta_\wedge(\varphi) \otimes \iota + \iota \otimes \theta_\wedge(\psi))(\pi_1 z \otimes 1 + 1 \otimes \pi_2 z)\\ &= \varphi\pi_1 z \otimes 1 + 1 \otimes \psi\pi_2 z = i_1\varphi\pi_1 z + i_2\psi\pi_2 z\\ &= (\varphi \oplus \psi)z = \theta_\wedge(\varphi \oplus \psi)(z)\end{aligned}$$

and so Formula (5.47) follows.

5.18. Direct Sums of Dual Spaces

Consider the dual pairs E, E^* and F, F^* of vector spaces. Then the induced scalar product in $E \oplus F$ and $E^* \oplus F^*$ is given by

$$\langle z^*, z\rangle = \langle x^*, x\rangle + \langle y^*, y\rangle \qquad z = (x, y),\ z^* = (x^*, y^*).$$

The multiplication operator in the algebra $\wedge(E \oplus F)$ is given by

$$\mu(a \otimes b) = \mu(a) \circ \omega^q \otimes \mu(b) \qquad a \in \wedge E,\ b \in \wedge^q F, \tag{5.48}$$

where ω^q denotes qth iterate of the canonical involution of the graded algebra $\wedge E$ (see Section 6.6 of *Linear Algebra*).

Dualizing (5.48) we obtain the relation

$$i(a \otimes b) = \omega^q \circ i(a) \otimes i(b). \tag{5.49}$$

In particular it follows that

$$i(h \otimes 1) = i(h) \otimes \iota \qquad h \in E$$

and

$$i(1 \otimes k) = \omega \otimes i(k) \qquad k \in E$$

whence

$$i(h \oplus k) = i(h) \otimes \iota + \omega \otimes i(k). \tag{5.50}$$

Since the vector spaces $E \oplus F$ and $E^* \oplus F^*$ are dual, a scalar product $\langle\langle , \rangle\rangle$ is induced in $\wedge(E \oplus F)$ and $\wedge(E^* \oplus F^*)$. It will now be shown that this scalar product coincides with the induced scalar product if $\wedge(E \oplus F)$ and $\wedge(E^* \oplus F^*)$ are considered as the tensor products $\wedge E \hat{\otimes} \wedge F$ and $\wedge E^* \hat{\otimes} \wedge F^*$. In other words, it will be proved that

$$\langle\langle u^* \otimes v^*, u \otimes v \rangle\rangle = \langle u^*, u \rangle \langle v^*, v \rangle \tag{5.51}$$

for all $u \in \wedge E$, $v \in \wedge F$, $u^* \in \wedge E^*$, $v^* \in \wedge F^*$. Without loss of generality we may assume that all the elements in (5.51) are homogeneous,

$$u \in \wedge^p E, \qquad v \in \wedge^r F, \qquad u^* \in \wedge^q E^*, \qquad v^* \in \wedge^s F^*.$$

If $p + r \neq q + s$, both sides of (5.51) are zero and hence only the case $p + r = q + s$ remains to be considered. Then we have

$$\langle\langle u^* \otimes v^*, u \otimes v \rangle\rangle = i(u \otimes v)(u^* \otimes v^*) = \omega^r i(u)u^* \otimes i(v)v^*.$$

If $p = q$ and $r = s$, it follows that

$$i(u)u^* = \langle u^*, u \rangle \quad \text{and} \quad i(v)v^* = \langle v^*, v \rangle,$$

while if $p \neq q$, it follows that either $p > q$ or $r > s$, so that both sides of (5.51) are again zero.

5.19. The Diagonal Mapping

Consider now the case $F = E$ and let the *diagonal mapping* $\Delta: E \to E \oplus E$ be defined by

$$\Delta = i_1 + i_2.$$

Then the product $u^* \wedge v^*$ of two elements u^* and v^* of $\wedge E^*$ can be written in the form

$$u^* \wedge v^* = \Delta^\wedge(u^* \otimes v^*). \tag{5.52}$$

In fact, if j_1 and j_2 denote the canonical injections of E^* into $E^* \oplus E^*$, Formula (5.42) yields

$$u^* \otimes v^* = (j_1)_\wedge u^* \wedge (j_2)_\wedge v^* = (\pi_1)^\wedge u^* \wedge (\pi_2)^\wedge v^*.$$

Applying $\Delta^\wedge$ we obtain

$$\begin{aligned} \Delta^\wedge(u^* \otimes v^*) &= \Delta^\wedge(\pi_1)^\wedge u^* \wedge \Delta^\wedge(\pi_2)^\wedge v^* \\ &= (\pi_1 \circ \Delta)^\wedge u^* \wedge (\pi_2 \circ \Delta)^\wedge v^* = \iota^\wedge u^* \wedge \iota^\wedge v^* \\ &= u^* \wedge v^*. \end{aligned}$$

Formula (5.52) shows that $\Delta^\wedge$ is the structure map of the algebra $\wedge E^*$ (see Section 2.1).

5.20. Direct Sums of Several Vector Spaces

Now consider the direct sum of r vector spaces

$$E = \bigoplus_{\rho=1}^{r} E_\rho.$$

Then an r-linear mapping

$$\psi : \wedge E_1 \times \cdots \times \wedge E_r \to \wedge E$$

is defined by

$$\psi(u_1, \ldots, u_r) = (i_1)_\wedge u_1 \wedge \cdots \wedge (i_r)_\wedge u_r \qquad u_\rho \in E_\rho,\ i_\rho : E_\rho \to E.$$

The r-linear mapping ψ induces a linear map

$$f : \wedge E_1 \mathbin{\hat{\otimes}} \cdots \mathbin{\hat{\otimes}} \wedge E_r \to \wedge E$$

such that

$$f(u_1 \otimes \cdots \otimes u_r) = (i_1)_\wedge u_1 \wedge \cdots \wedge (i_r)_\wedge u_r. \tag{5.53}$$

The same argument as in the case $r = 2$ shows that f is a homomorphism and, in fact, an isomorphism of the graded algebra $\wedge E_1 \mathbin{\hat{\otimes}} \cdots \mathbin{\hat{\otimes}} \wedge E_r$ onto $\wedge E$. Formula (5.53) shows that f is homogeneous of degree zero. Hence we may write

$$\wedge(E_1 \oplus \cdots \oplus E_r) \cong \wedge E_1 \mathbin{\hat{\otimes}} \cdots \mathbin{\hat{\otimes}} \wedge E_r$$

and

$$u_1 \otimes \cdots \otimes u_r = (i_1)_\wedge u_1 \wedge \cdots \wedge (i_r)_\wedge u_r.$$

Comparing the homogeneous subspaces of degree p in the relation we obtain

$$[\wedge(E_1 \oplus \cdots \oplus E_r)]_p \cong \sum_{p_1 + \cdots + p_r = p} \wedge^{p_1} E_1 \mathbin{\hat{\otimes}} \cdots \mathbin{\hat{\otimes}} \wedge^{p_r} E_r.$$

5.21. Exterior Algebra Over a Graded Vector Space

Let $E = \sum_{i=1}^{r} E_i$ be a graded vector space, where the subspaces E_i are homogeneous of degree k_i. Then there exists precisely one gradation in the algebra $\wedge E$ such that the injection $i : E \to \wedge E$ is homogeneous of degree zero.

The uniqueness follows immediately from the fact that the algebra $\wedge E$ is generated by the vectors of E and the unit element 1 (which is necessarily of degree zero). To prove the existence of such a gradation consider first the algebra $\wedge E_i$. By assigning the degree pk_i to the subspace $\wedge^p E_i$ we make $\wedge E_i$ into a graded algebra. Now writing

$$\wedge E = \wedge E_1 \mathbin{\hat{\otimes}} \cdots \mathbin{\hat{\otimes}} \wedge E_r,$$

we recall that the gradations of the $\wedge E_i$ induce a gradation in the algebra $\wedge E$. Clearly the injection $i: E \to \wedge E$ is homogeneous of degree zero and so the proof is complete.

The algebra $\wedge E$ together with the above gradation is called the *graded tensor algebra over the graded vector space E*. The subspace of homogeneous elements of degree k is given by

$$(\wedge E)_k = \sum_{(p)} \wedge^{p_1} E_1 \hat{\otimes} \cdots \hat{\otimes} \wedge^{p_r} E_r$$

where the sum is extended over all r-tuples $(p_1, \ldots, p_r)$ subject to

$$\sum_{i=1}^{r} p_i k_i = k.$$

Suppose now that the vector space E has finite dimension and that the gradation is positive. Then the Poincaré polynomial of the graded space $\wedge E_i$ is given by

$$P_i(t) = (1 + t^{k_i})^{n_i} \qquad n_i = \dim E_i, \; i = 1, \ldots, r.$$

Since the space $\wedge E$ is a tensor product of the spaces $\wedge E_i$ we obtain for the Poincaré polynomial $P(t)$ of $\wedge E$ (in view of Section 2.6) the expression

$$P(t) = (1 + t^{k_1})^{n_1} \cdots (1 + t^{k_r})^{n_r}.$$

Problems

1. Let E be a vector space and $\wedge E$ be an exterior algebra over E. Show that

$$u \wedge v = \pi_\wedge(u \otimes v) \qquad u, v \in \wedge E$$

where π_1 and π_2 are the canonical projections of $E \oplus E$ onto E and $\pi = \pi_1 + \pi_2$.

2. Let $E = E_1 + E_2$ be a decomposition of E and set $E_{12} = E_1 \cap E_2$.
(a) Establish a natural isomorphism

$$\psi: E_1/E_{12} \oplus E_2/E_{12} \xrightarrow{\cong} E/E_{12}.$$

(b) Consider the canonical projections

$$\rho_1: E_1 \to E_1/E_{12} \qquad \rho_2: E_2 \to E_2/E_{12} \qquad \rho: E \to E/E_{12}$$

and let $\varphi: E_1 \oplus E_2 \to E$ be the linear map given by

$$\varphi(x_1, x_2) = x_1 + x_2 \qquad x_1 \in E_1, x_2 \in E_2.$$

Show that the diagram

$$\begin{array}{ccc} \wedge E_1 \hat{\otimes} \wedge E_2 & \longrightarrow & \wedge E \\ \Big\downarrow {\scriptstyle (\rho_1)_\wedge \otimes (\rho_2)_\wedge} & & \Big\downarrow {\scriptstyle \rho_\wedge} \\ \wedge(E_1/E_{12}) \hat{\otimes} \wedge(E_2/E_{12}) & \xrightarrow[\cong]{} & \wedge(E/E_{12}) \end{array}$$

is commutative.

Ideals in $\wedge E$

5.22. Graded Ideals

Suppose that

$$I = \sum_p I_p, \qquad I_p = I \cap \wedge^p E,$$

is a graded left ideal in the algebra $\wedge E$. Then we have, for every p-vector $u \in \wedge^p E$ and any element $v = \sum_q v_q \in I$,

$$v \wedge u = \sum_q v_q \wedge u = \sum_q (-1)^{pq} u \wedge v_q \in I$$

and so I is a two-sided ideal. The same argument shows that every graded right ideal is two-sided.

Now let $a \in \wedge^p E$ be an arbitrary homogeneous element, and consider the graded subspace I_a of $\wedge E$ consisting of the elements $u \wedge a$, $u \in \wedge E$. Clearly I_a is a graded left ideal in $\wedge E$, and hence it is a two-sided ideal. Since $a \in I_a$, it follows that I_a is the smallest (graded) ideal in $\wedge E$ containing a, i.e.,

$$I_a = \bigcap_{a \in I} I.$$

I_a is called the ideal *generated by a*. A homogeneous element $a \neq 0$ is called a *divisor* of an element $u \in \wedge E$ if there exists an element $v \in \wedge E$ such that $u = a \wedge v$ or equivalently, if $u \in I_a$.

More generally, every homogeneous subspace $A \subset \wedge^p E$ generates a graded ideal I_A defined by

$$I_A = \left\{ \sum_i u_i \wedge a_i ; u_i \in \wedge E, a_i \in A \right\}. \tag{5.54}$$

I_A is the intersection of all graded ideals containing A. If B is a subspace of A it is clear that $I_B \subset I_A$. Now consider two homogeneous subspaces $A \subset \wedge^p E$ and $B \subset \wedge^q E$. It will be shown that $I_A = I_B$ if and only if $A = B$ (and hence $p = q$). Clearly $A = B$ implies that $I_A = I_B$. Conversely, assume that $I_A = I_B$. Then every element $b \in B$ can be written in the form

$$b = \sum_i u_i \wedge a_i \qquad u_i \in \wedge^{q-p} E, \qquad a_i \in A$$

and hence it follows that $q \geq p$. The same argument shows that $p \geq q$, whence $p = q$. Consequently, the u_i are scalars, and so $b \in A$, i.e., $B \subset A$. Similarly we obtain that $A \subset B$, whence $A = B$.

As a special case of this result we have

$$I_a = I_b \qquad a \in \wedge^p E, b \in \wedge^q E$$

if and only if $a = \lambda b$, $\lambda \neq 0$.

The ideal $I_{\wedge^p E}$, $p \geq 0$, will be denoted by I^p. It follows from (5.54) that

$$I^p = \sum_{j \geq p} \wedge^j E.$$

The ideals I^p form a *filtration* of the algebra $\wedge E$, i.e.,

$$\wedge E = I^0 \supset I^1 \supset I^2 \supset \cdots.$$

The ideal $I^1 = I_E$ is often denoted by $\wedge^+ E$,

$$\wedge^+ E = \sum_{j > 0} \wedge^j E.$$

If E is of dimension n we have

$$I^n = \wedge^n E$$

and

$$I^p = 0 \quad \text{if } p > n.$$

5.23. Direct Decompositions

Let

$$E = E_1 \oplus E_2$$

be a direct decomposition of E. Then we have (considering I_{E_1} as an ideal in $\wedge E$)

$$\boxed{I_{E_1} = \wedge^+ E_1 \hat{\otimes} \wedge E_2.} \tag{5.55}$$

In fact, since $\wedge^+ E_1 \subset I_{E_1}$, it follows that $\wedge^+ E_1 \hat{\otimes} \wedge E_2 \subset I_{E_1}$. Conversely, let $y \wedge v$ ($y \in E_1$, $v \in \wedge E$) be a linear generator of I_{E_1}. Writing

$$v = \sum_i a_i \otimes b_i \qquad a_i \in \wedge E_1,\ b_i \in \wedge E_2,$$

we obtain

$$y \wedge v \in \wedge^+ E_1 \hat{\otimes} \wedge E_2$$

and so

$$I_{E_1} \subset \wedge^+ E_1 \hat{\otimes} \wedge E_2.$$

Writing $\wedge E$ in the form

$$\begin{aligned} \wedge E = \wedge E_1 \hat{\otimes} \wedge E_2 &= (\Gamma \hat{\otimes} \wedge E_2) \oplus (\wedge^+ E_1 \hat{\otimes} \wedge E_2) \\ &= \wedge E_2 \oplus (\wedge^+ E_1 \hat{\otimes} \wedge E_2), \end{aligned}$$

we obtain, in view of (5.55), the relation

$$\boxed{\wedge E = I_{E_1} \oplus \wedge E_2.} \tag{5.56}$$

5.24. Linear Maps

Let $\varphi: E \to F$ be a linear map and consider the induced homomorphism $\varphi_\wedge: \wedge E \to \wedge F$. Generalizing the result of Section 5.8, we shall prove that

$$\boxed{\ker \varphi_\wedge = I_{\ker \varphi}.} \tag{5.57}$$

Let $E' \subset E$ be a subspace such that

$$E = \ker \varphi \oplus E'.$$

Then we can write $\varphi = 0 \oplus \varphi'$, where φ' denotes the restriction of φ to $E' \subset E$. Since φ' is injective so is $(\varphi'_\wedge)_\wedge$ and hence (see Section 5.16)

$$\ker(\varphi_\wedge) = \ker 0_\wedge \otimes \wedge E' = \wedge^+ \ker \varphi \otimes \wedge E'.$$

In view of (5.55), we have

$$\wedge^+ \ker \varphi \mathbin{\hat{\otimes}} \wedge E' = I_{\ker \varphi}.$$

Combining these relations, we obtain (5.57).

5.25. Invertible Elements, Maximum and Minimum Ideals

Proposition 5.25.1. *An element $z = \sum_i z_i$, $z_i \in \wedge^i E$, of $\wedge E$ is invertible if and only if $z_0 \neq 0$. z is nilpotent if and only if $z_0 = 0$.*

Proof. Since nilpotency and invertibility are mutually exclusive properties, it is sufficient to show that z is nilpotent (respectively invertible) if $z \in \wedge^+ E$ (respectively $z \notin \wedge^+ E$).

If $z \in \wedge^+ E$, then $z \in \wedge^+ F$, where F is a finite-dimensional subspace of E. It follows that $z^m = 0$ for $m > \dim F$ and so z is nilpotent. If $z \notin \wedge^+ E$, then, for some $\lambda \neq 0$,

$$\lambda z = 1 - a \qquad a \in \wedge^+ E.$$

Now consider the identity

$$(1 - a) \wedge (1 + a + 2!a^2 + \cdots + k!a^k) = 1 - (k + 1)!a^{k+1}.$$

Since a is nilpotent, it follows from this relation that $1 - a$ has an inverse and hence z has an inverse as well. □

Corollary I. *If $z \in \wedge E$ is invertible, then z^{-1} is a polynomial in z.*

Corollary II. *Every proper ideal in $\wedge E$ is contained in $\wedge^+ E$ and so $\wedge E$ has a maximum ideal, namely $\wedge^+ E$.*

Proof. If $I \not\subset \wedge^+ E$ is an ideal in $\wedge E$, then I contains invertible elements. Hence, $1 \in I$ and so $I = \wedge E$. □

Proposition 5.25.2. *Let E be an n-dimensional vector space and let e be a basis vector for $\wedge^n E$. Then for every element $u \neq 0$ of $\wedge E$, there exists an element $v \in \wedge E$ such that*

$$u \wedge v = e \quad \text{and} \quad v \wedge u = \pm e.$$

Proof. Let $u = \sum_i u_i$, $u_i \in \wedge^i E$ and assume that $u_r \neq 0$ and $u_i = 0$ for $i < r$. Choose a basis $\{e_\nu\}$ $(\nu = 1, \dots, n)$ of E such that $e = e_1 \wedge \cdots \wedge e_n$. Then

$$u_r = \sum_{<} \lambda^{\nu_1, \dots, \nu_r} e_{\nu_1} \wedge \cdots \wedge e_{\nu_r}.$$

Without loss of generality we may assume that $\lambda^{1, \dots, r} \neq 0$. Multiplying u by

$$v = (\lambda^{1, \dots, r})^{-1} e_{r+1} \wedge \cdots \wedge e_n,$$

we obtain

$$u \wedge v = u_r \wedge v = e_1 \wedge \cdots \wedge e_n = e$$

and

$$v \wedge u = v \wedge u_r = e_{r+1} \wedge \cdots \wedge e_n \wedge e_1 \wedge \cdots \wedge e_r = (-1)^{r(n-r)} e,$$

which proves the proposition. □

Corollary. *If E has finite dimension, then every (two-sided) nontrivial ideal I in $\wedge E$ contains $I^n = \wedge^n E$ and so $\wedge E$ has a minimum ideal, namely $\wedge^n E$. Conversely, if E is a vector space such that $\wedge E$ has a minimum ideal, then E has finite dimension.*

Proof. Let $I \neq 0$ be an ideal in $\wedge E$ and $u \neq 0$ be an arbitrary element in I. Then by the above proposition there is an element $v \in \wedge E$ such that $u \wedge v = e$ whence $I^n \subset I$. To prove the second part consider the ideals $I^q = \sum_{p \geq q} \wedge^p E$, $q \geq 0$. If E has infinite dimension, it follows that $\bigcap_q I^q = 0$ and so $\wedge E$ has no minimum ideal. □

5.26. The Annihilator

Let $u \in \wedge E$ be a homogeneous element. Then a graded ideal $N(u)$ in the algebra $\wedge E$ is determined by

$$N(u) = \ker \mu(u). \tag{5.58}$$

$N(u)$ is called the *annihilator* of u. It follows from the definition that

$$N(0) = \wedge E, \qquad N(1) = 0$$

and that

$$N(u) \subset N(v) \quad \text{whenever } u \text{ divides } v.$$

More generally, if $U \in \wedge E$ is a homogeneous subspace of $\wedge E$, the space

$$N(U) = \bigcap_{u \in U} N(u) \tag{5.59}$$

is called the *annihilator* of U. As an intersection of graded ideals $N(U)$ is itself a graded ideal. For $U = E$ we obtain that

$$N(E) = \begin{cases} 0 & \text{if } \dim E = \infty, \\ \wedge^n E & \text{if } \dim E = n. \end{cases}$$

It follows from the definition that $N(V) \supset N(U)$ whenever $V \subset U$.

Now consider the special case $U = \wedge^p F$, where F is a subspace of E. $N(\wedge^p F)$ consists of the elements $u \in \wedge E$ satisfying

$$y_1 \wedge \cdots \wedge y_p \wedge u = 0 \qquad y_i \in F.$$

It follows from the definition of $N(\wedge^p F)$ that

$$N(F) \subset N(\wedge^2 F) \subset \cdots \subset N(\wedge^p F) \subset \cdots. \tag{5.60}$$

Proposition 5.26.1. *Let F be an m-dimensional subspace of E. Then the annihilator $N(\wedge^{m-p+1} F)$ coincides with the ideal generated by $\wedge^p F$,*

$$\boxed{N(\wedge^{m-p+1} F) = I_{\wedge^p F} \qquad 0 \le p \le m + 1.} \tag{5.61}$$

PROOF. If $z_1 \wedge \cdots \wedge z_p$, $z_i \in F$, is a generator of $I_{\wedge^p F}$ and $y_j \in F (j = p, \ldots, m)$ are arbitrary vectors, then

$$\mu(z_1 \wedge \cdots \wedge z_p)(y_p \wedge \cdots \wedge y_m) = z_1 \wedge \cdots \wedge z_p \wedge y_p \wedge \cdots \wedge y_m = 0,$$

whence $I_{\wedge^p F} \subset N(\wedge^{m-p+1} F)$. To prove that

$$N(\wedge^{m-p+1} F) \subset I_{\wedge^p F}$$

we show first that every element $u \in I_{\wedge^p F}$ can be written as

$$u = u' + \sum_i a_i \otimes b_i \qquad u' \in I_{\wedge^{p+1} F},\ a_i \in \wedge^p F,\ b_i \in H, \tag{5.62}$$

where H is a complementary subspace to F in E. Without loss of generality we may assume that u is of the form

$$u = a \wedge v \qquad a \in \wedge^p F,\ v \in \wedge E.$$

Since $E = F \oplus H$, we have

$$v = \sum_j v_j \otimes b_j \qquad v_j \in F, b_j \in \wedge H,$$

whence

$$u = a \wedge \sum_j v_j \otimes b_j = \sum_j (a \wedge v_j) \otimes b_j.$$

Let u' be the sum of all terms in this equation for which v_j has positive degree. Then we have

$$u = u' + a \otimes \sum_j b_j \qquad a \in \wedge^p F, b_j \in \wedge H, u' \in I_{\wedge^{p+1}F},$$

which proves (5.62).

Now we prove (5.61) by induction on p. For $p = 0$ we have

$$N(\wedge^{m+1}F) = N(0) = \wedge E \quad \text{and} \quad I_{\wedge^0 F} = \wedge E$$

and so (5.61) is correct. Now assume that (5.61) holds for the integer p and let $u \in N(\wedge^{m-p}F)$ be an arbitrary element. In view of (5.60) we have $u \in N(\wedge^{m-p+1}F)$ and hence by the induction hypothesis, $u \in I_{\wedge^p F}$. In view of (5.62), we can write

$$u = u' + \sum_i a_i \otimes b_i \qquad u' \in I_{\wedge^{p+1}F}, a_i \in \wedge^p F, b_i \in \wedge H. \tag{5.63}$$

Now let $\{e_\mu\}$ $(\mu = 1, \ldots, m)$ be a basis for F. Then (5.63) can be written in the form

$$u = u' + \sum_< e_{\mu_1} \wedge \cdots \wedge e_{\mu_p} \otimes c^{\mu_1, \ldots, \mu_p} \qquad c^{\mu_1, \ldots, \mu_p} \in \wedge H. \tag{5.64}$$

Choose a fixed p-tuple $(\mu_1, \ldots, \mu_p)$ and let $(\mu_{p+1}, \ldots, \mu_m)$ be the complementary $(m - p)$-tuple. Multiplying (5.64) by $e_{\mu_{p+1}} \wedge \cdots \wedge e_{\mu_m}$ and observing that $u \in N(\wedge^{m-p}F)$ and $u' \in I_{\wedge^{p+1}F} \subset N(\wedge^{m-p}F)$, we obtain

$$0 = e \otimes c^{\mu_1, \ldots, \mu_p} \qquad e \in \wedge^m F, e \neq 0.$$

It follows that $c^{\mu_1, \ldots, \mu_p} = 0$ and so $u = u' \in I_{\wedge^{p+1}F}$. This completes the proof. □

Applying Proposition 5.26.1 for the special cases $p = 1$ and $p = m$, we obtain immediately the following

Corollary. *Let $u = x_1 \wedge \cdots \wedge x_m$ be a nonzero decomposable m-vector and $X \subset E$ be the subspace generated by the vectors x_i $(i = 1, \ldots, m)$. Then*

$$N(u) = I_X \qquad (p = 1)$$

and

$$N(X) = I_u \qquad (p = m).$$

In particular, u is divisible by a vector $y \neq 0$ if and only if $y \wedge u = 0$.

5.27.

As an application of the results of Section 5.26 we prove

Proposition 5.27.1. *Let E be an n-dimensional vector space and $\varphi: E \to \wedge E$ a linear map. Then*

$$x \wedge \varphi x = 0 \qquad x \in E \tag{5.65}$$

if and only if

$$\varphi x = x \wedge v \qquad x \in E \tag{5.66}$$

for some fixed element $v \in \wedge E$. The element v is uniquely determined mod $\wedge^n E$.

PROOF. It is trivial that (5.66) implies (5.65). Conversely, suppose (5.65) holds. Then it follows that

$$x \wedge \varphi y + y \wedge \varphi x = 0 \qquad x, y \in E.$$

Now we proceed by induction with respect to n. For $n = 0$ the proposition is trivial. Assume now that it is correct for $\dim E = n - 1$. Choose an arbitrary vector $a \neq 0$, $a \in E$. Since $a \wedge \varphi a = 0$, it follows from (5.61) with $m = p = 1$ that there is an element $c \in \wedge E$ such that

$$\varphi a = a \wedge c.$$

Now define a mapping $\sigma: E \to \wedge E$ by

$$\sigma x = \varphi x - x \wedge c. \tag{5.67}$$

Then we have

$$\sigma a = \varphi a - a \wedge c = 0, \tag{5.68}$$

$$x \wedge \sigma x = x \wedge \varphi x = 0, \tag{5.69}$$

and

$$a \wedge \sigma x = a \wedge \varphi x - a \wedge x \wedge c = -x \wedge \varphi a + x \wedge a \wedge c = 0. \tag{5.70}$$

Now consider the 1-dimensional subspace E_1 generated by a and write $E = E_1 \oplus F$. Since

$$\wedge E = \wedge E_1 \hat{\otimes} \wedge F,$$

it follows that

$$\sigma x = 1 \otimes \sigma_0 x + a \otimes \sigma_1 x \qquad x \in E,$$

where σ_0 and σ_1 are well-defined linear maps. Multiplying by a we obtain in view of (5.70)

$$0 = a \otimes \sigma_0 x \qquad x \in E,$$

whence $\sigma_0 x = 0$ and so

$$\sigma x = a \otimes \sigma_1 x \qquad x \in E. \tag{5.71}$$

Now let $y \in F$ be an arbitrary vector. Then relations (5.71) and (5.69) yield

$$-(1 \wedge a) \otimes (y \wedge \sigma_1 y) = (1 \otimes y) \wedge (a \otimes \sigma_1 y) = y \wedge \sigma y = 0$$

whence

$$y \wedge \sigma_1 y = 0 \qquad y \in F.$$

Now it follows from the induction hypothesis that

$$\sigma_1 y = y \wedge v_1 \qquad y \in F, \tag{5.72}$$

where $v_1 \in \wedge F$ is a fixed element. Since every vector $x \in E$ can be written in the form $x = \lambda a + y$, $\lambda \in \Gamma$, $y \in F$, we obtain, in view of (5.68), (5.71) and (5.72),

$$\begin{aligned} \sigma x = \lambda \sigma a + \sigma y = \sigma y = a \otimes \sigma_1 y &= a \otimes (y \wedge v_1) \\ = -y \wedge (a \wedge v_1) &= -x \wedge (a \wedge v_1) \end{aligned} \tag{5.73}$$

Combining (5.67) and (5.73) we find that

$$\varphi x = \sigma x + x \wedge c = x \wedge (c - a \wedge v_1)$$

and so the induction is closed.

To prove the uniqueness part of the proposition assume that $u_1 \in \wedge E$ and $u_2 \in \wedge E$ are two elements such that

$$\varphi x = x \wedge u_1 \quad \text{and} \quad \varphi x = x \wedge u_2.$$

Then

$$x \wedge (u_2 - u_1) = 0 \qquad x \in E$$

and so $u_2 - u_1 \in N(E) = \wedge^n E$. □

Corollary. *Let $\varphi : E \to \wedge^{k+1} E$ be a linear map such that*

$$x \wedge \varphi x = 0 \qquad x \in E.$$

Then there exists an element $u \in \wedge^k E$ such that

$$\varphi x = x \wedge u \qquad x \in E.$$

If $\varphi \neq 0$, then u is uniquely determined.

PROOF. In view of the above proposition, we can write

$$\varphi x = x \wedge v,$$

where $v \in \wedge E$. Applying the projection $\rho_{k+1}: \wedge E \to \wedge^{k+1} E$ to this relation, we obtain

$$\varphi x = \rho_{k+1}(x \wedge v) = x \wedge \rho_k v = x \wedge u \qquad u = \rho_k v$$

and so the first part of the corollary follows. Suppose now that

$$\varphi x = x \wedge u_1 \quad \text{and} \quad \varphi x = x \wedge u_2 \qquad u_1, u_2 \in \wedge^k E.$$

Then $u_2 - u_1 \in \wedge^n E$. On the other hand we have that $u_2 - u_1 \in \wedge^k E$ whence $u_2 - u_1 \in \wedge^n E \cap \wedge^k E$. But, if $\varphi \neq 0$, it follows that $k + 1 \leq n$ and so we obtain $u_2 - u_1 = 0$. □

If φ is a *homogeneous* linear map, Proposition 5.27.1 can be extended to spaces of infinite dimension.

Proposition 5.27.2. *Let $\varphi: E \to \wedge E$ be a homogeneous mapping of degree k such that $x \wedge \varphi x = 0$. Then there exists an element $u \in \wedge^k E$ such that $\varphi x = x \wedge u$. If $\varphi \neq 0$ the element u is uniquely determined by φ.*

PROOF. We prove first the uniqueness of u. Assume that

$$\varphi x = x \wedge u_1 \quad \text{and} \quad \varphi x = x \wedge u_2.$$

Then

$$x \wedge (u_2 - u_1) = 0 \qquad x \in E$$

and so

$$u_2 - u_1 \in N(E).$$

If the dimension of E is infinite we have $N(E) = 0$, whence $u_2 = u_1$. If $\dim E = n$, we have $N(E) = \wedge^n E$. If $\varphi \neq 0$, it follows that $k \neq n$ and so $u_2 - u_1 = 0$.

To show the existence of u we state first the following

Lemma. *Let $\varphi: E \to \wedge^{k+1} E$ be a linear map such that for a subspace F of finite dimension*

$$y \wedge \varphi y = 0 \qquad y \in F.$$

Then there exists an element $v \in \wedge^k E$ such that

$$\varphi y = y \wedge v \qquad y \in F.$$

The above lemma is proved by induction on $\dim F$ in the same way as Proposition 5.27.1.

Now consider a linear map $\varphi: E \to \wedge^{k+1} E$ satisfying $x \wedge \varphi x = 0$. We may assume that $\varphi \neq 0$. Choose $a \in E$ such that $\varphi a \neq 0$. Let H be a subspace of dimension $k + 1$ such that $a \in H$. Then, by the above lemma, there exists an element $u \in \wedge^k E$ such that

$$\varphi y = y \wedge u \qquad y \in H. \tag{5.74}$$

Now we show that

$$\varphi x = x \wedge u \quad \text{for every } x \in E.$$

Let $x \in E$ be an arbitrary vector and consider the subspace $H_1 \subset E$ generated by x and H. In view of the above lemma there exists an element $v \in \wedge^k E$ such that

$$\varphi z = z \wedge v \qquad z \in H_1. \tag{5.75}$$

From (5.74) and (5.75) we obtain

$$y \wedge (u - v) = 0 \qquad y \in H,$$

whence $u - v \in N(H)$. In view of (5.61) we have $N(H) = I_{\wedge^{k+1} H}$. On the other hand $u - v$ is of degree k and so $u - v = 0$. □

Proposition 5.27.2 permits us to give explicitly the forms of a derivation of odd degree and an antiderivation of even degree in $\wedge E$.

Let θ be a homogeneous derivation of degree k, where k is odd. Then we have $\theta = \Omega(\varphi)$ where φ denotes the restriction of θ to E (see Section 5.11). In view of (5.32), we have $x \wedge \varphi x = 0$ and hence, by the Corollary to Proposition 5.27.1, there exists an element $a \in \wedge^k E$ such that

$$\varphi x = a \wedge x \qquad x \in E.$$

Now Formula (5.33) yields

$$\theta u = \begin{cases} a \wedge u & u \in \wedge^p E, \ p \text{ odd}, \\ 0 & u \in \wedge^p E, \ p \text{ even}. \end{cases}$$

If θ is an antiderivation of even degree in $\wedge E$, we have again $x \wedge \varphi x = 0$, where φ denotes the restriction of θ to E. Hence φ can be written in the form

$$\varphi x = a \wedge x \qquad a \in \wedge^k E.$$

Now (5.34) gives

$$\theta u = \begin{cases} a \wedge u & u \in \wedge^p E, \ p \text{ odd} \\ 0 & u \in \wedge^p E, \ p \text{ even}. \end{cases}$$

Problems

1. Let I be a right ideal in $\wedge E$. Assume that $a \wedge b \in I$ and $b \notin \wedge^+ E$. Prove that $a \in I$.

2. Show that the generator of a graded principal ideal is homogeneous. Conclude that there exist nongraded principal ideals in $\wedge E$.

3. Construct a left ideal in $\wedge E$ which is not a right ideal.

4. Given a subspace $F \subset E$ prove that the algebra $\wedge(E/F)$ is isomorphic to the algebra $\wedge E/I_F$.

5. Let $E = E_1 + E_2$ be a decomposition of E and consider the homomorphism

$$\varphi_\wedge : \wedge E_1 \hat{\otimes} \wedge E_2 \to \wedge E,$$

which is induced by the linear map

$$\varphi: E_1 \oplus E_2 \to E_1 + E_2.$$

Determine the kernel of $\varphi_\wedge$.

6. Let φ be linear transformation of a finite-dimensional vector space E and denote the Fitting null- and 1-components of φ by $F_0(\varphi)$ and $F_1(\varphi)$. Prove that

$$F_0(\varphi_\wedge) = I_{F_0(\varphi)} \quad \text{and} \quad F_1(\varphi_\wedge) = \wedge F_1(\varphi).$$

Hint: See Problem 5, Section 2, Chapter XIII of *Linear Algebra*.

Ideals and Duality

5.28.

In this paragraph E, E^* will denote a pair of dual finite-dimensional vector spaces. Let I be a graded ideal in $\wedge E$. Then the orthogonal complement $I^\perp$ is stable under the operations $i(h)$, $h \in E$. In fact, if $u^* \in I^\perp$ is an arbitrary element we have, for every $u \in I$,

$$\langle i(h)u^*, u\rangle = \langle u^*, h \wedge u\rangle = 0,$$

and so $i(h)u^* \in I^\perp$. Conversely, if I is a graded subspace of $\wedge E$ such that $I^\perp$ is stable under every $i(h)$, then I is an ideal. Let $u \in I$ and $h \in E$ be arbitrary. Then we have, for every $u^* \in I^\perp$,

$$\langle u^*, h \wedge u\rangle = \langle i(h)u^*, u\rangle = 0$$

whence

$$h \wedge u \in (I^\perp)^\perp = I.$$

Now let F be a subspace of E and $F^\perp$ be the orthogonal complement. Then the subalgebra $\wedge(F^\perp)$ and the ideal I_F are orthogonal complements with respect to the scalar product in $\wedge E$ and $\wedge E^*$,

$$\boxed{\wedge(F^\perp) = (I_F)^\perp.} \tag{5.76}$$

In fact, consider the canonical projection

$$\pi: E \to E/F$$

and the injection

$$j: E^* \leftarrow F^\perp.$$

In view of Section 2.23 of *Linear Algebra*, the spaces E/F and $F^\perp$ are dual with respect to the scalar product defined by

$$\langle y^*, \pi x\rangle = \langle jy^*, x\rangle \qquad y^* \in F^\perp, x \in E. \tag{5.77}$$

This shows that the mappings π and j are dual. Consequently, the induced homomorphisms

$$\pi_\wedge : \wedge E \to \wedge(E/F)$$

and

$$j_\wedge : \wedge E^* \leftarrow \wedge(F^\perp)$$

are dual as well. This implies that

$$\operatorname{Im} j_\wedge = (\ker \pi_\wedge)^\perp.$$

Since $\ker \pi = F$ we have, in view of (5.57)

$$\ker \pi_\wedge = I_F.$$

On the other hand, it follows from Section 5.8 and the definition of j that

$$\operatorname{Im} j_\wedge = \wedge \operatorname{Im} j = \wedge(F^\perp).$$

Combining the above relations we obtain (5.76).

As an immediate consequence of (5.76) we have the formula

$$(\wedge F)^\perp = I_{F^\perp}, \tag{5.78}$$

which is obtained by applying (5.76) to the subspace $F^\perp \subset E^*$ and taking orthogonal complements on both sides.

Proposition 5.28.1. *Let $F \subset E$, $F^* \subset E^*$ be two dual subspaces. Then the ideals I_F and I_{F^*} are dual as well.*

PROOF. Let $F_1 = (F^*)^\perp$ and $F_1^* = F^\perp$. Then we have the direct decompositions

$$E = F \oplus F_1$$

and

$$E^* = F^* \oplus F_1^*.$$

This yields, in view of (5.56)

$$\wedge E = I_F \oplus \wedge F_1$$

and

$$\wedge E^* = I_{F^*} \oplus \wedge F_1^*.$$

Since $\wedge F_1^* = \wedge(F^\perp) = (I_F)^\perp$ and $\wedge F_1 = \wedge(F^{*\perp}) = I_{F^*}$, it follows that the ideals I_F and I_{F^*} are dual. □

Proposition 5.28.2. *Let $F \subset E$ be a subspace. Then the subalgebra $\wedge F$ is stable under the operations $i(h^*)$, $h^* \in E^*$. Conversely, if A is a subalgebra of $\wedge E$ stable under $i(h^*)$, $h^* \in E^*$, then there exists a subspace $F \subset E$ such that $A = \wedge F$.*

PROOF. Let $y_1 \wedge \cdots \wedge y_p$, $y_i \in F$, be any decomposable element of $\wedge F$. Then, for every $h^* \in E^*$,

$$i(h^*)(y_1 \wedge \cdots \wedge y_p) = \sum_j (-1)^{j+1} \langle h^*, y_j \rangle y_1 \wedge \cdots \wedge \hat{y}_j \wedge \cdots \wedge y_p \in \wedge F.$$

Hence $\wedge F$ is stable under $i(h^*)$. Conversely, assume that A is a subalgebra of $\wedge E$ which is stable under every $i(h^*)$. Then A is stable under every operator $i(u^*)$, $u^* \in \wedge E$. If $A = 0$ the proposition is trivial and so we may assume that $A \neq 0$. We show first that $1 \in A$. Let $u \neq 0$ be an element of A. We write u in the form

$$u = \sum_{p=0}^{r} u_p \qquad u_p \in \wedge^p E, u_r \neq 0.$$

Since $u_r \neq 0$, there exists an r-vector $u^* \in \wedge^r E^*$ such that $\langle u^*, u_r \rangle = 1$. Applying $i(u^*)$, we obtain

$$i(u^*)u = \langle u^*, u_r \rangle = 1.$$

Since $i(u^*)\, u \in A$ and $\langle u^*, u_r \rangle = 1$, it follows that $1 \in A$.

Now consider the subspace $F = A \cap E$. It will be shown that $A = \wedge F$. Clearly, $\wedge F \subset A$. Let

$$u = \sum_{p=0}^{n} u_p \qquad u_p \in \wedge^p E, n = \dim E,$$

be an arbitrary element of A and assume by induction that

$$u_\nu \in \wedge F \quad \text{for } \nu > q.$$

Since $\wedge^{n+1} E = 0$, this relation is correct for $q = n$. Define v by

$$v = u - \sum_{\nu=q+1}^{n} u_\nu. \tag{5.79}$$

Since $u \in A$ and $u_\nu \in \wedge F \subset A$, for $\nu \geq q + 1$, we have $v \in A$. Now let $u^* \in \wedge^{q-1} E^*$ be an arbitrary element. Applying $i(u^*)$ to (5.79) we obtain

$$\begin{aligned} i(u^*)v &= i(u^*)u_{q-1} + i(u^*)u_q \\ &= \langle u^*, u_{q-1} \rangle + i(u^*)u_q. \end{aligned} \tag{5.80}$$

Since A is stable under $i(u^*)$, we have $i(u^*)v \in A$. Since $1 \in A$, it follows from (5.80) that $i(u^*)u_q \in A$. On the other hand $i(u^*)u_q$ is of degree 1 and hence

$$i(u^*)u_q \in A \cap E = F.$$

Now let $y^* \in F^\perp$ be arbitrary. Then

$$\langle y^* \wedge u^*, u_q \rangle = \pm \langle y^*, i(u^*)u_q \rangle = 0.$$

Hence, u_q is orthogonal to the ideal $I_{F^\perp}$. Now it follows from (5.78) that $u_q \in \wedge F$ and so the induction is closed. □

PROBLEMS

1. Let I be a subspace of $\wedge E$. Prove that I is a left ideal if and only if the orthogonal complement $I^\perp$ is stable under $i(h)$ for every $h \in E$.

2. Given a subspace $F \subset E$ prove that

$$\wedge F = \bigcap_{u^* \in I_{F^\perp}} \ker i(u^*).$$

3. Let $h \in E$ be a fixed vector. Define operators $\partial: \wedge E \to \wedge E$ and $\delta: \wedge E^* \to \wedge E^*$ by

$$\partial u = \mu(h)u \qquad u \in \wedge E$$

and

$$\delta u^* = i(h)u^* \qquad u^* \in \wedge E^*.$$

(a) Show that $(\wedge E, \partial)$ is a graded differential space and $(\wedge E^*, \delta)$ is a graded differential algebra.

(b) Prove that

$$H(\wedge E) = 0 \quad \text{and} \quad H(\wedge E^*) = 0.$$

4. Let E, E^* be a pair of dual vector spaces and consider a graded subspace $U \subset \wedge E$. Show that U is an ideal in $\wedge E$ if and only if $U^\perp$ is stable under the operations $i(h)$, $h \in E$.

The Algebra of Skew-Symmetric Functions

5.29. Skew-Symmetric Functions

Let E be an n-dimensional vector space and consider the space $T^p(E)$ of p-linear functions in E (see Section 3.18). Then, if $\Phi \in T^p(E)$ and $\sigma \in S_p$, an element $\sigma\Phi \in T^p(E)$ is defined by

$$(\sigma\Phi)(x_1, \ldots, x_p) = \Phi(x_{\sigma(1)}, \ldots, x_{\sigma(p)}).$$

The function Φ is called *skew-symmetric*, if

$$\sigma\Phi = \varepsilon_\sigma \Phi.$$

The skew-symmetric functions form a subspace of $T^p(E)$ which will be denoted by $A^p(E)$.

Every p-linear function Φ in E determines a skew-symmetric p-linear function $A\Phi$ given by

$$A\Phi = \frac{1}{p!} \sum_\sigma \varepsilon_\sigma(\sigma\Phi).$$

$A\Phi$ is called the *skew-symmetric part of* Φ. In particular, if Φ is skew-symmetric, then $A\Phi = \Phi$.

Thus the operator $A: T^p(E) \to T^p(E)$ is a projection operator. It is called the *antisymmetry operator*.

Now let $\Phi \in T^p(E)$ and $\Psi \in T^q(E)$ and consider the $(p + q)$-linear function $\Phi \cdot \Psi$ (see Section 3.18). A simple calculation shows that

$$A(\Phi \cdot \Psi) = A(A\Phi \cdot \Psi) = A(\Phi \cdot A\Psi). \tag{5.81}$$

This formula implies that

$$A(\Phi \cdot \Psi) = A(A\Phi \cdot A\Psi).$$

Thus the skew-symmetric part of a product depends only on the skew-symmetric parts of the factors.

5.30. The Algebra $A^{\bullet}(E)$

The *Grassmann product* of two skew-symmetric functions $\Phi \in A^p(E)$ and $\Psi \in A^q(E)$ is the $(p + q)$-linear skew-symmetric function $\Phi \wedge \Psi$ given by

$$\Phi \wedge \Psi = \frac{(p+q)!}{p!q!} A(\Phi \cdot \Psi).$$

Explicitly,

$$(\Phi \wedge \Psi)(x_1, \ldots, x_{p+q}) = \frac{1}{p!q!} \sum_\sigma \varepsilon_\sigma \Phi(x_{\sigma(1)}, \ldots, x_{\sigma(p)}) \Psi(x_{\sigma(p+1)}, \ldots, x_{\sigma(p+q)}).$$

Proposition 5.30.1. *The Grassmann product has the following properties*:

$$\Phi \wedge \Psi = (-1)^{pq} \Psi \wedge \Phi \qquad \Phi \in A^p(E), \Psi \in A^q(E), \tag{A}$$

$$(\Phi \wedge \Psi) \wedge X = \Phi \wedge (\Psi \wedge X) \qquad \Phi, \Psi, X \in A(E). \tag{B}$$

PROOF. (A) Observe that

$$\Phi \cdot \Psi = \sigma(\Psi \cdot \Phi),$$

where σ is the permutation

$$(1, \ldots, p, p+1, \ldots, p+q) \to (p+1, \ldots, p+q, 1, \ldots, p).$$

Applying A to this equation and observing that $\varepsilon_\sigma = (-1)^{pq}$ we obtain (A).

(B) Let $\Phi \in A^p(E)$, $\Psi \in A^q(E)$, and $X \in A^r(E)$. Then

$$\begin{aligned}(\Phi \wedge \Psi) \wedge X &= \frac{(p+q+r)!}{p!q!r!} A[(\Phi \cdot \Psi) \cdot X] \\ &= \frac{(p+q+r)!}{p!q!r!} A[A(\Phi \cdot \Psi) \cdot X] \\ &= \frac{(p+q+r)!}{p!q!r!} A(\Phi \cdot \Psi \cdot X).\end{aligned}$$

Similarly,

$$\Phi \wedge (\Psi \wedge \mathrm{X}) = \frac{(p+q+r)!}{p!q!r!} A(\Phi \cdot \Psi \cdot \mathrm{X})$$

and so **(B)** follows.

The $\wedge$-product makes the direct sum

$$A^{\cdot}(E) = \sum_{p=0}^{n} A^{p}(E)$$

into a graded associative algebra called the *Grassmann algebra over E.*

5.31. Homomorphisms and Derivations

A linear map $\varphi: E \to F$ induces a homomorphism $\varphi^{A}: A^{\cdot}(E) \leftarrow A^{\cdot}(F)$ given by

$$(\varphi^{A}\Psi)(x_1, \ldots, x_p) = \Psi(\varphi x_1, \ldots, \varphi x_p) \qquad \Psi \in A^{p}(F).$$

Next, let φ be a linear transformation of E. Then φ determines a derivation $\theta^{A}(\varphi)$ in $A^{\cdot}(E)$ given by

$$(\theta^{A}(\varphi)\Phi)(x_1, \ldots, x_p) = \sum_{\nu=1}^{p} \Phi(x_1, \ldots, \varphi x_{\nu}, \ldots, x_p).$$

To show that this is indeed a derivation consider the derivation $\theta^{T}(\varphi)$ induced by φ in the algebra $T^{\cdot}(E)$ (see Section 3.18). It follows from the definitions that

$$\theta^{A}(\varphi)\Phi = \theta^{T}(\varphi)\Phi \qquad \Phi \in A^{\cdot}(E).$$

Moreover, the operator $\theta^{T}(\varphi)$ satisfies the relations

$$\theta^{T}(\varphi) \circ \sigma = \sigma \circ \theta^{T}(\varphi)$$

whence

$$\theta^{A}(\varphi) \circ A = A \circ \theta^{T}(\varphi).$$

It follows that

$$\begin{aligned}
\theta^{A}(\varphi)(\Phi \wedge \Psi) &= \frac{(p+q)!}{p!q!} \theta^{A}(\varphi)A(\Phi \cdot \Psi) \\
&= \frac{(p+q)!}{p!q!} A\theta^{T}(\varphi)(\Phi \cdot \Psi) \\
&= \frac{(p+q)!}{p!q!} A(\theta^{T}(\varphi) \cdot \Psi + \Phi \cdot \theta^{T}(\varphi)\Psi) \\
&= \theta^{T}(\varphi)\Phi \wedge \Psi + \Phi \wedge \theta^{T}(\varphi)\Psi \\
&= \theta^{A}(\varphi)\Phi \wedge \Psi + \Phi \wedge \theta^{A}(\varphi)\Psi.
\end{aligned}$$

5.32. The Operator $i_A(h)$

Recall from Section 3.19 the definition of the linear operator $i_A(h): T^{\bullet}(E) \to T^{\bullet}(E)$. It follows that if $\Phi \in A^{\bullet}(E)$, then

$$i_A(h)\Phi = i_1(h)\Phi.$$

Thus,

$$(i_A(h)\Phi)(x_1, \ldots, x_{p-1}) = \Phi(h, x_1, \ldots, x_{p-1}) \qquad \Phi \in A^p(E).$$

$i_A(h)$ is called the *substitution operator* in the algebra $A^{\bullet}(E)$.

Lemma. *Let A denote the antisymmetry operator. Then*

$$i_A(h) \circ A = A \circ i_A(h).$$

PROOF. Let $\Phi \in T^p(E)$. It is sufficient to consider the case

$$\Phi = f_1 \cdot \ldots \cdot f_p \qquad f_i \in E^*.$$

Then we have

$$A\Phi = \frac{1}{p!} \sum_{\sigma} \varepsilon_\sigma f_{\sigma(1)} \cdot \ldots \cdot f_{\sigma(p)}$$

and thus,

$$\begin{aligned} i_A(h)(A\Phi) &= i_1(h)(A\Phi) \\ &= \frac{1}{p!} i_1(h) \sum_{\sigma} \varepsilon_\sigma f_{\sigma(1)} \cdot \ldots \cdot f_{\sigma(p)} \\ &= \frac{1}{p!} \sum_{\sigma} \varepsilon_\sigma f_{\sigma(1)}(h) f_{\sigma(2)} \cdot \ldots \cdot f_{\sigma(p)} \\ &= \frac{1}{p!} \sum_{\mu=1}^{p} \sum_{\sigma(1)=\mu} \varepsilon_\sigma f_\mu(h) f_{\sigma(2)} \cdot \ldots \cdot f_{\sigma(p)} \\ &= \frac{1}{p!} \sum_{\mu=1}^{p} f_\mu(h) A_\mu, \end{aligned} \tag{5.82}$$

where

$$A_\mu = \sum_{\sigma(1)=\mu} \varepsilon_\sigma f_{\sigma(2)} \cdot \ldots \cdot f_{\sigma(p)}.$$

Now we show that

$$A_\mu = (-1)^{\mu-1}(p-1)! A(f_1 \otimes \cdots \otimes \hat{f}_\mu \cdot \ldots \cdot f_p).$$

In fact, fix μ and let $\sigma_\mu \in S_p$ be the permutation defined by

$$\sigma_\mu(1) = 2, \ldots, \quad \sigma_\mu(\mu - 1) = \mu, \quad \sigma_\mu(\mu) = 1, \quad \sigma_\mu(\mu + 1) = \mu + 1, \quad \cdots, \quad \sigma_\mu(p) = p.$$

Then every permutation $\sigma \in S_p$ determines a permutation $\tau_\sigma \in S_p$ given by

$$\tau_\sigma = \sigma \circ \sigma_\mu .$$

Since

$$\varepsilon(\sigma_\mu) = (-1)^{\mu-1},$$

$$\varepsilon(\tau_\sigma) = (-1)^{\mu-1}\varepsilon(\sigma).$$

Thus we obtain

$$\begin{aligned} A_\mu &= \sum_{\tau(\mu)=\mu} (-1)^{\mu-1}\varepsilon(\tau) f_{\tau(1)} \cdot \ldots \cdot \hat{f}_{\tau(\mu)} \cdot \ldots \cdot f_{\tau(p)} \\ &= (-1)^{\mu-1}(p-1)! A(f_1 \cdot \ldots \cdot \hat{f}_\mu \cdot \ldots \cdot f_p). \end{aligned} \tag{5.83}$$

Equations (5.82) and (5.83) yield

$$\begin{aligned} i_A(h)(A\Phi) &= \frac{(p-1)!}{p!} \sum_\mu (-1)^{\mu-1} f_\mu(h) A(f_1 \cdot \ldots \cdot \hat{f}_\mu \cdot \ldots \cdot f_p) \\ &= \frac{1}{p} A\left[\sum_\mu (-1)^{\mu-1} f_\mu(h) f_1 \cdot \ldots \cdot \hat{f}_\mu \cdot \ldots \cdot f_p\right] \\ &= \frac{1}{p} A \sum_\mu (-1)^{\mu-1} i_\mu(h)(f_1 \cdot \ldots \cdot f_p) \\ &= A i_A(h)\Phi \end{aligned}$$

and so the lemma is established. □

Proposition 5.32.1. *The operator $i_A(h)$ is an antiderivation in the algebra $A^\bullet(E)$,*

$$i_A(h)(\Phi \wedge \Psi) = i_A(h)\Phi \wedge \Psi + (-1)^p \Phi \wedge i_A(h)\Psi \qquad \Phi \in A^p(E),\ \Psi \in A^q(E).$$

PROOF. Since

$$\Phi \wedge \Psi = \frac{(p+q)!}{p!q!} A(\Phi \cdot \Psi),$$

we have to show that

$$i_A(h)A(\Phi \cdot \Psi) = \frac{p}{p+q} A(i_A(h)\Phi \cdot \Psi) + (-1)^p \frac{q}{p+q} A(\Phi \cdot i_A(h)\Psi).$$

By the lemma,

$$i_A(h)A(\Phi \cdot \Psi) = \frac{1}{p+q} \sum_{\mu=1}^{p+q} (-1)^{\mu-1} A i_\mu(h)(\Phi \cdot \Psi).$$

Now,

$$i_\mu(h)(\Phi \cdot \Psi) = \begin{cases} i_\mu(h)\Phi \cdot \Psi & \mu \le p, \\ \Phi \cdot i_{\mu-p}(h)\Psi & \mu \ge p+1, \end{cases}$$

(see Formula 3.18). Thus we obtain

$$\begin{aligned} i(h)A(\Phi \cdot \Psi) &= \frac{1}{p+q} \sum_{\mu=1}^{p} (-1)^{\mu-1} A(i_\mu(h)\Phi \cdot \Psi) \\ &\quad + \frac{1}{p+q} \sum_{\mu=p+1}^{p+q} (-1)^{\mu-1} A(\Phi \cdot i_{\mu-p}(h)\Psi) \\ &= \frac{1}{p+q} A\left(\sum_{\mu=1}^{p} (-1)^{\mu-1} i_\mu(h)\Phi \cdot \Psi\right) \\ &\quad + \frac{1}{p+q} A\left(\Phi \cdot \sum_{\mu=p+1}^{p+q} (-1)^{\mu-1} i_{\mu-p}(h)\Psi\right) \\ &= \frac{p}{p+q} A(i_A(h)\Phi \cdot \Psi) + (-1)^p \frac{q}{p+q} A(\Phi \cdot i_A(h)\Psi), \end{aligned}$$

which completes the proof. □

5.33. The Isomorphism $\wedge E^* \xrightarrow{\cong} T^\cdot(E)$

We show that the Grassmann algebra over E is isomorphic to the exterior algebra over E^*. In fact, consider the isomorphism $\alpha: \otimes^p E^* \xrightarrow{\cong} T^p(E)$ (see Section 3.20). It is easily checked that the diagram

$$\begin{array}{ccc} \otimes^p E^* & \xrightarrow[\cong]{\alpha} & T^p(E) \\ {\scriptstyle \pi^A}\big\downarrow & & \big\downarrow{\scriptstyle A} \\ \otimes^p E^* & \xrightarrow[\cong]{\alpha} & T^p(E) \end{array}$$

commutes, where π^A denotes the alternator (see Section 4.2). Since

$$\operatorname{Im} \pi^A = X^p(E^*) \quad \text{and} \quad \operatorname{Im} A = A^p(E),$$

it follows that α restricts to a linear isomorphism

$$\alpha: X^p(E^*) \xrightarrow{\cong} A^p(E).$$

Next observe that, in view of the commutative diagram above,

$$\begin{aligned} \alpha(u \cap v) &= \alpha\pi^A(u \otimes v) = A\alpha(u \otimes v) \\ &= A(\alpha(u) \cdot \alpha(v)) = \frac{p!q!}{(p+q)!} \alpha(u) \wedge \alpha(v) \qquad u \in X^p(E^*), v \in X^q(E^*). \end{aligned}$$

Thus the map

$$\beta: X(E^*) \to A^{\boldsymbol{\cdot}}(E)$$

defined by

$$\beta(u) = p!\alpha(u) \qquad u \in X^p(E^*)$$

is an algebra isomorphism.

Composing this isomorphism with the algebra isomorphism

$$\eta: \wedge E^* \xrightarrow{\cong} X(E^*),$$

obtained in Section 5.3, we obtain an algebra isomorphism

$$\wedge E^* \underset{\cong}{\xrightarrow{\eta}} X(E^*) \underset{\cong}{\xrightarrow{\beta}} A^{\boldsymbol{\cdot}}(E).$$

Under this isomorphism,

(1) The homomorphism $\varphi^\wedge: \wedge E^* \leftarrow \wedge F^*$ corresponds to the homomorphism $\varphi^A: A(E) \leftarrow A(F)$ (see Sections 5.9 and 5.31).
(2) The derivation $\theta^\wedge(\varphi)$ corresponds to the derivation $\theta^A(\varphi)$ (see Sections 5.10 and 5.31).
(3) The operator $i(h)$ corresponds to the substitution operator $i_A(h)$ (see Sections 5.14 and 5.32).

5.34. The Algebra $A_{\boldsymbol{\cdot}}(E)$

Denote by $A_p(E)$ ($p \geq 1$) the space of skew-symmetric p-linear mappings in the dual space E^* and set $A_0(E) = \Gamma$. Then we have the $\wedge$-multiplication between $A_p(E)$ and $A_q(E)$ (see Section 5.30). It makes the direct sum

$$A_{\boldsymbol{\cdot}}(E) = \sum_{p=0}^{n} A_p(E)$$

into an associative algebra which is isomorphic to the algebra $\wedge E$.

Now we show that the scalar product between $T^p(E)$ and $T_p(E)$ defined in Section 3.22 restricts to a scalar product between $A^p(E)$ and $A_p(E)$.

In fact, it is easy to check the relation

$$\langle \Phi, \sigma\Psi \rangle = \langle \sigma^{-1}\Phi, \Psi \rangle \qquad \Phi \in A^p(E), \Psi \in A_p(E).$$

It follows that

$$\langle \Phi, A\Psi \rangle = \langle A\Phi, \Psi \rangle.$$

Now suppose that $\Phi_1 \in A^p(E)$ is an element such that $\langle \Phi_1, \Psi_1 \rangle = 0$ for every $\Psi_1 \in A_p(E)$. Then we have for every $\Psi \in T_p(E)$

$$\langle \Phi_1, \Psi \rangle = \langle A\Phi_1, \Psi \rangle = \langle \Phi_1, A\Psi \rangle = 0$$

whence $\Phi_1 = 0$. In the same way it follows that if $\Psi_1 \in A_p(E)$ is an element such that $\langle \Phi_1, \Psi_1 \rangle = 0$ for every $\Phi_1 \in A^p(E)$, then $\Psi_1 = 0$.

Thus a scalar product $\langle\,,\,\rangle_A$ between the spaces $A^p(E)$ and $A_p(E)$ is defined by

$$\langle \Phi, \Psi\rangle_A = \frac{1}{p!}\langle \Phi, \Psi\rangle \qquad \Phi \in A^p(E),\ \Psi \in A_p(E).$$

In particular,

$$\langle \Phi, x_1 \wedge \cdots \wedge x_p\rangle_A = \Phi(x_1, \ldots, x_p) \qquad x_\nu \in E$$

and

$$\langle f_1 \wedge \cdots \wedge f_p, \Psi\rangle_A = \Psi(f_1, \ldots, f_p) \qquad f_\nu \in E^*.$$

In fact,

$$\begin{aligned}\langle \Phi, x_1 \wedge \cdots \wedge x_p\rangle_A &= \frac{1}{p!}\langle \Phi, x_1 \wedge \cdots \wedge x_p\rangle\\ &= \langle \Phi, \pi_A(x_1 \otimes \cdots \otimes x_p)\rangle = \langle A\Phi, x_1 \otimes \cdots \otimes x_p\rangle\\ &= \langle \Phi, x_1 \otimes \cdots \otimes x_p\rangle = \Phi(x_1, \ldots, x_p).\end{aligned}$$

The second formula is obtained in the same way.

Finally we show that

$$\langle f_1 \wedge \cdots \wedge f_p, x_1 \wedge \cdots \wedge x_p\rangle_A = \det(f_i(x_j)) \qquad f_i \in E^*,\ x_i \in E.$$

In fact,

$$\begin{aligned}\langle f_1 \wedge \cdots \wedge f_p, x_1 \wedge \cdots \wedge x_p\rangle_A &= \frac{1}{p!}\langle f_1 \wedge \cdots \wedge f_p, x_1 \wedge \cdots \wedge x_p\rangle\\ &= \langle f_1 \wedge \cdots \wedge f_p, A(x_1 \otimes \cdots \otimes x_p)\rangle\\ &= \sum_\sigma \varepsilon_\sigma f_1(x_{\sigma(p)}) \cdots f_p(x_{\sigma(p)})\\ &= \det(f_i(x_j)).\end{aligned}$$

6 Mixed Exterior Algebra

Throughout this chapter E^*, E will denote a pair of dual vector spaces over a field of characteristic zero.

The Algebra $\wedge(E^*, E)$

6.1. Skew-Symmetric Maps of Type (p, q)

A skew-symmetric map of type (p, q) from E^*, E into a vector space H is a $(p + q)$-linear mapping

$$\psi: \underbrace{E^* \times \cdots \times E^*}_{p} \times \underbrace{E \times \cdots \times E}_{q} \to H$$

which satisfies

$$\psi(x^*_{\sigma(1)}, \ldots, x^*_{\sigma(p)}; x_{\tau(1)}, \ldots, x_{\tau(q)}) = \varepsilon_\sigma \varepsilon_\tau \psi(x^*_1, \ldots, x^*_p; x_1, \ldots, x_q)$$

for all permutations $\sigma \in S_p$ and $\tau \in S_q$.

Proposition 6.1.1. *Every skew-symmetric map* ψ *of type* (p, q) *determines a unique linear map*

$$f: \wedge^p E^* \otimes \wedge^q E \to H$$

such that

$$f[(x^*_1 \wedge \cdots \wedge x^*_p) \otimes (x_1 \wedge \cdots \wedge x_q)] = \psi(x^*_1, \ldots, x^*_p; x_1, \ldots, x_q).$$

PROOF. The uniqueness follows from the fact that the products

$$(x^*_1 \wedge \cdots \wedge x^*_p) \otimes (x_1 \wedge \cdots \wedge x_q)$$

span the space $(\wedge^p E^*) \otimes (\wedge^q E)$.

To prove existence define a linear map

$$g:(\otimes^p E^*) \otimes (\otimes^q E) \to H$$

by

$$g(x_1^* \otimes \cdots \otimes x_p^* \otimes x_1 \otimes \cdots \otimes x_q) = \psi(x_1^*, \ldots, x_p^*; x_1, \ldots, x_q)$$

(see Section 1.20) and consider the bilinear mapping

$$\beta:(\otimes^p E^*) \times (\otimes^q E) \to H$$

given by

$$\beta(u, v) = g(u \otimes v).$$

The skew-symmetry of ψ implies that $\beta(u, v)$ depends only on the vectors $\pi_1 u$ and $\pi_2 v$ where

$$\pi_1 : \otimes^p E^* \to \wedge^p E^* \quad \text{and} \quad \pi_2 : \otimes^q E \to \wedge^q E$$

are the canonical projections (see Section 5.3). Thus a bilinear mapping

$$\gamma: \wedge^p E^* \times \wedge^q E \to H$$

is defined by

$$\gamma(\pi_1 u, \pi_2 v) = \beta(u, v).$$

This map, in turn, induces a linear map

$$f:(\wedge^p E^*) \otimes (\wedge^q E) \to H.$$

It follows that

$$\begin{aligned} f[(x_1^* \wedge \cdots \wedge x_p^*) \otimes (x_1 \wedge \cdots \wedge x_q)] &= \gamma\,[\pi_1(x_1^* \otimes \cdots \otimes x_p^*), \pi_2(x_1 \wedge \cdots \wedge x_q)] \\ &= \beta(x_1^* \otimes \cdots \otimes x_p^*, x_1 \otimes \cdots \otimes x_q) \\ &= g(x_1^* \otimes \cdots \otimes x_p^* \otimes x_1 \otimes \cdots \otimes x_q) \\ &= \psi(x_1^*, \ldots, x_p^*; x_1, \ldots, x_q) \end{aligned}$$

and so the proof is complete. □

6.2. The Algebra $\wedge(E^*, E)$

The *mixed exterior algebra* over the pair E^*, E, denoted by $\wedge(E^*, E)$ is defined to be the canonical tensor product of the algebras $\wedge E^*$ and $\wedge E$ (see Section 2.2),

$$\wedge(E^*, E) = \wedge E^* \otimes \wedge E.$$

The multiplication in this algebra will be denoted by $\cdot$ and is determined by the equation

$$(u^* \otimes u) \cdot (v^* \otimes v) = (u^* \wedge v^*) \otimes (u \wedge v) \qquad u^*, v^* \in \wedge E^*, u, v \in \wedge E.$$

Thus $\wedge(E^*, E)$ is a graded associative algebra with unit element $1 \otimes 1$. It is generated by the elements $1 \otimes 1$, $x^* \otimes 1$ and $1 \otimes x$ with $x^* \in E^*$ and $x \in E$.

If $w \in \wedge(E^*, E)$, we shall define w^k ($k \geq 0$) by

$$w^k = \frac{1}{k!}\underbrace{w \cdot \cdots \cdot w}_{k} \qquad k \geq 1$$

and

$$w^0 = 1.$$

Now consider the bigradation of $\wedge(E^*, E)$ by the subspaces

$$\wedge_q^p(E^*, E) = \wedge^p E^* \otimes \wedge^q E.$$

Clearly,

$$w_1 \cdot w_2 = (-1)^{p_1 p_2 + q_1 q_2} w_2 \cdot w_1 \qquad w_1 \in \wedge_{q_1}^{p_1}(E^*, E), w_2 \in \wedge_{q_2}^{p_2}(E^*, E). \quad (6.1)$$

Since

$$p_1 p_2 + q_1 q_2 \equiv (p_1 + q_1)p_2 + (p_2 + q_2)q_1 \pmod 2,$$

it follows that $p_1 p_2 + q_1 q_2$ is even whenever $p_1 + q_1$ and $p_2 + q_2$ are and that

$$w_1 \cdot w_2 = w_2 \cdot w_1 \quad \text{if } p_1 + q_1 \text{ and } p_2 + q_2 \text{ are even.}$$

In this case we have the binomial formula

$$(w_1 + w_2)^k = \sum_{i+j=k} w_1^i \cdot w_2^j.$$

The *scalar product* between $\wedge E^*$ and $\wedge E$ defined by

$$\langle x_1^* \wedge \cdots \wedge x_p^*, x_1 \wedge \cdots \wedge x_q \rangle = \begin{cases} 0 & \text{if } p \neq q, \\ \det(\langle x_i^*, x_j \rangle) & \text{if } p = q, \end{cases}$$

(see Section 5.6) induces an inner product in $\wedge(E^*, E)$ via

$$\langle u^* \otimes u, v^* \otimes v \rangle = \langle u^*, v \rangle \langle v^*, u \rangle \qquad u^*, v^* \in \wedge E^*, u, v \in \wedge E. \quad (6.2)$$

(The symmetry and the nondegeneracy are easily checked.)

Now fix an element $z \in \wedge(E^*, E)$ and, denote by $\mu(z)$ the left multiplication by z,

$$\mu(z)w = z \cdot w \qquad w \in \wedge(E^*, E);$$

let $i(z)$ be the dual operator,

$$\langle i(z)w_1, w_2 \rangle = \langle w_1, \mu(z)w_2 \rangle \qquad w_1, w_2 \in \wedge(E^*, E).$$

Then, if $z \in \wedge_q^p(E^*, E)$ and $w \in \wedge_s^r(E^*, E)$,

$$i(z)w \in \wedge_{s-p}^{r-q}(E^*, E) \quad \text{if } r \geq p \text{ and } s \geq q$$

and

$$i(z)w = 0 \quad \text{otherwise.}$$

It follows from the definition that

$$\mu(u^* \otimes u) = \mu(u^*) \otimes \mu(u) \qquad u^* \in \wedge E^*, u \in \wedge E,$$

where the operators on the right-hand side are the left multiplications in $\wedge E^*$ and $\wedge E$ respectively. In particular,

$$\mu(1 \otimes u) = \iota \otimes \mu(u) \qquad u \in \wedge E$$

and

$$\mu(u^* \otimes 1) = \mu(u^*) \otimes \iota \qquad u^* \in \wedge E^*.$$

Dualizing the relation

$$\mu(z_1 \cdot z_2) = \mu(z_1) \circ \mu(z_2) \qquad z_1, z_2 \in \wedge(E^*, E),$$

we obtain

$$i(z_1 \cdot z_2) = i(z_2) \circ i(z_1). \tag{6.3}$$

Note that if $z \in \wedge_q^p(E^*, E)$ and $w \in \wedge_p^q(E^*, E)$, then $i(z)w$ is the element in $\wedge_0^0(E^*, E) = \Gamma$ given by

$$i(z)w = \langle z, w \rangle.$$

Next, consider the flip operators

$$Q_E: \wedge(E^*, E) \to \wedge(E, E^*) \quad \text{and} \quad Q_{E^*}: \wedge(E, E^*) \to \wedge(E^*, E)$$

given by

$$Q_E(u^* \otimes u) = u \otimes u^* \qquad u^* \in \wedge E^*, u \in \wedge E$$

and

$$Q_{E^*}(u \otimes u^*) = u^* \otimes u \qquad u^* \in \wedge E^*, u \in \wedge E.$$

They are algebra isomorphisms as well as isometries with respect to the inner products in $\wedge(E^*, E)$ and in $\wedge(E, E^*)$.

The subspace

$$\Delta E = \sum_{p \geq 0} \Delta_p E, \quad \text{where } \Delta_p E = \wedge_p^p(E^*, E),$$

of $\wedge(E^*, E)$ is obviously a subalgebra. It is called the *diagonal subalgebra.* Formula (6.1) implies that the diagonal subalgebra is commutative. Moreover, ΔE is stable under the operators $i(z)$ if $z \in \Delta E$. Finally, the restriction of the inner product in $\wedge(E^*, E)$ to ΔE is nondegenerate.

Next, let F, F^* be a second pair of dual vector spaces and let

$$\varphi: E \to F, \qquad \varphi^*: E^* \leftarrow F^*$$

and

$$\psi: E \leftarrow F, \qquad \psi^*: E^* \to F^*$$

be a pair of dual maps. Then we have the induced algebra homomorphisms

$$\psi^\wedge \otimes \varphi_\wedge : \wedge(E^*, E) \to \wedge(F^*, F)$$

and

$$\varphi^\wedge \otimes \psi_\wedge : \wedge(E^*, E) \leftarrow \wedge(F^*, F).$$

It is easy to check that these maps are dual.

Since $\psi^\wedge \otimes \varphi_\wedge$ is an algebra homomorphism, we have the relation

$$(\psi^\wedge \otimes \varphi_\wedge)(z \cdot w) = (\psi^\wedge \otimes \varphi_\wedge)z \cdot (\psi^\wedge \otimes \varphi_\wedge)w \qquad z, w \in \wedge(E^*, E)$$

or equivalently,

$$(\psi^\wedge \otimes \varphi_\wedge) \circ \mu(z) = \mu[\psi^\wedge \otimes \varphi_\wedge)z] \circ (\psi^\wedge \otimes \varphi_\wedge).$$

Dualizing, we obtain

$$i(z) \circ (\varphi^\wedge \otimes \psi_\wedge) = (\varphi^\wedge \otimes \psi_\wedge) \circ i[(\psi^\wedge \otimes \varphi_\wedge)z]. \tag{6.4}$$

Finally, note that a pair of dual isomorphisms $\varphi: E \xrightarrow{\cong} F$, $\varphi^*: E^* \xleftarrow{\cong} F^*$ induces an algebra isomorphism

$$\alpha_\varphi : \wedge(E^*, E) \xrightarrow{\cong} \wedge(F^*, F)$$

given by

$$\alpha_\varphi = (\varphi^\wedge)^{-1} \otimes \varphi_\wedge .$$

Next, consider the linear map

$$T_E: \wedge(E^*, E) \to L(\wedge E; \wedge E)$$

defined by

$$T_E(a^* \otimes b)v = \langle a^*, v\rangle b \qquad v \in \wedge E \tag{6.5}$$

(see Section 1.26). Since

$$\langle a^*, v\rangle = 0 \quad \text{if } a^* \in \wedge^p E^*, v \in \wedge^q E, p \neq q,$$

T_E restricts to linear maps

$$\wedge^p_q(E^*, E) \to L(\wedge^p E; \wedge^q E).$$

The dual of the linear transformation

$$T_E(a^* \otimes b) \colon \wedge E \to \wedge E$$

is the linear transformation $T_E(b \otimes a^*)$. Finally, recall from Section 1.26 that T_E is a linear isomorphism if E has finite dimension.

6.3. The Box Product of Linear Transformations

Let $\varphi_i (i = 1, \ldots, p)$ be linear transformations of E. Then a linear transformation

$$\varphi_1 \square \cdots \square \varphi_p \colon \wedge^p E \to \wedge^p E$$

is given by

$$(\varphi_1 \square \cdots \square \varphi_p)(x_1 \wedge \cdots \wedge x_p) = \sum_\sigma \varepsilon_\sigma \varphi_1 x_{\sigma(1)} \wedge \cdots \wedge \varphi_p x_{\sigma(p)}.$$

It is called the *box product* of the φ_i. In particular,

$$(\varphi_1 \square \varphi_2)(x_1 \wedge x_2) = \varphi_1 x_1 \wedge \varphi_2 x_2 - \varphi_1 x_2 \wedge \varphi_2 x_1.$$

The box product formula can be written in the form

$$(\varphi_1 \square \cdots \square \varphi_p)(x_1 \wedge \cdots \wedge x_p) = \sum_\sigma \varphi_{\sigma(1)} x_1 \wedge \cdots \wedge \varphi_{\sigma(p)} x_p$$

This show that the box product is symmetric,

In fact, let $\tau \in S_p$ be any permutation. Then we have

$$(\varphi_{\tau(1)} \square \cdots \square \varphi_{\tau(p)})(x_1 \wedge \cdots \wedge x_p) = \sum_\sigma \varphi_{\sigma\tau(1)} x_1 \wedge \cdots \wedge \varphi_{\sigma\tau(p)} x_p.$$

Setting $\sigma\tau = \rho$, we obtain

$$\begin{aligned}(\varphi_{\tau(1)} \square \cdots \square \varphi_{\tau(p)})(x_1 \wedge \cdots \wedge x_p) &= \sum_\rho \varphi_{\rho(1)} x_1 \wedge \cdots \wedge \varphi_{\rho(p)} x_p \\ &= (\varphi_1 \square \cdots \square \varphi_p)(x_1 \wedge \cdots \wedge x_p).\end{aligned}$$

It follows from the definition of the box product that

$$\frac{1}{p!}\underbrace{(\varphi \square \cdots \square \varphi)}_{p} = \wedge^p \varphi. \tag{6.6}$$

Proposition 6.3.1. *The operator T_E satisfies the relation*

$$T_E(z_1 \cdots z_p) = T_E(z_1) \square \cdots \square T_E(z_p) \qquad z_\nu \in E^* \otimes E.$$

PROOF. It is sufficient to consider the case $z_i = y_i^* \otimes y_i$, where $y_i^* \in E^*$ and $y_i \in E$ $(i = 1, \ldots, p)$. Then we have, for $x_i \in E$ $(i = 1, \ldots, p)$,

$$\begin{aligned}
T_E(z_1 \cdots z_p)(x_1 \wedge \cdots \wedge x_p) \\
&= T_E(y_1^* \wedge \cdots \wedge y_p^* \otimes y_1 \wedge \cdots \wedge y_p)(x_1 \wedge \cdots \wedge x_p) \\
&= \langle y_1^* \wedge \cdots \wedge y_p^*, x_1 \wedge \cdots \wedge x_p\rangle (y_1 \wedge \cdots \wedge y_p) \\
&= \sum_\sigma \varepsilon_\sigma \langle y_1^*, x_{\sigma(1)}\rangle \cdots \langle y_p^*, x_{\sigma(p)}\rangle (y_1 \wedge \cdots \wedge y_p) \\
&= \sum_\sigma \varepsilon_\sigma \langle y_1^*, x_{\sigma(1)}\rangle y_1 \wedge \cdots \wedge \langle y_p^*, x_{\sigma(p)}\rangle y_p \\
&= \sum_\sigma \varepsilon_\sigma T(z_1)x_{\sigma(1)} \wedge \cdots \wedge T(z_p)x_{\sigma(p)} \\
&= (T_E(z_1) \,\square \cdots \square\, T_E(z_p))(x_1 \wedge \cdots \wedge x_p).
\end{aligned}$$

□

Corollary. *Let* $z \in E^* \otimes E$. *Then*

$$T_E(z^p) = \wedge^p(T_E z).$$

The Composition Product

6.4

We now define a second multiplication in the space $\wedge E^* \otimes \wedge E$ which will be denoted by $\circ$ and called the *composition product*. Given $u^*, v^* \in \wedge E^*$ and $u, v \in \wedge E$, set

$$(u^* \otimes u) \circ (v^* \otimes v) = \langle u^*, v\rangle v^* \otimes u. \tag{6.7}$$

(cf. Section 1.26).

Then we have the relation

$$T_E(w_1 \circ w_2) = T_E(w_1) \circ T_E(w_2) \qquad w_1, w_2 \in \wedge(E^*, E), \tag{6.8}$$

where T_E is the operator defined by (6.5) and the right-hand side is the composition of the linear transformations $T_E(w_1)$ and $T_E(w_2)$. Note that if $\dim E < \infty$, then T_E is a linear isomorphism (cf. Section 1.26) and so the composition product is determined by relation (6.8) in this case.

It follows easily from the definition that the composition algebra is associative. In particular,

$$\begin{aligned}
(u_1^* \otimes u_1) \circ (u_2^* \otimes u_2) \circ \cdots \circ (u_p^* \otimes u_p) \\
= \langle u_1^*, u_2\rangle \langle u_2^*, u_3\rangle \cdots \langle u_{p-1}^*, u_p\rangle u_p^* \otimes u_1.
\end{aligned} \tag{6.9}$$

The kth power of an element w in the composition algebra will be denoted by $w^{\textcircled{k}}$,

$$w^{\textcircled{k}} = \underbrace{w \circ \cdots \circ w}_{k}.$$

Note that the composition algebra has no unit element unless *dim* $E < \infty$ (see Section 1.26).

It follows from the definition that

$$\wedge_q^p(E^*, E) \circ \wedge_s^r(E^*, E) = 0 \quad \text{if } p \neq s \tag{6.10}$$

and

$$\wedge_q^p(E^*, E) \circ \wedge_p^r(E^*, E) \subset \wedge_q^r(E^*, E). \tag{6.11}$$

In particular,

$$\wedge_p^p(E^*, E) \circ \wedge_p^p(E^*, E) \subset \wedge_p^{p-p}(E^*, E)$$

and so the subspaces $\Delta_p E = \wedge_p^p(E^*, E)$ are subalgebras. Moreover, T_E restricts to homomorphisms

$$T_p: \Delta_p(E) \to L(\wedge^p E; \wedge^p E) \quad \text{for each } p.$$

Formula (6.11) shows that the composition product is *not* homogeneous with respect to the usual gradation in $\wedge(E^*, E)$. However, if we introduce a new gradation (called the *cross-gradation*) by setting

$$\text{Deg } w = p - q \quad w \in \wedge_q^p(E^*, E),$$

then we have, for any two homogeneous elements w_1 and w_2 for which $w_1 \circ w_2 \neq 0$,

$$\text{Deg}(w_1 \circ w_2) = \text{Deg } w_1 + \text{Deg } w_2.$$

Thus, the composition product preserves the cross-gradation.

6.5

Lemma I. *Let* $w_1, w_2 \in \wedge(E^*, E)$. *Then*

$$w_1 \circ w_2 = [\iota \otimes T_E(w_1)]w_2 = [T_E(w_2)^* \otimes \iota]w_1.$$

PROOF. We may assume that $w_1 = u^* \otimes u$ and $w_2 = v^* \otimes v$. Then formula (6.7) yields

$$[\iota \otimes T_E(w_1)]w_2 = v^* \otimes T_E(w_1)v = \langle u^*, v\rangle(v^* \otimes u) = w_1 \circ w_2.$$

The second relation is established in the same way. □

Lemma II. *Let* $w \in \wedge_p^p(E^*, E)$ *and* $z \in E^* \otimes E$. *Then*

$$(T_E(z)^\wedge \otimes \iota)w = w \circ z^p$$

and

$$(\iota \otimes T_E(z)_\wedge)w = z^p \circ w.$$

PROOF. Applying Lemma I with $w_1 = w$ and $w_2 = z^p$, we obtain

$$w \circ z^p = [T_E(z^p)^* \otimes \iota]w.$$

By the corollary to Proposition 6.3.1, $T_E(z^p) = \wedge^p(T_E z)$. Thus we obtain, since $w \in \wedge_p^p(E^*, E)$,

$$w \circ z^p = [(\wedge^p(T_E z))^* \otimes \iota]w = [T_E(z)^\wedge \otimes \iota]w.$$

The second formula follows by a similar argument. □

The following proposition states a relation between the multiplications in the mixed exterior algebra and the composition algebra.

Proposition 6.5.1. *Let* $z_\nu \in E^* \otimes E$ $(\nu = 1, \ldots, p)$, $p \geq 2$ *and* $z \in E^* \otimes E$. *Then,*

$$i(z)(z_1 \cdots z_p) = \sum_\nu \langle z, z_\nu \rangle z_1 \cdots \hat{z}_\nu \cdots z_p.$$
$$- \sum_{\nu<\mu} (z_\mu \circ z \circ z_\nu + z_\nu \circ z \circ z_\mu) \cdot z_1 \cdots \hat{z}_\nu \cdots \hat{z}_\mu \cdots z_p.$$

PROOF. We may assume that $z_\nu = y_\nu^* \otimes y_\nu$, where $y_\nu^* \in E^*$ and $y_\nu \in E$ and that $z = x^* \otimes x$. Then we have (see Corollary II to Proposition 5.14.1)

$$\begin{aligned} i(z)(z_1 \cdots z_p) &= [i(x) \otimes i(x^*)](y_1^* \wedge \cdots \wedge y_p^* \otimes y_1 \wedge \cdots \wedge y_p) \\ &= \sum_{\nu,\mu} (-1)^{\mu+\nu} \langle y_\mu^*, x \rangle \langle x^*, y_\nu \rangle y_1^* \wedge \cdots \wedge \hat{y}_\mu^* \wedge \cdots \wedge y_p^* \\ &\quad \otimes y_1 \wedge \cdots \wedge \hat{y}_\nu \wedge \cdots \wedge y_p. \end{aligned}$$

Now write this sum in three parts with $\nu = \mu$, $\nu < \mu$ *and* $\nu > \mu$ to obtain

$$\begin{aligned} &i(z)(z_1 \cdots z_p) \\ &= \sum_\nu \langle y_\nu^*, x \rangle \langle x^*, y_\nu \rangle y_1^* \wedge \cdots \wedge \widehat{y_\nu^*} \wedge \cdots \wedge y_p^* \otimes y_1 \wedge \cdots \wedge \hat{y}_\nu \wedge \cdots \wedge y_p \\ &\quad - \sum_{\nu<\mu} \langle y_\mu^*, x \rangle \langle x^*, y_\nu \rangle (y_\nu^* \otimes y_\mu)[y_1^* \wedge \cdots \wedge \widehat{y_\nu^*} \wedge \cdots \wedge \widehat{y_\mu^*} \wedge \cdots \wedge y_p^*] \\ &\quad \otimes [y_1 \wedge \cdots \wedge \hat{y}_\nu \wedge \cdots \wedge \hat{y}_\mu \wedge \cdots \wedge y_p] \\ &\quad - \sum_{\mu<\nu} \langle y_\mu^*, x \rangle \langle x^*, y_\nu \rangle (y_\nu^* \otimes y_\mu)[y_1^* \wedge \cdots \wedge \widehat{y_\mu^*} \wedge \cdots \wedge \widehat{y_\nu^*} \wedge \cdots \wedge y_p^*] \\ &\quad \otimes [y_1 \wedge \cdots \wedge \hat{y}_\mu \wedge \cdots \wedge \hat{y}_\nu \wedge \cdots \wedge y_p]. \end{aligned}$$

Since, in view of (6.9)

$$\langle y_\mu^*, x\rangle\langle x^*, y_\nu\rangle(y_\nu^* \otimes y_\mu) = z_\mu \circ z \circ z_\nu$$

it follows that

$$\begin{aligned} i(z)(z_1 \cdots z_p) &= \sum_\nu \langle z_\nu, z\rangle z_1 \cdots \hat{z}_\nu \cdots z_p \\ &\quad - \sum_{\nu<\mu} (z_\mu \circ z \circ z_\nu) \cdot z_1 \cdots \hat{z}_\nu \cdots \hat{z}_\mu \cdots z_p \\ &\quad - \sum_{\mu<\nu} (z_\mu \circ z \circ z_\nu) \cdot z_1 \cdots \hat{z}_\mu \cdots \hat{z}_\nu \cdots z_p. \end{aligned}$$

Interchanging the roles of μ and ν in the last term and combining it with the preceding term now completes the proof. □

Corollary. *Let* $z \in E^* \otimes E$ *and* $w \in E^* \otimes E$. *Then*

$$i(z)w^p = \langle z, w\rangle w^{p-1} - (w \circ z \circ w) \cdot w^{p-2} \qquad p \geqq 2.$$

Poincaré Duality

In Sections 6.6–6.16 E denotes an n-dimensional vector space over a field of characteristic zero.

6.6. The Isomorphism T_E

Consider the linear isomorphism T_E from $\wedge(E^*, E)$ to $L(\wedge E; \wedge E)$ given by

$$T_E(a^* \otimes b)u = \langle a^*, u\rangle b \qquad u \in \wedge E$$

(see Section 6.5). It restricts for each pair (p, q) to an isomorphism

$$T_E : \wedge_q^p(E^*, E) \xrightarrow{\cong} L(\wedge^p E; \wedge^q E).$$

This isomorphism satisfies the relation

$$\operatorname{tr}(T_E w_1 \circ T_E w_2) = \langle w_1, w_2\rangle \qquad w_1 \in T_q^p(E^*, E),\ w_2 \in T_p^q(E^*, E). \quad (6.12)$$

In fact, let $w_1 = a^* \otimes b$, $a^* \in \wedge^p E^*$, $b \in \wedge^q E$, $w_2 \in b^* \otimes a$, $b^* \in \wedge^q E^*$, and $a \in \wedge^p E$. Then

$$\langle w_1, w_2\rangle = \langle a^*, a\rangle\langle b^*, b\rangle.$$

On the other hand, Formulas (6.7) and (6.8) yield

$$\operatorname{tr}(T_E(a^* \otimes b) \circ T_E(b^* \otimes a)) = \langle a^*, a\rangle\langle b^*, b\rangle$$

and so Formula (6.12) is established.

6.7. The Unit Tensors

Define tensors $t_E \in \wedge(E^*, E)$ and $t_p \in \wedge_p^p(E^*, E)$ by

$$t_E = T_E^{-1}(\iota_{\wedge E})$$

and

$$t_p = T_E^{-1}(\iota_{\wedge^p E}) \qquad (p = 0, \ldots, n).$$

Then t_E is the unit element of the composition algebra while t_p is the unit element of the subalgebra $\Delta_p(E)$. The tensor t_p will be called the *unit tensor of degree p*. We shall denote t_1 simply by t. Thus,

$$t = T_E^{-1}(\iota_E).$$

The tensor t_n can be written in the form

$$t_n = e^* \otimes e,$$

where e^*, e is a pair of dual basis vectors of $\wedge^n E^*$ and $\wedge^n E$. Clearly,

$$t_E = \sum_{p=0}^{n} t_p. \tag{6.13}$$

The unit tensor t_p coincides with the pth power of t in the algebra $\wedge(E^*, E)$,

$$t_p = t^p \qquad (p = 0, \ldots, n). \tag{6.14}$$

In fact, the corollary to Proposition 6.3.1 yields

$$T_E(t^p) = \wedge^p T_E(t) = \wedge^p \iota_E = \iota_{\wedge^p E} \tag{6.15}$$

and so (6.14) follows.

Next observe that Formula (6.12) applied with $w_1 = t_p$ and $w_2 \in \wedge_p^p(E^*, E)$ yields

$$\langle t_p, w \rangle = \operatorname{tr}(T_E(t_p) \circ T_E(w)) = \operatorname{tr} T_E(w).$$

Thus,

$$\langle t_p, w \rangle = \operatorname{tr} T_E(w) \qquad w \in \wedge_p^p(E^*, E). \tag{6.16}$$

In particular,

$$\langle t_p, u^* \otimes u \rangle = \langle u^*, u \rangle \qquad u^* \in \wedge^p E^*, u \in \wedge^p E. \tag{6.17}$$

Proposition 6.7.1. *The unit tensors satisfy the relation*

$$t_p \cdot t_q = \binom{p+q}{q} t_{p+q} \qquad p \geq 0, q \geq 0 \tag{6.18}$$

and

$$i(t_q)t_p = \binom{n-p+q}{q} t_{p-q} \qquad p \geq q. \tag{6.19}$$

Proof. The first part follows directly from Formula (6.14). To prove the second part consider first the case $q = 1$. Then the corollary of Proposition 6.5.1 (applied with $z = w = t$) yields

$$i(t)t^p = \langle t, t\rangle t^{p-1} - (t \circ t \circ t) \cdot t^{p-2}.$$

Since, by (6.16), $\langle t, t\rangle = n$ and $t \circ t = t$, we obtain

$$\begin{aligned} i(t)t^p &= nt^{p-1} - (p-1)t^{p-1} \\ &= (n - p + 1)t^{p-1}. \end{aligned}$$

Thus (6.19) is correct for $q = 1$. Now the general formula follows via induction on q. □

i. $$\langle t_p, t_p\rangle = \binom{n}{p} \qquad (p = 0, \ldots, n)$$

ii. $$i(t_q)t_n = t_{n-q} \qquad (q = 0, \ldots, n).$$

6.8. The Poincaré Isomorphism

Choose a basis vector e of $\wedge^n E$ and let e^* be the unique basis vector of $\wedge^n E^*$ such that $\langle e^*, e\rangle = 1$. Then, as noted previously,

$$e^* \otimes e = t_n.$$

Now define linear maps

$$D_e : \wedge E \to \wedge E^*$$

and

$$D^e : \wedge E^* \to \wedge E$$

by setting

$$D_e u = i(u)e^* \qquad u \in \wedge E$$

and

$$D^e u^* = i(u^*)e \qquad u^* \in \wedge E^*.$$

In particular, $D_e(1) = e^*$, $D_e(e) = 1$ and $D^e(1) = e$, $D^e(e^*) = 1$. Note that, for any nonzero scalar λ,

$$D_{\lambda e} = \lambda^{-1} D_e \quad \text{and} \quad D^{\lambda e} = \lambda D^e.$$

As immediate consequences of the definitions we have

$$D_e(u \wedge v) = i(v) D_e(u) \qquad u, v \in \wedge E$$

and

$$D^e(u^* \wedge v^*) = i(v^*) D^e(u^*) \qquad u^*, v^* \in \wedge E^*.$$

Moreover, the operators D_e and D^e restrict to linear maps

$$D_p\colon \wedge^p E \to \wedge^{n-p} E^* \qquad p = 0, \ldots, n$$

and

$$D^p\colon \wedge^p E^* \to \wedge^{n-p} E \qquad p = 0, \ldots, n.$$

Theorem 6.8.1

1. *D_e and D^e preserve the scalar products*

$$\langle D_e u, D^e u^* \rangle = \langle u^*, u \rangle.$$

2. *The duals D_p^* and $(D^p)^*$ are given by*

$$D_p^* = (-1)^{p(n-p)} D_{n-p} \qquad p = 0, \ldots, n$$

and

$$(D^p)^* = (-1)^{p(n-p)} D^{n-p} \qquad p = 0, \ldots, n.$$

3.
$$D^p \circ D_{n-p} = (-1)^{p(n-p)} \iota \qquad p = 0, \ldots, n$$

and

$$D_p \circ D^{n-p} = (-1)^{p(n-p)} \iota \qquad p = 0, \ldots, n.$$

In particular, the maps D_e, D^e, D_p, and D^p are all linear isomorphisms.

PROOF

1. It is sufficient to consider the case $u \in \wedge^p E$, $u^* \in \wedge^p E^*$. Then we have, in view of (6.17),

$$\begin{aligned}\langle D_e u, D^e u^* \rangle &= \langle i(u)e^*, i(u^*)e \rangle \\ &= \langle t_{n-p}, i(u)e^* \otimes i(u^*)e \rangle = \langle t_{n-p}, i(u^* \otimes u) t_n \rangle.\end{aligned}$$

Since the diagonal subalgebra is commutative,

$$\langle t_{n-p}, i(u^* \otimes u) t_n \rangle = \langle u^* \otimes u, i(t_{n-p}) t_n \rangle.$$

Thus we obtain, in view of the Corollary to Proposition 6.7.1,

$$\langle D_e u, D^e u^* \rangle = \langle u^* \otimes u, t_p \rangle = \langle u^*, u \rangle.$$

2. Let $u \in \wedge^p E$ and $v \in \wedge^{n-p} E$. Then

$$\begin{aligned}\langle D_e u, v\rangle &= \langle i(u)e^*, v\rangle = i(v)i(u)e^* \\ &= (-1)^{p(n-p)} i(u)i(v)e^* = (-1)^{p(n-p)}\langle D_{n-p} v, u\rangle.\end{aligned}$$

This proves the first relation. The second relation is obtained in the same way.

3. Let $u, v \in \wedge^{n-p} E$. Then, by (2) and (1),

$$\langle (D^p \circ D_{n-p})u, v\rangle = (-1)^{p(n-p)}\langle D_{n-p} u, D^{n-p} v\rangle = (-1)^{p(n-p)}\langle u, v\rangle$$

which gives the first part of (3). The second part is established in the same way □

6.9. Naturality

Let $\varphi: E \xrightarrow{\cong} F$ be a linear isomorphism from E to a second vector space F. Choose a basis vector e of $\wedge^n E$ and set $f = \varphi_\wedge e$. Note that the vector $f^* = (\varphi^\wedge)^{-1} e^*$ satisfies $\langle f^*, f\rangle = 1$.

We show that the diagrams

$$\begin{array}{ccc} \wedge E & \xrightarrow[\cong]{\varphi_\wedge} & \wedge F \\ {\scriptstyle D_e}\downarrow{\scriptstyle\cong} & & {\scriptstyle\cong}\downarrow{\scriptstyle D_f} \\ \wedge E^* & \xrightarrow[\cong]{(\varphi^\wedge)^{-1}} & \wedge F^* \end{array} \qquad \begin{array}{ccc} \wedge E^* & \xrightarrow[\cong]{(\varphi^\wedge)^{-1}} & \wedge F^* \\ {\scriptstyle D^e}\downarrow{\scriptstyle\cong} & & {\scriptstyle\cong}\downarrow{\scriptstyle D^f} \\ \wedge E & \xrightarrow[\varphi_\wedge]{\cong} & \wedge F \end{array}$$

commute.

In fact, Formula 5.40 yields

$$(\varphi^\wedge \circ D_f \circ \varphi_\wedge)u = i(u)\varphi^\wedge f^* = D_e u$$

and so the first diagram commutes. The commutativity of the second diagram follows in the same way.

6.10. The Isomorphism D_E

In this section we introduce a canonical linear isomorphism

$$D_E: \wedge(E^*, E) \xrightarrow{\cong} \wedge(E^*, E)$$

(not depending on the choice of a basis vector of $\wedge^n E$). In fact, let

$$D_E w = i(w) t_n \qquad w \in \wedge(E^*, E).$$

If e^*, e is a pair of dual basis vectors of $\wedge^n E$ and $\wedge^n E^*$, we have,

$$D_E(u^* \otimes u) = i(u^* \otimes u)(e^* \otimes e) = D_e u \otimes D^e u^*. \tag{6.20}$$

Thus, the operators D_e, D^e, and D_E are related by

$$D_E = (D_e \otimes D^e) \circ Q_E, \tag{6.21}$$

where Q_E denotes the flip map defined in Section 6.2.

The corollary to Proposition 6.7.1 shows that

$$D_E t_p = t_{n-p} \qquad (p = 0, \dots, n).$$

In particular,

$$D_E(1) = t_n \qquad \text{and} \qquad D_E(t_n) = 1.$$

Moreover, it is immediate from (6.3) that

$$D_E(w_1 \cdot w_2) = i(w_2) D_E w_1 \qquad w_1, w_2 \in \wedge(E^*, E). \tag{6.22}$$

To state the analogue of Theorem 6.8.1 we introduce the involution Ω_E of $\wedge(E^*, E)$ given by

$$\Omega_E(z) = (-1)^{p(n-p)+q(n-q)} z \qquad z \in \wedge_q^p(E^*, E).$$

Then we have

Theorem 6.10.1

1. *D_E is an isometry.*
2. *The dual operator is given by*

$$D_E^* = \Omega_E \circ D_E.$$

3. $D_E^2 = \Omega_E$.

PROOF

1. By Theorem 6.8.1, Part (1) and Formula (6.20) we have

$$\begin{aligned} \langle D_E(u^* \otimes u), D_E(v^* \otimes v)\rangle &= \langle D_e u \otimes D^e u^*, D_e v \otimes D^e v^*\rangle \\ &= \langle D_e u, D^e v^*\rangle \langle D_e v, D^e u^*\rangle \\ &= \langle u^* \otimes u, v^* \otimes v\rangle. \end{aligned}$$

2. This follows from Theorem 6.8.1, Part (2) and (6.20).
3. This is a consequence of (1) and (2). □

Finally, observe that the isomorphism D_E restricts to a linear automorphism D_Δ of the space ΔE.

Theorem 6.10.1 implies that

$$D_\Delta^* = D_\Delta$$

and

$$D_\Delta^2 = \iota.$$

6.11. Naturality

Let $\varphi : E \xrightarrow{\cong} F$ be a linear isomorphism and consider the induced isomorphism $\alpha_\varphi : \wedge(E^*, E) \xrightarrow{\cong} \wedge(F^*, F)$ (see Section 6.2).

Then the diagram

$$\begin{array}{ccc} \wedge(E^*, E) & \xrightarrow[\cong]{\alpha_\varphi} & \wedge(F^*, F) \\ {\scriptstyle D_E} \downarrow \cong & & \cong \downarrow {\scriptstyle D_F} \\ \wedge(E^*, E) & \xrightarrow[\cong]{\alpha_\varphi} & \wedge(F^*, F) \end{array}$$

commutes. This follows from the naturality of D_e and D^e, Formula (6.21) and the relation

$$Q_F \circ \alpha_\varphi = \tilde{\alpha}_\varphi \circ Q_E,$$

where $\tilde{\alpha}_\varphi = \varphi_\wedge \otimes (\varphi^\wedge)^{-1}$.

6.12. The Intersection Product

We introduce a second product structure, the *intersection product*, in $\wedge E$ by setting

$$u \cap v = D^e[(D^e)^{-1}u \wedge (D^e)^{-1}v] \qquad u, v \in \wedge E.$$

Thus if $u \in \wedge^p E$ and $v \in \wedge^q E$, then $u \cap v \in \wedge^{p+q-n} E$. In particular, $u \cap v = 0$ if $p + q > 2n$ or $p + q < n$.

It is immediate from the definition that

$$u \cap v = (-1)^{(n-p)(n-q)} v \cap u \qquad u \in \wedge^p E,\ v \in \wedge^q E.$$

Moreover, if $e \in \wedge^n E$ is the element used to define D^e, then

$$u \cap e = e \cap u = u \qquad u \in \wedge E.$$

The intersection product makes $\wedge E$ into an algebra called the *intersection algebra*.

Now we show that D_e is an isomorphism from the intersection algebra to the exterior algebra,

$$D_e(u \cap v) = D_e u \wedge D_e v \qquad u, v \in \wedge E.$$

In fact, let $u \in \wedge^p E$ and $v \in \wedge^q E$. Then Theorem 6.8.1, Part (3) gives

$$\begin{aligned} D_e(u \cap v) &= (-1)^{p(n-p)+q(n-q)} (D^{n-p})^{-1} u \wedge (D^{n-q})^{-1} v \\ &= D_p u \wedge D_q v = D_e u \wedge D_e v. \end{aligned}$$

Now let $u \in \wedge^p E$ and $v \in \wedge^{n-p} E$. Then $u \cap v$ is a scalar. It will be denoted by $J(u, v)$. Theorem 6.8.1, Part (3) implies that

$$\begin{aligned} J(u, v) &= (u \cap v)\langle e^*, e\rangle = \langle D_e(u \cap v), e\rangle \\ &= \langle D_e u \wedge D_e v, e\rangle = \langle D_e v, i(D_e u)e\rangle \\ &= \langle D_e v, D^e D_e u\rangle = (-1)^{p(n-p)}\langle D_e v, u\rangle \\ &= (-1)^{p(n-p)}\langle e^*, v \wedge u\rangle = \langle e^*, u \wedge v\rangle. \end{aligned}$$

Thus,

$$J(u, v) = \langle e^*, u \wedge v\rangle.$$

To obtain a geometric interpretation of $J(u, v)$ let E be a *real* vector space. Orient E via the determinant function Δ given by

$$\Delta(x_1, \ldots, x_n) = \langle e^*, x_1 \wedge \cdots \wedge x_n\rangle \qquad x_\nu \in E.$$

Now consider two oriented subspaces E_1 and E_2 of dimensions p and $n - p$. Choose positive bases

$$a_1, \ldots, a_p \quad \text{and} \quad a_{p+1}, \ldots, a_n$$

of E_1 and E_2 respectively and set

$$u = a_1 \wedge \cdots \wedge a_p, \qquad v = a_{p+1} \wedge \cdots \wedge a_n.$$

Then

$$J(u, v) = \Delta(a_1, \ldots, a_n).$$

This shows that

i. $J(u, v) \neq 0$ if and only if $E_1 + E_2 = E$; that is, if and only if $E = E_1 \oplus E_2$.

ii. Suppose that $E = E_1 \oplus E_2$ holds. Then the orientation of E coincides with the orientations induced by E_1 and E_2 (see Section 4.29 of *Linear Algebra*) if and only if $J(u, v) > 0$.

6.13. The Duals of the Basis Elements

Now let $\{e_\nu\}$, $\{e^{*\nu}\}$ be a pair of dual bases. Then the dual images of the basis vectors $e_{\nu_1} \wedge \cdots \wedge e_{\nu_p}$ $(\nu_1 < \cdots < \nu_p)$ of $\wedge^p E$ are given by

$$D_e(e_{\nu_1} \wedge \cdots \wedge e_{\nu_p}) = (-1)^{\sum_{i=1}^{p}(\nu_i - i)} e^{*\nu_{p+1}} \wedge \cdots \wedge e^{*\nu_n}, \tag{6.23}$$

where $(\nu_{p+1}, \ldots, \nu_n)$ is the complementary $(n - p)$-tuple.

To prove (6.23) we write e^* in the form

$$e^* = \varepsilon_\sigma e^{*\nu_1} \wedge \cdots \wedge e^{*\nu_n},$$

where σ denotes the permutation $(1, \ldots, n) \to (\nu_1, \ldots, \nu_n)$.

Then we have

$$D_e(e_{\nu_1} \wedge \cdots \wedge e_{\nu_p}) = \varepsilon_\sigma i(e_{\nu_1} \wedge \cdots \wedge e_{\nu_p})(e^{*\nu_1} \wedge \cdots \wedge e^{*\nu_n})$$
$$= \varepsilon_\sigma i(e_{\nu_p}) \cdots i(e_{\nu_1})(e^{*\nu_1} \wedge \cdots \wedge e^{*\nu_n})$$
$$= \varepsilon_\sigma e^{*\nu_{p+1}} \wedge \cdots \wedge e^{*\nu_n}.$$

The relations $\nu_1 < \cdots < \nu_p$ and $\nu_{p+1} < \cdots < \nu_n$ imply that

$$\varepsilon_\sigma = (-1)^{\sum_{i=1}^{p}(\nu_i - i)}$$

and so Formula (6.23) is proved. In the same way it follows that

$$D^e(e^{*\nu_1} \wedge \cdots \wedge e^{*\nu_n}) = (-1)^{\sum_{i=1}^{p}(\nu_i - i)} e_{\nu_{p+1}} \wedge \cdots \wedge e_{\nu_n}.$$

In view of Theorem 6.8.1, Part (3) and (6.23) we obtain

$$D^e(e^{*\nu_{p+1}} \wedge \cdots \wedge e^{*\nu_n}) = D^e D_e (-1)^{\sum_{i=1}^{p}(\nu_i - i)} e_{\nu_1} \wedge \cdots \wedge e_{\nu_p}$$
$$= (-1)^{p(n-p)}(-1)^{\sum_{i=1}^{p}(\nu_i - i)} e_{\nu_1} \wedge \cdots \wedge e_{\nu_p},$$

i.e.,

$$D^e(e^{*\nu_{p+1}} \wedge \cdots \wedge e^{*\nu_n}) = (-1)^{p(n-p)}(-1)^{\sum_{i=1}^{p}(\nu_i - i)} e_{\nu_1} \wedge \cdots \wedge e_{\nu_p}. \quad (6.24)$$

(6.23) yields, in the case $p = n - 1$,

$$D_e(e_1 \wedge \cdots \wedge \hat{e}_i \wedge \cdots \wedge e_n) = (-1)^{n-i} e^{*i}. \quad (6.25)$$

6.14. The External Product

Consider the $(n - 1)$-linear skew-symmetric mapping of $\underbrace{E \times \cdots \times E}_{n-1}$ into E^* defined by

$$[x_1, \ldots, x_{n-1}] = D_e(x_1 \wedge \cdots \wedge x_{n-1}).$$

The vector $[x_1, \ldots, x_{n-1}]$ is called the *external product* of the vectors x_i $(i = 1, \ldots, n - 1)$. Clearly the definition of the *external product* depends on the choice of the basis vector $e^* \in \wedge^n E^*$. It follows that

$$\langle [x_1, \ldots, x_{n-1}], x_i \rangle = \langle e^*, x_1 \wedge \cdots \wedge x_{n-1} \wedge x_i \rangle = 0 \quad (i = 1, \ldots, n - 1)$$

showing that the external product is orthogonal to all factors. Now consider the external product of $n - 1$ vectors $x^{*\nu} \in E^*$,

$$[x^{*1}, \ldots, x^{*n-1}] = D^e(x^{*1} \wedge \cdots \wedge x^{*n-1}).$$

From Theorem 6.8.1, Part (1) we obtain

$$\langle [x_1, \ldots, x_{n-1}], [x^{*1}, \ldots, x^{*n-1}] \rangle = \langle x^{*1} \wedge \cdots \wedge x^{*n-1}, x_1 \wedge \cdots \wedge x_{n-1} \rangle$$
$$= \det(\langle x^{*i}, x_j \rangle).$$

This yields the *Lagrange identity*

$$\langle[x_1, \ldots, x_{n-1}], [x^{*1}, \ldots, x^{*n-1}]\rangle = \det(\langle x^{*i}, x_j\rangle) \qquad 0 \le i, j \le n-1.$$

For the external product of $(n-1)$ basis vectors e_ν we obtain, from (6.25)

$$[e_1, \ldots, \hat{e}_i, \ldots, e_z] = (-1)^{n-i} e^{*i} \qquad i = 1, \ldots, n.$$

This formula gives, in the case $n = 3$,

$$[e_1, e_2] = e^{*3}, \qquad [e_2, e_3] = e^{*1}, \qquad [e_3, e_1] = e^{*2}.$$

6.15. Euclidean Spaces

Suppose now that E is an n-dimensional oriented Euclidean space. Then all the spaces $\wedge^p E$ $(1 \le p \le n)$ are Euclidean and there exists precisely one unit vector $e \in \wedge^n E$ which represents the given orientation. Since E is self-dual we may set $E^* = E$. Then the Poincaré isomorphisms coincide and are given by

$$D_e u = i(u)e \qquad u \in \wedge E.$$

D_e maps $\wedge^p E$ onto $\wedge^{n-p} E$ $(0 \le p \le n)$. Theorem 6.8.1, Part (1) implies that

$$(D_e u, D_e v) = (u, v) \qquad u, v \in \wedge E$$

and so D_e is an isometric mapping. From Theorem 6.8.1, Part (2) we obtain

$$\tilde{D}_p = (-1)^{p(n-p)} D_{n-p}, \tag{6.26}$$

where $\tilde{D}_p$ denotes the adjoint of D_p.

Assume now that n is even, $n = 2m$. Then the operator D_m defines a linear transformation of $\wedge^m E$,

$$D_m \colon \wedge^m E \to \wedge^m E.$$

The above formula yields

$$\tilde{D}_m = (-1)^m D_m.$$

It follows that the transformation D_m is selfadjoint or skew depending on whether m is even or odd.

Problems

1. Define the operation of $GL(E)$ on $\wedge(E^*, E)$ by

$$\alpha z = [(\alpha^\wedge)^{-1} \otimes \alpha_\wedge] z \qquad \alpha \in GL(E).$$

Prove the following relations:

a. $\alpha(z_1 + z_2) = \alpha z_1 + \alpha z_2$
b. $\alpha(z_1 \cdot z_2) = \alpha z_1 \cdot \alpha z_2$
c. $(\alpha\beta)z = \alpha(\beta z)$
d. $\iota z = z$ (ι being the identity map).

Show that

$$\alpha t^p = t^p \qquad (0 \le p \le n),$$

where t is the unit tensor.

2. Verify the relations

$$i(1 \otimes u)t^p = u \qquad u \in \wedge^p E$$

$$i(u^* \otimes 1)t^p = u^* \otimes 1 \qquad u^* \in \wedge^p E^*$$

$$\langle t^{p+q}, t^p \wedge w\rangle = \binom{n-q}{p}\langle t^q, w\rangle \qquad w \in \wedge(E^*, E)$$

3. Verify the formulas

$$D_e i(v^*)u = (-1)^{q(p-q)} v^* \wedge D_e u \qquad u \in \wedge^p E,\ v^* \in \wedge^q E^*$$

$$D^e i(v)u^* = (-1)^{p(q-p)} v \wedge D^e u^* \qquad v \in \wedge^p E,\ u^* \in \wedge^q E^*$$

$$D_E i(z)w = (-1)^{p(s-p)+q(r-q)} z \cdot D_E w \qquad z \in \wedge^p_q(E^*, E),\ w \in \wedge^r_s(E^*, E).$$

4. Prove that

$$D_\Delta \mu(z) = i(z) D_\Delta \qquad z \in \Delta(E^*, E).$$

5. Let E and F be vector spaces of dimension n and let $\alpha: E \to F$ be a linear isomorphism.

   a. Denote by D^e and D^f the Poincaré isomorphisms

   $$D^e u^* = i(u^*)e, \qquad D^f v^* = i(v^*)f,$$

   where e is a basis vector of $\wedge^n E$ and $f = \alpha_\wedge e$. Show that

   $$D^f \circ (\alpha^\wedge)^{-1} = \alpha_\wedge \circ D^e.$$

   b. If $F = E$ prove that

   $$D_E(\alpha z) = \alpha D_E(z) \qquad z \in \wedge(E^*, E)$$

   (for the definition of αz see Problem 1).

6. Let E be an oriented Euclidean plane. Show that the linear map

$$D_e: E \to E$$

is a rotation with rotation angle $+\pi/2$.

7. Let E be an oriented 3-dimensional Euclidean space. Prove that

$$x \times y = D_e(x \wedge y) \qquad x, y \in E$$

(see Section 7.16 of *Linear Algebra*). Use the above relation to prove the formulas

$$(x_1 \times y_1, x_2 \times y_2) = (x_1, x_2)(y_1, y_2) - (x_1, y_2)(x_2, y_1)$$

and

$$(x \times y) \times z = y(x, z) - x(y, z).$$

8. Let $\{e_\nu\}$, $\{e^{*\nu}\}$ be a pair of dual bases of E and E^*. Prove that

$$D_e(x_1 \wedge \cdots \wedge x_p) = \sum_{\nu_1 < \cdots < \nu_p} p!(-1)^{\sum_{i=1}^{p}(\nu_i - i)} \langle e^{*\nu_1}, x_1\rangle \cdots \langle e^{*\nu_p}, x_p\rangle e^{*\nu_{p+1}} \wedge \cdots \wedge e^{*\nu_n}.$$

In the above sum, $(\nu_{p+1}, \ldots, \nu_n)$ denotes the ordered complement $(\nu_1, \ldots, \nu_p)$ in the natural order.

9. Show that the restriction of the isomorphism D_E to the subspaces $\wedge_0^n(E^*, E)$ and $\wedge_n^0(E^*, E)$ is the identity.

10. Show that the components of $D_E z$ are given by (see Problem 8)

$$(D_E z)_{\nu_{p+1}, \ldots, \nu_n}^{\mu_{q+1}, \ldots, \mu_n} = (-1)^{\sum_{i=1}^{p}(\nu_i - i) + \sum_{j=1}^{q}(\mu_j - j)} (z)_{\mu_1, \ldots, \mu_q}^{\nu_1, \ldots, \nu_p}$$

$$\mu_{q+1} < \cdots < \mu_n, \; \nu_{p+1} < \cdots < \nu_n,$$

where $(\nu_1, \ldots, \nu_p)$ is the complementary p-tuple of $(\nu_{p+1}, \ldots, \nu_n)$ and $(\mu_1, \ldots, \mu_q)$ is the complementary q-tuple of $(\mu_{q+1}, \ldots, \mu_n)$ in the natural order.

11. Let E^* and E be two dual 3-dimensional spaces and $a \in E$, $a \neq 0$ and $b^* \in E^*$ be two given vectors. Prove that there exists a vector $x \in E$ such that $[a, x] = b^*$ if and only if $\langle b^*, a\rangle = 0$, and prove that, if x_0 is a particular solution of $[a, x] = b^*$, then the general solution is given by $x = x_0 + \lambda a$, $\lambda \in \Gamma$.

12. Consider an $(n - 1)$-linear skew-symmetric map φ of E into a vector space F. Prove that there exists exactly one linear map $\chi: E^* \to F$ such that

$$\varphi(x_1, \ldots, x_{n-1}) = \chi[x_1, \ldots, x_{n-1}] \qquad x_\nu \in E \; (\nu = 1, \ldots, n-1).$$

13. If α is a linear automorphism of E, prove that

$$[\alpha x_1, \ldots, \alpha x_{n-1}] = \det \alpha (\alpha^{-1})^*[x_1, \ldots, x_{n-1}] \qquad x_\nu \in E \; (\nu = 1, \ldots, n-1).$$

14. Let e_ν $(\nu = 1, \ldots, n)$ be a basis of E such that $e_1 \wedge \cdots \wedge e_n = e$. Given $n - 1$ vectors

$$x_i = \sum_\nu \xi_i^\nu e_\nu \qquad (i = 1, \ldots, n-1),$$

show that the vector $y^* = [x_1, \ldots, x_{n-1}]$ has components

$$\eta_\nu = (-1)^{n-\nu} \begin{vmatrix} \xi_1^1 & \cdots & \hat{\xi}_1^\nu & \cdots & \xi_1^n \\ \vdots & & & & \vdots \\ \xi_{n-1}^1 & \cdots & \hat{\xi}_{n-1}^\nu & \cdots & \xi_{n-1}^n \end{vmatrix} \qquad (\nu = 1, \ldots, n).$$

15. Using the formula in Problem 14 derive the relation

$$\begin{vmatrix} \sum_{\nu=1}^n \xi_1^\nu \eta_\nu^1 & \cdots & \sum_{\nu=1}^n \xi_1^\nu \eta_\nu^{n-1} \\ \vdots & & \vdots \\ \sum_{\nu=1}^n \xi_{n-1}^\nu \eta_\nu^1 & \cdots & \sum_{\nu=1}^n \xi_{n-1}^\nu \eta_\nu^{n-1} \end{vmatrix} = \sum_{\nu=1}^n \begin{vmatrix} \xi_1^1 & \cdots & \hat{\xi}_1^\nu & \cdots & \xi_1^n \\ \vdots & & & & \vdots \\ \xi_{n-1}^1 & \cdots & \hat{\xi}_{n-1}^\nu & \cdots & \xi_{n-1}^n \end{vmatrix} \begin{vmatrix} \eta_1^1 & \cdots & \hat{\eta}_1^\nu & \cdots & \eta_1^n \\ \vdots & & & & \vdots \\ \eta_{n-1}^1 & \cdots & \hat{\eta}_{n-1}^\nu & \cdots & \eta_{n-1}^n \end{vmatrix}$$

from the Lagrange identity.

16. Find the minimum polynomial, the characteristic polynomial, the trace, and the determinant of D_Δ. *Hint:* In case $\dim E = 2m$ consider the restriction of D_Δ to the subspace $\wedge_m^m(E^*, E)$.

17. Suppose E is a Euclidean space of dimension $n = 2m$ and set $E^* = E$. Prove that

$$\wedge_m^m(E, E) = X^2(\wedge^m E) \oplus Y^2(\wedge^m E)$$

is a decomposition of $\wedge_m^m(E, E)$ into orthogonal spaces stable under D_Δ (see Sections 4.2 and 4.10 with E replaced by $\wedge^m E$).

Applications to Linear Transformations

In Section 6.16 E and E^* denote a dual pair of n-dimensional vector spaces.

If $\varphi: E \to E$ is a linear transformation, then the induced homomorphism $\varphi_\wedge : \wedge E \to \wedge E$ will also be denoted by $\wedge \varphi$ and the restriction of $\varphi_\wedge$ to $\wedge^p E$ will be denoted by $\wedge^p \varphi$.

6.16. The Isomorphism T

Let $t \in E^* \otimes E$ be the unit tensor for E and E^* (see Section 1.26). Then t determines a linear map

$$T: L(E; E) \to E^* \otimes E$$

defined by

$$T(\varphi) = (\iota \otimes \varphi)t \qquad \varphi \in L(E; E).$$

(Notice that this isomorphism T is the inverse of the T in Section 1.26.) In particular, we have

$$T(\iota) = t.$$

Now let $a \in E$ and $b^* \in E^*$ be arbitrary vectors. Then it follows that

$$\begin{aligned}\langle T(\varphi), b^* \otimes a \rangle &= \langle (\varphi \otimes \iota)t, b^* \otimes a \rangle = \langle t, \varphi^* b^* \otimes a \rangle \\ &= \langle \varphi^* b^*, a \rangle = \langle b^*, \varphi a \rangle,\end{aligned}$$

whence

$$\langle T(\varphi), b^* \otimes a \rangle = \langle b^*, \varphi a \rangle \qquad \varphi \in L(E; E),\ a \in E,\ b^* \in E^*. \tag{6.27}$$

Using this relation, we obtain

$$\langle T(\varphi), t \rangle = \sum_\nu \langle e^{*\nu}, \varphi e_\nu \rangle = \operatorname{tr} \varphi,$$

i.e.,

$$\langle T(\varphi), t \rangle = \operatorname{tr} \varphi \qquad \varphi \in L(E; E). \tag{6.28}$$

Similarly, a linear map

$$\tilde{T}: L(E^*; E^*) \to E^* \otimes E$$

is defined by

$$\tilde{T}(\chi) = (\chi \otimes \iota)t \qquad \chi \in L(E^*; E^*).$$

An argument similar to that given above shows that

$$\langle \tilde{T}(\chi), b^* \otimes a \rangle = \langle \chi b^*, a \rangle = \langle b^*, \chi^* a \rangle \qquad \chi \in L(E^*; E^*), a \in E, b^* \in E^*.$$

Comparing this relation with (6.27) we find that

$$T(\varphi) = \tilde{T}(\varphi^*) \qquad \varphi \in L(E; E). \tag{6.29}$$

Now let $\psi \in L(E; E)$ be a second linear transformation. Then we have

$$T(\psi\varphi) = (\iota \otimes \psi\varphi)t = (\iota \otimes \psi)[(\iota \otimes \varphi)t] = (\iota \otimes \psi)T(\varphi),$$

whence

$$T(\psi \circ \varphi) = (\psi \otimes \iota)T(\varphi). \tag{6.30}$$

With the aid of (6.28) and (6.30) we shall prove that T and $\tilde{T}$ are linear isomorphisms. In fact, for any $\varphi \in L(E; E)$, $\psi^* \in L(E^*; E^*)$ we have

$$\begin{aligned}\langle T(\varphi), \tilde{T}(\psi^*) \rangle &= \langle T(\varphi), (\psi^* \otimes \iota)t \rangle = \langle (\iota \otimes \psi)T(\varphi), t \rangle \\ &= \langle T(\psi \circ \varphi), t \rangle = \operatorname{tr}(\psi \circ \varphi),\end{aligned}$$

i.e.,

$$\langle T(\varphi), \tilde{T}(\psi^*) \rangle = \operatorname{tr}(\psi \circ \varphi) \qquad \varphi, \psi \in L(E; E).$$

Since the bilinear function $(\psi, \varphi) \to \operatorname{tr}(\psi \circ \varphi)$ is nondegenerate, it follows that T and $\tilde{T}$ are linear isomorphisms.

The Skew Tensor Product of $\wedge E^*$ and $\wedge E$

6.17. The Algebra $\wedge E^* \hat{\otimes} \wedge E$

Recall that so far we have defined two multiplications in the space $\wedge E^* \otimes \wedge E$, the canonical tensor product of the algebras $\wedge E^*$ and $\wedge E$ (see Section 6.1) and the composition product (see Sections 6.4 and 6.5). We shall now introduce a third algebra structure in this space, namely the skew tensor product of the graded algebras $\wedge E^*$ and $\wedge E$. The corresponding multiplication will be denoted by $\wedge$.

Thus,

$$(u^* \otimes u) \wedge (v^* \otimes v) = (-1)^{qr}(u^* \wedge v^*) \otimes (u \wedge v)$$
$$u^* \in \wedge E^*, u \in \wedge^q E, v^* \in \wedge^r E, v \in \wedge E.$$

$\wedge E^* \hat{\otimes} \wedge E$ is an associative algebra with unit element $1 \otimes 1$. It is generated by the elements $1 \otimes 1$, $x^* \otimes 1$ and $1 \otimes x$ where $x^* \in E^*$ and $x \in E$. The formula above implies that

$$w_1 \wedge w_2 = (-1)^{(p+q)(r+s)} w_2 \wedge w_1$$
$$w_1 \in \wedge^p E^* \otimes \wedge^q E, w_2 \in \wedge^r E^* \otimes \wedge^s E. \quad (6.31)$$

as is easily verified. In particular, if $p + q$ or $r + s$ is even, then

$$w_1 \wedge w_2 = w_2 \wedge w_1. \quad (6.32)$$

For $w \in \wedge E^* \hat{\otimes} \wedge E$ we shall set

$$w^{\hat{k}} = \frac{1}{k!} \underbrace{w \wedge \cdots \wedge w}_{k} \qquad k \geq 1$$

and

$$w^{\hat{0}} = 1 \otimes 1.$$

Then (6.32) yields the binomial formula

$$(w_1 + w_2)^{\hat{k}} = \sum_{i+j=k} w_1^{\hat{i}} \wedge w_2^{\hat{j}} \qquad w_1, w_2 \in \wedge^p E^* \otimes \wedge^q E, pq \text{ even}.$$

Note that the products in the algebras $\wedge E^* \hat{\otimes} \wedge E$ and $\wedge E^* \otimes \wedge E$ are connected by the relation

$$(u^* \otimes u) \wedge (v^* \otimes v) = (-1)^{qr}(u^* \otimes u) \cdot (v^* \otimes v) \qquad u \in \wedge^q E, v^* \in \wedge^r E^*.$$

In particular,

$$z_1 \wedge \cdots \wedge z_k = (-1)^{k(k-1)/2} z_1 \cdots z_k \qquad z_i \in E^* \otimes E, \quad (6.33)$$

and so

$$z^{\hat{k}} = (-1)^{k(k-1)/2} z^k \qquad z \in E^* \otimes E.$$

6.18. The Inner Product in $E^* \hat{\otimes} E$

Consider the bilinear function $\langle\!\langle , \rangle\!\rangle$ in the space $\wedge E^* \otimes \wedge E$ defined by

$$\langle\!\langle u^* \otimes u, v^* \otimes v \rangle\!\rangle = \begin{cases} (-1)^{pq}\langle u^*, v\rangle\langle v^*, u\rangle & \text{if } p = s \text{ and } r = q, \\ 0 & \text{otherwise,} \end{cases}$$

where $u^* \in \wedge^p E^*$, $u \in \wedge^q E$, $v^* \in \wedge^r E^*$, $v \in \wedge^s E$. It is easily checked that this bilinear function is an inner product.

Next, recall from Section 5.15 the isomorphism $f: \wedge E^* \hat{\otimes} \wedge E \to \wedge(E^* \oplus E)$ given by $f(u^* \otimes v) = i_\wedge u^* \wedge j_\wedge v$ where $i: E^* \to E^* \oplus E$ and $j: E \to E^* \oplus E$ denote the inclusion maps.

Proposition 6.18.1. *Define an inner product in* $E^* \oplus E$ *by*

$$\langle\langle x^* \oplus x, y^* \oplus y \rangle\rangle = \langle x^*, y \rangle + \langle y^*, x \rangle.$$

Then f *is an isometry,*

$$\langle\langle f(u^* \otimes u), f(v^* \otimes v) \rangle\rangle = \langle\langle u^* \otimes u, v^* \otimes v \rangle\rangle.$$

PROOF. We may assume that

$$u^* = x_1^* \wedge \cdots \wedge x_p^*, \qquad u = x_1 \wedge \cdots \wedge x_q$$

and

$$v^* = y_1^* \wedge \cdots \wedge y_r^*, \qquad v = y_1 \wedge \cdots \wedge y_s.$$

Then we have

$$f(u^* \otimes u) = (ix_1^* \wedge \cdots \wedge ix_p^*) \wedge (jx_1 \wedge \cdots \wedge jx_q)$$
$$f(v^* \otimes v) = (iy_1^* \wedge \cdots \wedge iy_r^*) \wedge (jy_1 \wedge \cdots \wedge jy_s)$$

whence

$$\langle\langle f(u^* \otimes u), f(v^* \otimes v) \rangle\rangle = \langle\langle (ix_1^* \wedge \cdots \wedge ix_p^*) \wedge (jx_1 \wedge \cdots \wedge jx_q), (iy_1^* \wedge \cdots \wedge iy_r^*) \wedge (jy_1 \wedge \cdots \wedge jy_s) \rangle\rangle.$$

Thus,

$$\langle\langle f(u^* \otimes u), f(v^* \otimes v) \rangle\rangle = 0 \quad \text{if } p + r \neq q + s.$$

On the other hand, if $p + r = q + s$, then

$$\langle\langle f(u^* \otimes u), f(v^* \otimes v) \rangle\rangle = \det\begin{pmatrix} \langle\langle ix_\alpha^*, iy_\gamma^* \rangle\rangle & \langle\langle ix_\alpha^*, jy_\delta \rangle\rangle \\ \langle\langle jx_\beta, iy_\gamma^* \rangle\rangle & \langle\langle jx_\beta, jy_\delta \rangle\rangle \end{pmatrix}.$$

But, since

$$\langle\langle ix_\alpha^*, iy_\gamma^* \rangle\rangle = 0 \qquad \alpha = 1, \ldots, p,$$
$$\langle\langle ix_\alpha^*, jy_\delta \rangle\rangle = \langle\langle x_\alpha^*, y_\delta \rangle\rangle \qquad \beta = 1, \ldots, q,$$
$$\langle\langle jx_\beta, iy_\gamma^* \rangle\rangle = \langle\langle y_\gamma^*, x_\beta \rangle\rangle \qquad \gamma = 1, \ldots, r,$$
$$\langle\langle jx_\beta, jy_\delta \rangle\rangle = 0, \qquad \delta = 1, \ldots, s,$$

it follows that

$$\langle\langle f(u^* \otimes u), f(v^* \otimes v) \rangle\rangle = 0 \quad \text{unless } s = p \text{ and } r = q.$$

If $s = p$ and $r = q$, we obtain

$$\begin{aligned}\langle\!\langle f(u^* \otimes u), f(v^* \otimes v)\rangle\!\rangle &= (-1)^{pq} \det(\langle x_\alpha^*, y_\delta\rangle)\det(\langle y_\gamma^*, x_\beta\rangle)\\ &= (-1)^{pq}\langle u^*, v\rangle\langle v^*, u\rangle\\ &= \langle\!\langle u^* \otimes u, v^* \otimes v\rangle\!\rangle\end{aligned}$$

which completes the proof. □

Finally note that the inner product defined above and the inner product defined in Section 6.2 are obtained from each other by the formula

$$\langle\!\langle w_1, w_2\rangle\!\rangle = (-1)^{pq}\langle w_1, w_2\rangle \qquad w_1 \in \wedge_q^p(E^*, E),\ w_2 \in \wedge_p^q(E^*, E). \tag{6.34}$$

In particular,

$$\langle\!\langle z_1 \wedge \cdots \wedge z_p, w_1 \wedge \cdots \wedge w_p\rangle\!\rangle = (-1)^p\langle z_1 \wedge \cdots \wedge z_p, w_1 \wedge \cdots \wedge w_p\rangle$$
$$z_i \in E^* \otimes E,\ w_i \in E^* \otimes E. \tag{6.35}$$

7 Applications to Linear Transformations

In this chapter E continues to denote an n-dimensional vector space over a field of characteristic zero.

All the problems concerning this chapter are collected at the end of the chapter.

The Isomorphism D_L

7.1. Definition

Recall from Section 6.2 the isomorphism

$$T_E\colon \wedge(E^*, E) \to L(\wedge E;\ \wedge E).$$

Let

$$D_L\colon L(\wedge E;\ \wedge E) \xrightarrow{\cong} L(\wedge E;\ \wedge E)$$

denote the isomorphism defined by the commutative diagram

$$\begin{array}{ccc} \wedge(E^*, E) & \xrightarrow[\cong]{T_E} & L(\wedge E;\ \wedge E) \\ {\scriptstyle D_E}\big\downarrow{\scriptstyle \cong} & & {\scriptstyle \cong}\big\downarrow{\scriptstyle D_L} \\ \wedge(E^*, E) & \xrightarrow[\cong]{T_E} & L(\wedge E;\ \wedge E) \end{array}$$

where D_E is the Poincaré map. If $u^* \in \wedge E^*$ and $u \in \wedge E$, the linear transformation $D_L \circ T_E(u^* \otimes u)$ is explicitly given by

$$[D_L \circ T_E(u^* \otimes u)]v = \langle D_e u, v\rangle D^e u^*.$$

From the results of Section 6.10 and Formula (6.12) we obtain the following relations:

$$D_L^* = \Omega_L \circ D_L \tag{7.1}$$

$$D_L^2 = \Omega_L \tag{7.2}$$

$$D_L(\iota_{\wedge^p E}) = \iota_{\wedge^{n-p} E}$$

$$\operatorname{tr}(D_L\alpha \circ D_L\beta) = \operatorname{tr}(\alpha \circ \beta) \qquad \alpha, \beta \in L(\wedge E;\ \wedge E),$$

where Ω_L denotes the involution in $L(\wedge E;\ \wedge E)$ given by $\Omega_L = T_E \circ \Omega_E \circ T_E^{-1}$.

The last relation implies that

$$\operatorname{tr}(D_L \alpha) = \operatorname{tr} \alpha \qquad \alpha \in L(\wedge E; \wedge E).$$

Proposition 7.1.1. *Let e be a basis vector of $\wedge^n E$. Then we have for α, $\beta \in L(\wedge E; \wedge E)$*

1. $$D_L \alpha = D^e \circ \alpha^* \circ (D^e)^{-1}$$

and

2. $$D_L(\alpha \circ \beta) = D_L(\beta) \circ D_L(\alpha).$$

Proof. Let $\alpha = T_E(u^* \otimes u)$. Then we have, by Theorem 6.8.1,

$$\begin{aligned}[D^e \circ T_E(u^* \otimes u)^* \circ D_e^*](v) &= D^e(\langle D_e^* v, u\rangle u^*) = \langle D_e u, v\rangle D^e u^* \\ &= D_L T_E(u^* \otimes u)(v)\end{aligned}$$

and so (1) follows. (2) is an immediate consequence of (1). □

Next observe that the image of the diagonal subalgebra $\Delta(E)$ ander T_E is the space $\sum_{p=0}^n L(\wedge^p E; \wedge^p E)$. Thus D_L restricts to a linear automorphism D_L^Δ of $\sum_{p=0}^n L(\wedge^p E; \wedge^p E)$. Moreover, since Ω_L reduces to the identity in this space, Formulas (7.1) and (7.2) show that D_L^Δ is a selfadjoint involution.

Let φ be a linear transformation of E and consider the induced transformation $\iota \otimes \varphi_\wedge$ of $\wedge(E^*, E)$. To simplify notation we shall set

$$\iota \otimes \varphi_\wedge = \Phi.$$

Then

$$\varphi^\wedge \otimes \iota = \Phi^*.$$

Now Formula (6.4) yields the relations

$$\begin{aligned} i(z) \circ \Phi^* &= \Phi^* \circ i(\Phi z) \\ i(z) \circ \Phi &= \Phi \circ i(\Phi^* z). \end{aligned} \tag{7.3}$$

Moreover, if $\Phi = T_E(z)$, Lemma II of Section 6.3 implies that

$$\begin{aligned} z^p \circ w &= \Phi(w) \qquad & w \in \wedge(E^*, E) \\ w \circ z^p &= \Phi^*(w) \qquad & w \in \wedge(E^*, E). \end{aligned} \tag{7.4}$$

Setting $w = t_p$, we obtain

$$z^p = \Phi(t_p) = \Phi^*(t_p) \qquad (p = 0, \ldots, n). \tag{7.5}$$

7.2. The Determinant

Recall from Section 4.4 of *Linear Algebra* that the determinant of a linear transformation φ is defined by the equation

$$\Delta(\varphi x_1, \ldots, \varphi x_n) = \det \varphi\, \Delta(x_1, \ldots, x_n).$$

This can be written in the form

$$(\wedge^n \varphi^*)e^* = \det \varphi \cdot e^*$$

where e^* is a basis vector of $\wedge^n E^*$. Thus,

$$\det \varphi = \operatorname{tr}(\wedge^n \varphi^*) = \operatorname{tr}(\wedge^n \varphi).$$

Similarly,

$$\varphi^\wedge e^* = (\wedge^n \varphi)e = \det \varphi \cdot e,$$

where e is a basis vector of $\wedge^n E$.

Since $t_n = e^* \otimes e$, where e^*, e is a pair of dual basis vectors of $\wedge^n E^*$ and $\wedge^n E$, the relations above imply that

$$\Phi(t_n) = \Phi^*(t_n) = \det \varphi \cdot t_n.$$

Now Formula (7.5) shows that

$$z^n = \det \varphi \cdot t_n \qquad z \in E^* \otimes E.$$

Proposition 7.2.1. *Let $z \in E^* \otimes E$ and set $T_E(z) = \varphi$. Then for $0 \le p \le n$,*

1.
$$\Phi(D_E z^p) = \det \varphi \cdot t_{n-p}$$
$$\Phi^*(D_E z^p) = \det \varphi \cdot t_{n-p}$$

and

2.
$$z^p \circ D_E(z^{n-p}) = \det \varphi \cdot t_p$$
$$D_E(z^{n-p}) \circ z^p = \det \varphi \cdot t_p.$$

PROOF. Using (7.5), the second Formula (7.3) and Part (ii) of the Corollary to Proposition 6.7.1 we obtain

$$\begin{aligned}\Phi(D_E z^p) &= \Phi(i(z^p)t_n) = [\Phi \circ i(\Phi^* t_p)]t_n \\ &= (i(t_p) \circ \Phi)t_n = \det \varphi i(t_p)t_n = \det \varphi \cdot t_{n-p}\end{aligned}$$

which proves the first Relation (1). The second Relation (1) is obtained in the same way.

(2) In fact, the first Relation (7.4) and Part (1) of the proposition yield

$$z^p \circ D_E z^{n-p} = \Phi(D_E z^{n-p}) = \det \varphi \cdot t_p.$$

The second Formula (2) is established in a similar way. □

Corollary (Laplace formula). *Let φ be a linear transformation of E. Then*

1.
$$\wedge^p \varphi \circ D_L(\wedge^{n-p} \varphi) = \det \varphi \cdot \iota_{\wedge^p E}$$

and

2.
$$D_L(\wedge^p \varphi) \circ \wedge^{n-p} \varphi = \det \varphi \cdot \iota_{\wedge^{n-p} E}.$$

Proof. Apply T_E to Formula (2) in the proposition and observe that $T_E(z^p) = \wedge^p \varphi$. □

Proposition 7.2.2. *Let $z \in E^* \otimes E$ and set $T(z) = \varphi$. Then*

1. $$i(D_E z^p)z^q = \binom{2n-p-q}{n-p} \det \varphi \cdot z^{p+q-n}$$

and

2. $$D_E z^p \cdot D_E z^q = \binom{2n-p-q}{n-p} \det \varphi \cdot D_E z^{p+q-n}.$$

Proof. In fact, applying (7.5) the second Relation (7.3) and Part (1) of Proposition 7.2.1 we obtain

$$\begin{aligned} i(D_E z^p)z^q &= i(D_E z^p)\Phi(t_q) \\ &= \Phi i(\Phi^* D_E z^p) t_q = \det \varphi \cdot \Phi i(t_{n-p}) t_q \\ &= \binom{2n-p-q}{n-p} \det \varphi \cdot \Phi(t_{p+q-n}) \\ &= \binom{2n-p-q}{n-p} \det \varphi \cdot z^{p+q-n}. \end{aligned}$$

(2) Since the restriction of D_E to the diagonal algebra $\Delta(E)$ is an involution, we have, in view of (6.22),

$$\begin{aligned} D_E[i(D_E z^p)z^q] &= D_E[i(D_E z^p)D_E(D_E z^q)] \\ &= D_E[D_E(D_E z^q \cdot D_E z^p)] = D_E(z^q) \cdot D_E(z^p). \end{aligned}$$

Now (2) follows from (1), since Δ_E is commutative.

7.3. The Adjoint Tensor

Consider the symmetric $(n-1)$-linear mapping

$$\mathrm{Ad}: \underbrace{(E^* \otimes E) \times \cdots \times (E^* \otimes E)}_{n-1} \to E^* \otimes E$$

given by

$$\mathrm{Ad}(z_1, \ldots, z_{n-1}) = D_E(z_1 \cdots z_{n-1}) \qquad z_\nu \in E^* \otimes E,$$

and set

$$\mathrm{ad}(z) = \frac{1}{(n-1)!} \mathrm{Ad}(z, \ldots, z).$$

Thus,

$$\operatorname{ad}(z) = D_E(z^{n-1});$$

ad(z) is called the *adjoint tensor* of z. Observe that ad(z) does not depend linearly on z except in the case $n = 1$.

Since D_E is an involution in the diagonal subalgebra ΔE, we have the relation

$$D_E \operatorname{ad}(z) = z^{n-1} \qquad z \in E^* \otimes E.$$

Moreover, Proposition 7.2.1, (2) implies that

$$\operatorname{ad}(z) \circ z = z \circ \operatorname{ad}(z) = \det T_E(z)t.$$

Next, let $z \in E^* \otimes E$ and define tensors $B_p(z) \in E^* \otimes E$ ($p = 0, \ldots, n-1$) by the expansion

$$\operatorname{ad}(z + \lambda t) = \sum_{p=0}^{n-1} B_p(z)\lambda^{n-p-1}.$$

Proposition 7.3.1. *The tensors $B_p(z)$ are given by*

$$B_p(z) = i(z^p)t_{p+1} \qquad (p = 0, \ldots, n-1).$$

PROOF. Since, by the binomial formula (see Section 6.2)

$$(z + \lambda t)^{n-1} = \sum_{p=0}^{n-1} (z^p \cdot t_{n-1-p})\lambda^{n-1-p}$$

we have

$$\operatorname{ad}(z + \lambda t) = \sum_{p=0}^{n-1} D_E(z^p \cdot t_{n-1-p})\lambda^{n-1-p}.$$

But, since $\Delta(E)$ is commutative,

$$D_E(z^p \cdot t_{n-p-1}) = i(z^p \cdot t_{n-p-1})t_n = i(z^p)t_{p+1}$$

and so we obtain

$$\operatorname{ad}(z + \lambda t) = \sum_{p=0}^{n-1} i(z^p)t_{p+1}\lambda^{n-p-1}.$$

It follows that

$$B_p(z) = i(z^p)t_{p+1} \qquad (p = 0, \ldots, n-1). \qquad \square$$

Next we prove the *Jacobi identity*

$$D_E(\operatorname{ad} z)^p = (\det T_E(z))^{p-1}z^{n-p} \qquad 1 \le p \le n. \tag{7.6}$$

Proposition 7.2.2, (1) yields

$$i(\operatorname{ad} z)z^q = (n + 1 - q)\det T_E(z)z^{q-1}$$

whence, by induction,

$$i((\text{ad } z)^p)z^q = \binom{n+p-q}{p}\det(T_E(z))^p z^{q-p}.$$

In particular,

$$i(\text{ad } z)^{p-1})z^{n-1} = p(\det T_E z)^{p-1}z^{n-p} \qquad 1 \le p \le n.$$

On the other hand,

$$i[(\text{ad } z)^{p-1}]z^{n-1} = i[(\text{ad } z)^{p-1}]D_E \text{ ad } z$$

and thus, in view of Formula (6.22)

$$\begin{aligned} i((\text{ad } z)^{p-1})z^{n-1} &= D_E[(\text{ad } z)\cdot(\text{ad } z)^{p-1}] \\ &= pD_E(\text{ad } z)^p. \end{aligned}$$

Combining these relations we obtain the Jacobi identity.

Setting $p = n - 1$ in this identity yields the formula

$$\text{ad ad } z = (\det T_E(z))^{n-2}z \qquad z \in E^* \otimes E,\ n \ge 2.$$

7.4. The Classical Adjoint Transformation

Let $\varphi_1, \dots, \varphi_{n-1}$ be linear transformations of E. Fix a nonzero determinant function Δ in E. Then, for every n-tuple of vectors $(x_1, \dots, x_n)$, a linear transformation $\Omega_\Delta(x_1, \dots, x_n)$ of E is defined by

$$\Omega_\Delta(x_1, \dots, x_n)h = \sum_\sigma \varepsilon_\sigma \Delta(\varphi_1 x_{\sigma(1)}, \dots, \varphi_{n-1}x_{\sigma(n-1)}, h)x_{\sigma(n)}.$$

It is easy to check that $\Omega_\Delta(x_1, \dots, x_n)$ is skew-symmetric in the x_ν. Thus the correspondence

$$(x_1, \dots, x_n) \mapsto \Omega_\Delta(x_1, \dots, x_n)$$

defines a skew-symmetric n-linear mapping

$$\underbrace{E \times \cdots \times E}_{n} \to L(E; E).$$

Hence, there is a unique element in $L(E; E)$, denoted by $\text{Ad}(\varphi_1, \dots, \varphi_{n-1})$, such that

$$\Omega_\Delta(x_1, \dots, x_n) = \Delta(x_1, \dots, x_n)\text{Ad}(\varphi_1, \dots, \varphi_{n-1}).$$

Now the above definition reads

$$\text{Ad}(\varphi_1, \dots, \varphi_{n-1})h \cdot \Delta(x_1, \dots, x_n) = \sum_\sigma \varepsilon_\sigma \Delta(\varphi_1 x_{\sigma(1)}, \dots, \varphi_{n-1}x_{\sigma(n-1)}, h)x_{\sigma(n)}.$$

In particular, if $\{e_1, \ldots, e_n\}$ is a basis of E such that $\Delta(e_1, \ldots, e_n) = 1$, then we have, for $h \in E$,

$$\operatorname{Ad}(\varphi_1, \ldots, \varphi_{n-1})h = \sum_\sigma \varepsilon_\sigma \Delta(\varphi_1 e_{\sigma(1)}, \ldots, \varphi_{n-1} e_{\sigma(n-1)}, h) e_{\sigma(n)}.$$

We shall show that the function Ad is symmetric. In fact, let τ be a permutation of $(1, \ldots, n-1)$ and set

$$\Omega_\tau(x_1, \ldots, x_n)h = \sum_\sigma \varepsilon_\sigma \Delta(\varphi_{\tau(1)} x_{\sigma(1)}, \ldots, \varphi_{\tau(n-1)} x_{\sigma(n-1)}, h) x_{\sigma(n)}.$$

It has to be shown that $\Omega_\tau = \Omega_\Delta$. Extend τ to a permutation of $(1, \ldots, n)$ by setting $\tau(n) = n$ and let $\alpha = \sigma\tau^{-1}$. Then we have

$$\begin{aligned} \Omega_\tau(x_1, \ldots, x_n)h &= \sum_\alpha \varepsilon_\alpha \Delta(\varphi_1 x_{\alpha(1)}, \ldots, \varphi_{n-1} x_{\alpha(n-1)}, h) x_{\alpha(n)} \\ &= \Omega_\Delta(x_1, \ldots, x_n)h, \end{aligned}$$

whence $\Omega_\tau = \Omega_\Delta$.

For a single linear transformation φ we set

$$\operatorname{ad}(\varphi) = \frac{1}{(n-1)!} \operatorname{Ad}(\underbrace{\varphi, \ldots, \varphi}_{n-1}).$$

Since

$$\sum_\sigma \Delta(\underbrace{\varphi x_1, \ldots, h}_{\sigma(n)}, \ldots, \varphi x_n) x_{\sigma(n)} = (n-1)! \sum_{i=1}^n \Delta(\varphi x_1, \ldots, h, \ldots, \varphi x_n) x_i,$$

it follows that $\operatorname{ad}(\varphi)$ coincides with the classical adjoint of φ as defined in Section 4.6 of *Linear Algebra.*

Proposition 7.4.1. *The operators* D_L *and* Ad *are connected by the relation*

1. $$D_L(\varphi_1 \square \cdots \square \varphi_{n-1}) = \operatorname{Ad}(\varphi_1, \ldots, \varphi_{n-1}) \qquad \varphi_\nu \in L(E; E),\ \nu = 1, \ldots, n-1.$$

In particular

2. $$D_L(\wedge^{n-1}\varphi) = \operatorname{ad}(\varphi) \qquad \varphi \in L(E; E).$$

Proof. Without loss of generality we may assume that the φ_ν are of the form

$$\varphi_\nu = T_E(a_\nu^* \otimes a_\nu) \qquad a_\nu^* \in E^*,\ a_\nu \in E.$$

Fix a basis $\{e_1, \ldots, e_n\}$ of E such that $\Delta(e_1, \ldots, e_n) = 1$ and set

$$e_1 \wedge \cdots \wedge e_n = e \qquad e^{*1} \wedge \cdots \wedge e^{*n} = e^*.$$

Then we have for $h \in E$

$$\begin{aligned}\operatorname{Ad}(\varphi_1, \ldots, \varphi_{n-1})h &= \sum_\sigma \varepsilon_\sigma \Delta(\langle a_1^*, e_{\sigma(1)}\rangle a_1, \ldots, \langle a_{n-1}^*, e_{\sigma(n-1)}\rangle a_{n-1}, h)e_{\sigma(n)} \\ &= \sum_\sigma \varepsilon_\sigma \langle a_1^*, e_{\sigma(1)}\rangle \cdots \langle a_{n-1}^*, e_{\sigma(n-1)}\rangle \cdot \Delta(a_1, \ldots, a_{n-1}, h)e_{\sigma(n)}.\end{aligned}$$

It follows that for $h^* \in E^*$

$$\begin{aligned}\langle h^*, \operatorname{Ad}(\varphi_1, \ldots, \varphi_{n-1})h\rangle &= \sum_\sigma \varepsilon_\sigma \langle a_1^*, e_{\sigma(1)}\rangle \cdots \langle a_{n-1}^*, e_{\sigma(n-1)}\rangle \langle h^*, e_{\sigma(n)}\rangle \cdot \Delta(a_1, \ldots, a_{n-1}, h) \\ &= \langle a_1^* \wedge \cdots \wedge a_{n-1}^* \wedge h^*, e_1 \wedge \cdots \wedge e_n\rangle \cdot \Delta(a_1, \ldots, a_{n-1}, h) \\ &= \langle a_1^* \wedge \cdots \wedge a_{n-1}^* \wedge h^*, e\rangle \cdot \langle e^*, a_1 \wedge \cdots \wedge a_{n-1} \wedge h\rangle \\ &= \langle h^*, a\rangle\langle a^*, h\rangle,\end{aligned}$$

where

$$a = i(a_1^* \wedge \cdots \wedge a_{n-1}^*)e \quad \text{and} \quad a^* = i(a_1 \wedge \cdots \wedge a_{n-1})e^*.$$

Thus,

$$\langle h^*, \operatorname{Ad}(\varphi_1, \ldots, \varphi_{n-1})h\rangle = \langle h^*, a\rangle\langle a^*, h\rangle. \tag{7.7}$$

On the other hand,

$$\begin{aligned}D_L(\varphi_1 \,\Box \cdots \Box\, \varphi_{n-1}) &= D_L T_E[(a_1^* \otimes a_1) \cdots (a_{n-1}^* \otimes a_{n-1})] \\ &= D_L T_E[(a_1^* \wedge \cdots \wedge a_{n-1}^*) \otimes (a_1 \wedge \cdots \wedge a_{n-1})] \\ &= T_E D_E[(a_1^* \wedge \cdots \wedge a_{n-1}^*) \otimes (a_1 \wedge \cdots \wedge a_{n-1})] \\ &= T_E[i(a_1 \wedge \cdots \wedge a_{n-1})e^* \otimes i(a_1^* \wedge \cdots \wedge a_{n-1}^*)e] \\ &= T_E(a^* \otimes a).\end{aligned}$$

It follows that

$$\begin{aligned}\langle h^*, D_L(\varphi_1 \,\Box \cdots \Box\, \varphi_{n-1})h\rangle &= \langle h^*, T_E(a^* \otimes a)h\rangle \\ &= \langle h^*, a\rangle\langle a^*, h\rangle.\end{aligned} \tag{7.8}$$

Now the proposition follows from Relations (7.7) and (7.8). □

Corollary I. *Let φ be a linear transformation of E. Then*

$$\varphi \circ \operatorname{ad}(\varphi) = \det \varphi \cdot \iota$$

and

$$\operatorname{ad}(\varphi) \circ \varphi = \det \varphi \cdot \iota$$

In particular, ad(φ) *is a linear isomorphism if and only if φ is.*

PROOF. In fact, Proposition 7.4.1 and the Laplace formula (see Section 7.2) yield

$$\mathrm{ad}(\varphi) \circ \varphi = \varphi \circ \mathrm{ad}(\varphi) = \varphi \circ D_L(\wedge^{n-1}\varphi) = \det \varphi \cdot \iota. \qquad \square$$

Corollary II. *If φ and ψ are linear transformations, then*

$$\mathrm{ad}(\psi \circ \varphi) = \mathrm{ad}(\varphi) \circ \mathrm{ad}(\psi).$$

PROOF. In view of Proposition 7.4.1 and Proposition 7.1.1, (2),

$$\begin{aligned}\mathrm{ad}(\psi \circ \varphi) &= D_L \wedge^{n-1}(\psi \circ \varphi) = D_L(\wedge^{n-1}\psi \circ \wedge^{n-1}\varphi) \\ &= D_L(\wedge^{n-1}\varphi) \circ D_L(\wedge^{n-1}\psi) = \mathrm{ad}(\psi) \circ \mathrm{ad}(\varphi). \qquad \square\end{aligned}$$

Corollary III. *Let φ be a linear transformation of E. Then*

$$D_L(\wedge^p \,\mathrm{ad}\, \varphi) = (\det \varphi)^{p-1} \wedge^{n-p}\varphi \qquad (p = 1, \ldots, n).$$

In particular,

$$\mathrm{ad}\;\mathrm{ad}(\varphi) = (\det \varphi)^{n-2}\varphi \qquad n \geq 2.$$

PROOF. Apply the proposition and the Jacobi identity (see Section 7.3). $\square$

Corollary IV. *Let $a_\nu^* \in E^*$, $a_\nu \in E$ $(\nu = 1, \ldots, n-1)$ and set*

$$\varphi_\nu = T_E(a_\nu^* \otimes a_\nu).$$

Assume that either $\{a_\nu^*\}$ or $\{a_\nu\}$ is linearly dependent. Then

$$\mathrm{Ad}(\varphi_1, \ldots, \varphi_{n-1}) = 0.$$

Characteristic Coefficients

7.5. Definition

Consider for each $p \geq 1$ the p-linear function C_p in $L(E; E)$ given by

$$C_p(\varphi_1, \ldots, \varphi_p) = \mathrm{tr}(\varphi_1 \,\square \cdots \square\, \varphi_p)$$

and set

$$C_0 = 1.$$

Since the box product is commutative, the functions C_p are symmetric. Note that

$$C_1(\varphi) = \mathrm{tr}\, \varphi \qquad \varphi \in L(E; E).$$

The function C_n will be denoted by Det,

$$\operatorname{Det}(\varphi_1, \ldots, \varphi_n) = \operatorname{tr}(\varphi_1 \,\square \cdots \square\, \varphi_n).$$

Now consider the tensors $z_\nu \in E^* \otimes E$ given by

$$z_\nu = T_E^{-1}(\varphi_\nu) \qquad \nu = 1, \ldots, p.$$

Then Proposition 6.3.1 and Formula (6.16) imply that

$$C_p(\varphi_1, \ldots, \varphi_p) = \operatorname{tr}(T_E(z_1) \,\square \cdots \square\, T_E(z_p)) = \langle z_1 \cdots z_p, t_p \rangle. \tag{7.9}$$

In particular,

$$C_p = 0 \quad \text{if } p > n.$$

For a single linear transformation φ we set

$$C_p(\varphi) = \frac{1}{p!} C_p(\varphi, \ldots, \varphi) \qquad (p = 0, \ldots, n).$$

Thus, in view of (6.6),

$$C_p(\varphi) = \operatorname{tr}(\wedge^p \varphi) \qquad (p = 0, \ldots, n).$$

Clearly,

$$C_p(\lambda\varphi) = \lambda^p C_p(\varphi) \qquad \lambda \in \Gamma.$$

Relation (7.9) implies that

$$C_p(\varphi) = \langle (T_E^{-1}\varphi)^p, t_p \rangle \qquad (p = 0, \ldots, n). \tag{7.10}$$

Recall from Section 7.2 that

$$\det \varphi = \operatorname{tr}(\wedge^n \varphi).$$

Thus

$$C_n(\varphi) = \det \varphi.$$

Equivalently,

$$\frac{1}{n!} \operatorname{Det}(\varphi, \ldots, \varphi) = \det \varphi.$$

Replacing φ by $\varphi + \psi$ in this relation we obtain the formula

$$\det(\varphi + \psi) = \frac{1}{n!} \sum_{p+q=n} \operatorname{Det}(\underbrace{\varphi, \ldots, \varphi}_{p}, \underbrace{\psi, \ldots, \psi}_{q}).$$

Proposition 7.5.1. *The $C_p(\varphi)$ satisfy the relation*

$$\det(\varphi + \lambda\iota) = \sum_{p=0}^{n} C_{n-p}(\varphi)\lambda^p \qquad \lambda \in \Gamma.$$

Thus $(-1)^p C_{n-p}(\varphi)$ is the pth *characteristic coefficient of φ.*

PROOF. Let $z \in E^* \otimes E$ be the unique tensor such that $T_E(z) = \varphi$. Then we have

$$\varphi + \lambda\iota = T_E(z + \lambda\not{t})$$

whence

$$\det(\varphi + \lambda\iota) = \langle (z + \lambda\not{t})^n, \not{t}_n \rangle.$$

Expanding by the binomial formula we obtain

$$\det(\varphi + \lambda\iota) = \sum_{p=0}^{n} \langle \not{t}_p \cdot z^{n-p}, \not{t}_n \rangle \lambda^p.$$

But

$$\begin{aligned} \langle \not{t}_p \cdot z^{n-p}, \not{t}_n \rangle &= \langle z^{n-p}, i(\not{t}_p)\not{t}_n \rangle \\ &= \langle z^{n-p}, \not{t}_{n-p} \rangle = C_{n-p}(\varphi) \qquad (p = 0, \ldots, n). \end{aligned}$$

It follows that

$$\det(\varphi + \lambda\iota) = \sum_{p=0}^{n} C_{n-p}(\varphi)\lambda^p. \qquad \square$$

Corollary. *If φ and ψ are linear transformations, then the characteristic polynomials of $\psi \circ \varphi$ and $\varphi \circ \psi$ coincide.*

PROOF. In fact, since the trace of a composition does not depend on the order,

$$\begin{aligned} C_p(\psi \circ \varphi) &= \operatorname{tr} \wedge^p(\psi \circ \varphi) = \operatorname{tr}(\wedge^p\psi \circ \wedge^p\varphi) \\ &= \operatorname{tr}(\wedge^p\varphi \circ \wedge^n\psi) = C_p(\varphi \circ \psi) \qquad (p = 0, \ldots, n). \qquad \square \end{aligned}$$

7.6. The Linear Transformations $A_p(\varphi)$

Let φ be a linear transformation and write

$$\operatorname{ad}(\varphi + \lambda\iota) = \sum_{p=0}^{n-1} A_p(\varphi)\lambda^{n-p-1},$$

where $A_p(\varphi) \in L(E; E)$.

In particular,

$$A_{n-1}(\varphi) = \operatorname{ad}(\varphi).$$

Proposition 7.6.1. *The transformation $A_p(\varphi)$ is given by*

$$A_p(\varphi) = \sum_{\nu=0}^{p} (-1)^\nu C_{p-\nu}(\varphi)\varphi^\nu \qquad (p = 0, \ldots, n-1).$$

PROOF. From the formula (see Corollary I to Proposition 7.4.1)

$$(\varphi + \lambda\iota) \circ \mathrm{ad}(\varphi + \lambda\iota) = \det(\varphi + \lambda\iota) \cdot \iota$$

we obtain

$$\sum_{p=1}^{n} \varphi \cdot A_{p-1}(\varphi)\lambda^{n-p} + \sum_{p=0}^{n-1} A_p(\varphi)\lambda^{n-p} = \sum_{p=0}^{n} C_p(\varphi)\lambda^{n-p} \cdot \iota.$$

Comparing the coefficients of λ^{n-p} on both sides of this equation yields the recursion formulas

$$\varphi \circ A_{p-1}(\varphi) + A_p(\varphi) = C_p(\varphi) \cdot \iota \qquad (p = 0, \ldots, n),$$

where $A_{-1}(\varphi) = 0$ and $A_n(\varphi) = 0$. From these relations we obtain

$$A_p(\varphi) = \sum_{\nu=0}^{p} (-1)^\nu C_{p-\nu}(\varphi)\varphi^\nu \qquad (p = 0, \ldots, n). \tag{7.11}$$

□

In particular, for $p = n - 1$,

$$\mathrm{ad}(\varphi) = \sum_{\nu=0}^{n-1} (-1)^\nu C_{n-\nu-1}(\varphi)\varphi^\nu.$$

Cayley–Hamilton theorem. *Every linear transformation φ satisfies its characteristic equation; that is,*

$$\sum_{\nu=0}^{n} (-1)^\nu C_{n-\nu}(\varphi)\varphi^\nu = 0.$$

PROOF. Apply (7.11) for $p = n$ observing that $A_n(\varphi) = 0$. □

Proposition 7.6.2. *The trace of the linear transformation $A_p(\varphi)$ is given by*

$$\mathrm{tr}\, A_p(\varphi) = (n - p)C_p(\varphi) \qquad (p = 0, \ldots, n).$$

PROOF. Let $z \in E^* \otimes E$ be the unique tensor such that $T_E(z) = \varphi$. Consider the tensors $B_p(z)$ determined by the expansion

$$\mathrm{ad}(z + \lambda\iota) = \sum_{p=0}^{n-1} B_p(z)\lambda^{n-p-1}$$

(see Section 7.3). Since

$$T_E B_p(z) = A_p(\varphi) \qquad (p = 0, \ldots, n - 1),$$

$$\mathrm{tr}\, A_p(\varphi) = \langle t, B_p(z)\rangle \qquad (p = 0, \ldots, n - 1). \tag{7.12}$$

Next observe that, by Proposition 7.3.1,

$$B_p(z) = i(z^p)t_{p+1} \qquad (p = 0, \ldots, n - 1).$$

It follows from (7.10) that

$$\begin{aligned}\langle \ell, B_p(z)\rangle &= i(\ell)B_p(z) = i(\ell)i(z^p)\ell_{p+1}\\ &= i(z^p)i(\ell)\ell_{p+1} = (n-p)\langle z^p, \ell_p\rangle = (n-p)C_p(\varphi). \quad (7.13)\end{aligned}$$

Relations (7.12) and (7.13) yield

$$\operatorname{tr} A_p(\varphi) = (n-p)C_p(\varphi) \qquad (p = 0, \dots, n). \qquad \square$$

Finally, we show that the characteristic coefficients of the classical adjoint transformation are given by

$$C_p(\operatorname{ad}\varphi) = (\det\varphi)^{p-1}C_{n-p}(\varphi) \qquad (p = 1, \dots, n). \tag{7.14}$$

In particular,

$$\det \operatorname{ad}(\varphi) = (\det\varphi)^{n-1} \quad \text{and} \quad \operatorname{tr}\operatorname{ad}(\varphi) = C_{n-1}(\varphi).$$

In fact, taking the trace on both sides of the Jacobi identity (7.6) and observing that for $\alpha \in L(\wedge E; \wedge E)$tr $D_L\alpha = \operatorname{tr}\alpha$ (see Section 7.1) we obtain

$$\operatorname{tr} \wedge^p \operatorname{ad}(\varphi) = (\det\varphi)^{p-1} \cdot \operatorname{tr} \wedge^{n-p}\varphi \qquad (p = 1, \dots, n)$$

whence (7.14).

7.7. The Trace Coefficients

Let Tr_p denote the symmetric p-linear function in $L(E; E)$ given by

$$\mathrm{Tr}_p(\varphi_1, \dots, \varphi_p) = \frac{1}{p!}\sum_\sigma \operatorname{tr}(\varphi_{\sigma(1)} \circ \cdots \circ \varphi_{\sigma(p)}) \qquad p \geq 1$$

and set

$$\mathrm{Tr}_0 = n.$$

The pth *trace coefficient* of a linear transformation φ is defined by

$$\mathrm{Tr}_p(\varphi) = \mathrm{Tr}_p(\varphi, \dots, \varphi) \qquad p \geq 1$$

and

$$\mathrm{Tr}_0(\varphi) = n.$$

Thus

$$\mathrm{Tr}_p(\varphi) = \operatorname{tr}\varphi^p \qquad p \geq 1.$$

In particular,

$$\mathrm{Tr}_p(\iota) = n.$$

Note that, in contrast to the characteristic coefficients, the trace coefficients do not vanish in general for $p > n$.

Proposition 7.7.1. *The trace coefficients and the characteristic coefficients are connected by the relation*

$$C_p(\varphi) = \frac{1}{p}\sum_{\nu=0}^{p-1}(-1)^{p-\nu-1}C_\nu(\varphi)\mathrm{Tr}_{p-\nu}(\varphi) \qquad p \geqq 1.$$

Proof. Taking the trace in the formula in Proposition 7.6.1 and using Proposition 7.6.2 we obtain

$$(n-p)C_p(\varphi) = \sum_{\nu=0}^{p}(-1)^\nu C_{p-\nu}(\varphi)\mathrm{tr}\,\varphi^\nu$$

$$= nC_p(\varphi) + \sum_{\nu=1}^{p}(-1)^\nu C_{p-\nu}(\varphi)\mathrm{tr}\,\varphi^\nu.$$

It follows that

$$C_p(\varphi) = -\frac{1}{p}\sum_{\nu=1}^{p}(-1)^\nu C_{p-\nu}(\varphi)\mathrm{tr}\,\varphi^\nu$$

$$= -\frac{1}{p}\sum_{\nu=0}^{p-1}(-1)^{p-\nu}C_\nu(\varphi)\mathrm{tr}\,\varphi^{p-\nu}. \qquad \square$$

7.8. Application to the Elementary Symmetric Functions

Fix a basis $\{e_1, \ldots, e_n\}$ of E and consider the linear transformation φ given by

$$\varphi e_\nu = \lambda_\nu e_\nu \; (\nu = 1, \ldots, n).$$

A simple calculation shows that

$$C_p(\varphi) = \sigma_p(\lambda_1, \ldots, \lambda_n)$$

while

$$\mathrm{Tr}_p(\varphi) = s_p(\lambda_1, \ldots, \lambda_n),$$

where σ_p and s_p denote the symmetric polynomials given by

$$\sigma_p(\lambda_1, \ldots, \lambda_n) = \sum_{\nu_1 < \cdots < \nu_p} \lambda_{\nu_1} \cdots \lambda_{\nu_p} \qquad p \geq 1$$

and

$$s_p(\lambda_1, \ldots, \lambda_n) = \sum_{\nu=1}^{p} \lambda_\nu^p.$$

Now Proposition 7.7.1 yields the classical recursion formulas for the σ_p in terms of the s_p,

$$\sigma_p = \frac{1}{p}\sum_{\nu=0}^{p-1}(-1)^{p-\nu-1}\sigma_\nu s_{p-\nu} \qquad (p = 1, \ldots, n).$$

7.9. Complex Vector Spaces

Let E be an n-dimensional complex vector space and let $E_{\mathbb{R}}$ denote the $2n$-dimensional real vector space. Let Δ_E be a nonzero determinant function in E. Regard Δ_E as a $\mathbb{C}$-valued n-linear function in $E_{\mathbb{R}}$ and set

$$\Delta = (-i)^n \Delta_E \wedge \bar{\Delta}_E,$$

where $\bar{\Delta}_E$ is defined by

$$\bar{\Delta}_E(x_1, \ldots, x_n) = \overline{\Delta_E(x_1, \ldots, x_n)} \qquad x_\nu \in E_{\mathbb{R}}.$$

Then Δ is linear over $\mathbb{R}$ and skew-symmetric. To show that Δ is real-valued and nonzero (and hence a determinant function in $E_{\mathbb{R}}$) choose a basis $\{a_1, \ldots, a_n\}$ of E. Since $\Delta(z_1, \ldots, z_n) = 0$ whenever the vectors $\{z_\nu\}$ are linearly dependent (over $\mathbb{C}$) it follows that

$$\begin{aligned}
&\Delta(a_1, \ldots, a_n, ia_1, \ldots, ia_n) \\
&\quad = \frac{(-i)^n}{(n!)^2} \sum_{\sigma,\tau} \varepsilon_\sigma \varepsilon_\tau \Delta_E(\alpha_{\sigma(1)}, \ldots, a_{\sigma(n)}) \bar{\Delta}_E(ia_{\tau(1)}, \ldots, ia_{\tau(n)}) \\
&\quad = \Delta_E(a_1, \ldots, a_n)\bar{\Delta}_E(a_1, \ldots, a_n) \\
&\quad = |\Delta_E(a_1, \ldots, a_n)|^2.
\end{aligned}$$

Thus,

$$\Delta(a_1, \ldots, a_n, ia_1, \ldots, ia_n) > 0$$

and so Δ is a nonzero determinant function in $E_{\mathbb{R}}$. If Δ_E is replaced by $\lambda\Delta_E$, where λ is a nonzero complex number, then Δ changes into $|\lambda|^2\Delta$. This shows that the orientation of $E_{\mathbb{R}}$ determined by Δ_E is independent of Δ_E and so $E_{\mathbb{R}}$ carries a *natural orientation.*

Proposition 7.9.1. *Let φ be a linear transformation of E and let $\varphi_{\mathbb{R}}$ denote the corresponding linear transformation of $E_{\mathbb{R}}$. Let f and $f_{\mathbb{R}}$ denote the characteristic polynomials of φ and $\varphi_{\mathbb{R}}$. Then*

$$f_{\mathbb{R}}(t) = f(t) \cdot \overline{f(t)}.$$

PROOF. We show first that

$$\det \varphi_{\mathbb{R}} = \det \varphi \cdot \overline{\det \varphi}. \tag{7.15}$$

In fact, let Δ_E be a nonzero determinant function in E and set $\Delta = i^n\Delta_E \wedge \bar{\Delta}_E$. Then

$$\varphi_{\mathbb{R}}^*\Delta = (-i)^n \varphi^*\Delta_E \wedge \varphi^*\bar{\Delta}_E.$$

But

$$\varphi^*\Delta_E = \det \varphi \cdot \Delta_E, \qquad \varphi^*\bar{\Delta}_E = \overline{\det \varphi} \cdot \bar{\Delta}_E$$

and so we obtain

$$\varphi_{\mathbb{R}}^{*}\Delta = (-1)^n \det \varphi\, \overline{\det \varphi} \cdot \Delta_E \cdot \overline{\Delta}_E = |\det \varphi|^2 \Delta$$

whence (7.15).

Replacing $\varphi_{\mathbb{R}}$ by $\varphi_{\mathbb{R}} - t\iota$ in this relation yields

$$\det(\varphi_{\mathbb{R}} - t\iota) = \det(\varphi - t\iota)\overline{\det(\varphi - t\iota)}$$

whence

$$f_{\mathbb{R}}(t) = f(t) \cdot \overline{f(t)}. \qquad \square$$

Corollary. *The characteristic coefficients of* $\varphi_{\mathbb{R}}$ *are given by*

$$C_r(\varphi)_{\mathbb{R}} = \sum_{p+q=r} C_p(\varphi)\overline{C_q(\varphi)} \qquad (r = 0, \ldots, n).$$

Problems

In the problems below T is the map defined in Section 6.16.

1. Find all linear transformations φ of E such that $T(\varphi)$ is decomposable.

2. Let

$$T(\varphi) = \sum_{i=1}^{r} a_i \otimes a^{*i} \qquad a_i \in E,\ a^{*i} \in E^*$$

be a representation of the tensor $T(\varphi)$ such that the vectors a_i and a^{*i} $(i = 1, \ldots, r)$ are linearly independent. Show that $r = \operatorname{rank}(\varphi)$.

3. Let E, E^* be a pair of finite-dimensional dual vector spaces and consider the linear map

$$f: E^* \otimes E \to L(E; E)$$

defined by $f(b^* \otimes a)x = \langle b^*, x\rangle a$. Prove that

a. $f = T^*$
b. $f = T^{-1}$.

4. Verify the relation

$$\langle z^p, (\operatorname{ad} z)^p\rangle = \binom{n}{p}(\det T^{-1}(z))^p \qquad 1 \le p \le n,\ z \in E^* \otimes E.$$

5. Show that

$$\det \wedge^p \varphi \cdot \det \wedge^{n-p} \varphi = (\det \varphi)^{\binom{n}{p}} \qquad 0 \le p \le n.$$

6. Show that the coefficient of λ^{n-p} in the characteristic polynomial of an $n \times n$-matrix is $(-1)^{n-p}$ times the sum of all principal minors of order p.

7. Let E be a Euclidean space and set $E^* = E$. Write

$$L(E; E) = S(E; E) \oplus A(E; E)$$

where $S(E; E)$ denotes the space of selfadjoint transformations and $A(E; E)$ denotes the space of skew transformations (see *Linear Algebra* Problem 3, Chapter IX, Section 2). Prove that

$$T(S(E; E)) = Y^2(E)$$

and

$$T(A(E; E)) = X^2(E).$$

(see Sections 4.2 and 4.10).

8. Let E be a real vector space and consider the bilinear function Φ in $\Delta(E, E^*)$ defined by

$$\Phi(u, v) = \langle D_\Delta u, v\rangle \qquad u, v \in \Delta(E, E^*).$$

a. Prove that Φ is symmetric and has signature zero.

Hint: Make E into a Euclidean space and set $E^* = E$; then consider first $\wedge^m_m(E, E)$ in the case $\dim E = 2m$. See problems 7 above, Problems 3 and 4 after Section 6.14, Problem 4, Section 5, Chapter IX of *Linear Algebra*, and Problem 3, Section 2, Chapter IX of *Linear Algebra*.

b. Given two linear transformations $\varphi, \psi \in H(\wedge E; \wedge E)$ (see problem 13) prove that

$$\Phi(T(\varphi), T(\psi)) = \operatorname{tr}((D\varphi)^* \circ \psi)$$

and conclude that the signature of the bilinear function

$$F(\varphi, \psi) = \Phi(T(\varphi), T(\psi))$$

is zero.

c. If $\dim E = 2m$ and φ, ψ are any two linear transformations of E such that φ is regular, prove that

$$\Phi(T(\varphi)^m, T(\psi)^m) = \det \varphi \operatorname{tr} \wedge^m(\varphi^{-1} \circ \psi)$$

and

$$\operatorname{tr} \wedge^m \varphi = \det \varphi \operatorname{tr} \wedge^m \varphi^{-1}.$$

Conclude that

$$\alpha_m = \det \varphi \cdot \tilde{\alpha}_m,$$

where α_m and $\tilde{\alpha}_m$ are the mth characteristic coefficients of φ and φ^{-1} respectively.

d. If $n = 2$ show that,

$$\Phi(T(\varphi), T(\psi)) = \operatorname{tr} \varphi \operatorname{tr} \psi - \operatorname{tr}(\varphi \circ \psi).$$

9. Use one of the formulas (2) in Proposition 7.2.2 to derive the *classical Laplace expansion formula* for a determinant:

   Let $A = (\alpha_j^i)$ be an $n \times p$ matrix and fix a p-tuple $(\lambda_1, \ldots, \lambda_p)$ such that $\lambda_1 < \cdots < \lambda_p$. Let $(\lambda_{p+1}, \ldots, \lambda_n)$ denote the complementary $(n-p)$-tuple in increasing order. Then

$$\det A = \sum_{\nu_1 < \cdots < \nu_p} (-1)^{\Sigma_{i=1}^{n} \lambda_i + \nu_i} \det A_{\lambda_1, \ldots, \lambda_p}^{\nu_1, \ldots, \nu_p} \det A_{\lambda_{p+1}, \ldots, \lambda_n}^{\nu_{p+1}, \ldots, \nu_n},$$

   where $(\nu_{p+1}, \ldots, \nu_n)$ is the complementary $(n-p)$-tuple of $(\nu_1, \ldots, \nu_n)$. Here $A_{\beta_1, \ldots, \beta_q}^{\alpha_1, \ldots, \alpha_q}$ denotes the $q \times q$ matrix consisting of the rows $\beta_1, \ldots, \beta_q$ and the columns $\alpha_1, \ldots, \alpha_q$.

10. Show that the rank of the adjoint transformation is given by

$$r(\operatorname{ad} \varphi) = n \quad \text{if } r(\varphi) = n$$

$$r(\operatorname{ad} \varphi) = 1 \quad \text{if } r(\varphi) = n - 1$$

$$r(\operatorname{ad} \varphi) = 0 \quad \text{if } r(\varphi) \leq n - 2.$$

   Use these relations to prove the formula

$$\operatorname{ad} \operatorname{ad} \varphi = (\det \varphi)^{n-2} \varphi.$$

11. Given an $n \times n$ matrix $A = (\alpha_\nu^\mu)$, define the matrix $(\tilde{\alpha}_\nu^\mu)$ by

$$\tilde{\alpha}_\nu^\mu = \begin{vmatrix} \alpha_1^1 \cdots \hat{\alpha}_1^\mu \cdots \alpha_1^n \\ \vdots \qquad\qquad \vdots \\ \hat{\alpha}_\nu^1 \cdots \hat{\alpha}_\nu^\mu \cdots \hat{\alpha}_\nu^n \\ \vdots \qquad\qquad \vdots \\ \alpha_n^1 \cdots \hat{\alpha}_n^\mu \cdots \alpha_n^n \end{vmatrix} \qquad (\nu, \mu = 1, \ldots, n).$$

   Using the Jacobi identity in Section 7.3 prove the *classical Jacobian identities:*

$$\begin{vmatrix} \tilde{\alpha}_{\nu_1}^{\mu_1} & \cdots & \tilde{\alpha}_{\nu_p}^{\mu_1} \\ \vdots & & \vdots \\ \tilde{\alpha}_{\nu_1}^{\mu_p} & \cdots & \tilde{\alpha}_{\nu_p}^{\mu_p} \end{vmatrix} = \begin{vmatrix} \alpha_{\nu_{p+1}}^{\mu_{p+1}} & \cdots & \alpha_{\nu_n}^{\mu_{p+1}} \\ \vdots & & \vdots \\ \alpha_{\nu_{p+1}}^{\mu_n} & \cdots & \alpha_{\nu_n}^{\mu_n} \end{vmatrix} \qquad (\det A)^{p-1}$$

$$(\nu_1 < \cdots < \nu_p, \mu_1 < \cdots < \mu_p),$$

   where $(\mu_{p+1}, \ldots, \mu_n)$ and $(\nu_{p+1}, \ldots, \nu_n)$ are the complementary $(n-p)$-tuples of $(\mu_1, \ldots, \mu_p)$ and $(\nu_1, \ldots, \nu_p)$ respectively.

12. Use the formula

$$\langle D_e u, v \rangle = \langle e^*, u \wedge v \rangle \qquad u \in \wedge^p E, v \in \wedge^{n-p} E$$

   to derive the classical Laplace expansion formula (see Problem 9).

   Hint: Given an $n \times n$ matrix (α_ν^μ), define the vectors x_i by

$$x_i = \sum_\nu \alpha_i^\nu e_\nu \qquad i = 1, \ldots, n.$$

   Then apply the above formula with

$$u = x_1 \wedge \cdots \wedge x_p \quad \text{and} \quad v = x_{p+1} \wedge \cdots \wedge x_n.$$

13. Let E be a Euclidean space and consider the space $H(\wedge E; \wedge E)$ of homogeneous linear transformations $\wedge E \to \wedge E$ of degree zero. Then the Poincaré isomorphism induces a linear automorphism D_H of $H(\wedge E; \wedge E)$. Given an isometry $\varphi: E \to E$, prove that $\wedge \varphi$ is an eigenvector of D_H with eigenvalue $+1$.

14. Consider the isomorphism

$$*: H(\wedge E; \wedge E) \xrightarrow{\cong} H(\wedge E^*; \wedge E^*)$$

defined by $*(\varphi) = \varphi^*$. Prove that $*$ commutes with D_L,

$$D_L^{-1}(\varphi^*) = (D_L\varphi)^* \qquad \varphi \in H(\wedge E; \wedge E).$$

Skew and Skew-Hermitian Transformations 8

Throughout this chapter E denotes an n-dimensional vector space over a field of characteristic zero.

The Characteristic Coefficients of a Skew Linear Transformation

8.1. Definition

Let E be an n-dimensional inner product space see Section 1.25 and denote the inner product by $(,)$. It determines a linear isomorphism σ

$$\sigma : E \xrightarrow{\cong} E^*$$

via

$$\langle \sigma x, y \rangle = (x, y) \qquad x, y \in E.$$

Thus E can be regarded as self-dual. Hence every subspace $F \subset E$ determines a second subspace $F^{\perp}$ (the orthogonal complement of F). Its dimension is given by

$$\dim F^{\perp} = n - \dim F.$$

In particular, if the restriction of the inner product to F is nondegenerate, then $F \cap F^{\perp} = 0$ and so we have the direct decomposition

$$E = F \oplus F^{\perp}.$$

A basis $\{e_1, \ldots, e_n\}$ of E is called *orthogonal*, if

$$(e_\nu, e_\mu) = 0 \qquad \nu \neq \mu.$$

Note that $(e_\nu, e_\nu) \neq 0$ $(\nu = 1, \ldots, n)$, since the inner product is nondegenerate.

Every inner product space admits an orthogonal basis. In fact, consider the function

$$Q(x) = (x, x) \qquad x \in E.$$

Since

$$(x, y) = \tfrac{1}{2}(Q(x + y) - Q(x) - Q(y)),$$

it follows that $Q \neq 0$. Thus there is a vector e_1 such that $Q(e_1) \neq 0$. Let E_1 denote the subspace spanned by e_1. Then

$$E = E_1 \oplus E_1^\perp$$

and the restriction of the inner product in E to $E_1^\perp$ is again nondegenerate.

Now assume by induction that $E_1^\perp$ admits an orthogonal basis $\{e_2, \ldots, e_n\}$. Then $\{e_1, e_2, \ldots, e_n\}$ is an orthogonal basis of E.

Every linear transformation φ of E determines an *adjoint transformation* $\tilde{\varphi}$ by the equation

$$(\varphi x, y) = (x, \tilde{\varphi} y) \qquad x, y \in E.$$

If $\tilde{\varphi} = -\varphi$, then φ is called a *skew linear transformation*. The skew linear transformations form a subspace of $L(E; E)$ which will be denoted by $\mathrm{Sk}(E)$. Its dimension is given by

$$\dim \mathrm{Sk}(E) = \binom{n}{2}.$$

8.2. The Isomorphisms Φ_E and Ψ_E

Consider the linear operator $\Phi_E \colon \wedge^2 E \to \mathrm{Sk}(E)$ given by

$$\Phi_E(a \wedge b)x = (a, x)b - (b, x)a \qquad x \in E.$$

We show that Φ_E is an isomorphism. Since

$$\dim \mathrm{Sk}(E) = \binom{n}{2} = \dim \wedge^2 E,$$

it is sufficient to prove that Φ_E is injective.

Choose an orthogonal basis $\{e_1, \ldots, e_n\}$ of E and set

$$\Phi_E(e_i \wedge e_j) = \varphi_{ij} \qquad (i < j).$$

Then

$$\varphi_{ij}(x) = (e_i, x)e_j - (e_j, x)e_i \qquad x \in E.$$

Thus the relation

$$\sum_{i<j} \lambda^{ij}\varphi_{ij} = 0$$

implies that $\lambda^{ij} = 0$ and so Φ_E is injective.

Thus

$$\Phi_E \colon \wedge^2 E \xrightarrow{\cong} \mathrm{Sk}(E)$$

is a linear isomorphism. The inverse isomorphism will be denoted by Ψ_E.

Remark. In Section 8.3 we shall derive a formula which expresses the characteristic coefficients of a skew transformation φ in terms of the tensor $\Psi_E(\varphi)$ (see Theorem 8.3.1).

Next consider the projections

$$\pi_E \colon E \otimes E \to \wedge^2 E, \qquad \pi_{E^*} \colon E^* \otimes E^* \to \wedge^2 E^*$$

and

$$\pi \colon L_E \to \mathrm{Sk}(E),$$

where

$$\pi_E(a \otimes b) = a \wedge b, \qquad \pi_{E^*}(a^* \otimes b^*) = a^* \wedge b^*$$

and

$$\pi(\varphi) = \varphi - \tilde{\varphi} \qquad \varphi \in L_E.$$

We shall show that the isomorphisms Φ_E and $T_E \colon E^* \otimes E \xrightarrow{\cong} L_E$ (see Section (6.2) are connected by the following commutative diagrams:

$$\begin{array}{ccc} E^* \otimes E & \xrightarrow[T_E]{\cong} & L_E \\ {\scriptstyle \sigma^{-1}\otimes\iota}\downarrow{\scriptstyle \cong} & & \\ E \otimes E & & \downarrow{\scriptstyle \pi} \\ {\scriptstyle \pi_E}\downarrow & & \\ \wedge^2 E & \xrightarrow[\cong]{\Phi_E} & \mathrm{Sk}(E) \end{array} \qquad \begin{array}{ccccc} E^* \otimes E & & \xrightarrow[\cong]{T_E} & & L_E \\ {\scriptstyle \iota\otimes\sigma}\downarrow{\scriptstyle \cong} & & & & \\ E^* \otimes E^* & & & & \downarrow{\scriptstyle \pi} \\ {\scriptstyle \pi_{E^*}}\downarrow & & & & \\ \wedge^2 E^* & \xrightarrow[\cong]{\sigma_\wedge^{-1}} & \wedge^2 E & \xrightarrow[\cong]{\Phi_E} & \mathrm{Sk}(E) \end{array} \tag{8.1}$$

In fact, let $a^* \in E^*$ and $b \in E$. Then

$$\pi_E(\sigma^{-1} \otimes \iota)(a^* \otimes b) = \sigma^{-1}a^* \wedge b$$

and so

$$\Phi_E \pi_E(\sigma^{-1} \otimes \iota)(a^* \otimes b)x = (\sigma^{-1}a^*, x)b - (b, x)\sigma^{-1}a^*.$$

On the other hand, set $T_E(a^* \otimes b) = \varphi$. Then

$$\varphi x = \langle a^*, x\rangle b = (\sigma^{-1}a^*, x)b.$$

Hence

$$\tilde{\varphi}x = (b, x)\sigma^{-1}a^*$$

and so

$$(\varphi - \tilde{\varphi})x = (\sigma^{-1}a^*, x)b - (b, x)\sigma^{-1}a^*.$$

It follows that

$$\pi \circ T_E = \Phi_E \circ \pi_E \circ (\sigma^{-1} \otimes \iota)$$

and so the first diagram commutes. The commutativity of the second diagram follows from $\sigma_\wedge \circ \pi_E \circ (\sigma^{-1} \otimes \iota) = \pi_{E^*} \circ (\iota \otimes \sigma)$.

8.3. The Isomorphism τ

Let E^* be the dual space of E. Then the bilinear function $\langle\langle , \rangle\rangle$ defined by

$$\langle\langle x^* \oplus x, y^* \oplus y\rangle\rangle = \langle x^*, y\rangle + \langle y^*, x\rangle$$

(cf. Section 6.18) determines an inner product in the direct sum $E^* \oplus E$, as is easily checked. Observe that this inner product does not depend on the inner product in E.

Now consider the isomorphism $\sigma: E \xrightarrow{\cong} E^*$ induced by the inner product in E (see Section 8.1) and define a linear transformation τ of $E^* \oplus E$ by setting

$$\tau(x^* \oplus x) = (x^* + \sigma x) \oplus (-\sigma^{-1}x^* + x).$$

A simple calculation shows that

$$\tau^2(x^* \oplus x) = 2(\sigma x \oplus (-\sigma^{-1}x^*))$$

and so τ is a linear automorphism of $E^* \oplus E$.

Moreover, τ is selfadjoint with respect to the inner product $\langle\langle , \rangle\rangle$. In fact,

$$\begin{aligned}\langle\langle \tau(x^* \oplus x), y^* \oplus y\rangle\rangle &= \langle x^*, y\rangle + \langle \sigma x, y\rangle - \langle y^*, \sigma^{-1}x^*\rangle + \langle y^*, x\rangle \\ &= \langle x^*, y\rangle + \langle y^*, x\rangle + (x, y) - (y^*, x^*) \\ &= \langle\langle x^* \oplus x, \tau(y^* \oplus y)\rangle\rangle.\end{aligned}$$

Now extend τ to an algebra automorphism

$$\tau_\wedge : \wedge(E^* \oplus E) \xrightarrow{\cong} \wedge(E^* \oplus E).$$

Then $\tau_\wedge$ is selfadjoint with respect to the induced inner product in $\wedge(E^* \oplus E)$.

The automorphism $\tau_\wedge$ determines via the canonical isomorphism

$$f: \wedge E^* \hat{\otimes} \wedge E \xrightarrow{\cong} \wedge(E^* \oplus E)$$

(see Section 5.15) an algebra automorphism of $\wedge E^* \hat{\otimes} \wedge E$ which will also be denoted by $\tau_\wedge$. A straightforward computation shows that

$$\tau_\wedge(x^* \otimes 1) = x^* \otimes 1 - 1 \otimes \sigma^{-1}x^* \qquad x^* \in E^*$$

and

$$\tau_\wedge(1 \otimes x) = \sigma x \otimes 1 + 1 \otimes x \qquad x \in E.$$

These relations yield

$$\begin{aligned}\tau_\wedge(x^* \otimes x) = (x^* \wedge \sigma x) \otimes 1 - 1 \otimes (\sigma^{-1}x^* \wedge x)& \\ + x^* \otimes x + \sigma x \otimes \sigma^{-1}x^* \qquad x^* \in E^*, x \in E.&\end{aligned} \tag{8.2}$$

Let Q_E be the linear automorphism of $E^* \otimes E$ given by

$$Q_E(x^* \otimes x) = \sigma x \otimes \sigma^{-1}x^*.$$

Then Formula (8.2) yields

$$\begin{aligned}\tau_\wedge Q_E(x^* \otimes x) = (\sigma x \wedge x^*) \otimes 1 - 1 \otimes (x \wedge \sigma^{-1}x^*)& \\ + \sigma x \otimes \sigma^{-1}x^* + x^* \otimes x.&\end{aligned} \tag{8.3}$$

Adding (8.2) and (8.3) we obtain

$$\tau_\wedge[x^* \otimes x + Q_E(x^* \otimes x)] = 2(x^* \otimes x + Q_E(x^* \otimes x)) \qquad x^* \in E^*, x \in E.$$

Thus,

$$\tau_\wedge(u + Q_E u) = 2(u + Q_E u) \qquad u \in E^* \otimes E.$$

In particular, if $Q_E u = u$, then

$$\tau_\wedge u = 2u. \tag{8.4}$$

On the other hand, subtracting (8.3) from (8.2) yields

$$\begin{aligned}\tau_\wedge[x^* \otimes x - Q_E(x^* \otimes x)] &= 2[(x^* \wedge \sigma x) \otimes 1 - 1 \otimes (\sigma^{-1}x^* \wedge x)] \\ &= 2[\pi_{E^*}(x^* \otimes \sigma x) \otimes 1 - 1 \otimes \pi_E(\sigma^{-1}x^* \otimes x)],\end{aligned}$$

where

$$\pi_E : E \otimes E \to \wedge^2 E \quad \text{and} \quad \pi_{E^*} \circ E^* \otimes E^* \to \wedge^2 E^*$$

are the projections defined in Section 8.2. Thus we have the relation

$$\tau_\wedge(u - Q_E u) = 2[\pi_{E^*}(\iota \otimes \sigma)u \otimes 1 - 1 \otimes \pi_E(\sigma^{-1} \otimes \iota)u] \quad u \in E^* \otimes E. \tag{8.5}$$

In particular, if $Q_E u = -u$, this reduces to

$$\tau_\wedge u = \pi_{E^*}(\iota \otimes \sigma)u \otimes 1 - 1 \otimes \pi_E(\sigma^{-1} \otimes \iota)u. \tag{8.6}$$

Next, consider the linear isomorphism $T_E: E^* \otimes E \to L(E; E)$ (see Section 6.2). It is easily checked that the adjoint transformation of $T_E(u)$ is given by

$$\widetilde{T_E(u)} = T_E(Q_E u) \qquad u \in E^* \otimes E.$$

Thus $T_E(u)$ is skew if and only if u satisfies

$$Q_E(u) = -u.$$

Lemma. *Let φ be a skew linear transformation and set*

$$u = T_E^{-1}(\varphi), \qquad z = \Psi_E(\varphi).$$

Then

$$\tau_\wedge u = 2(\sigma_\wedge z \otimes 1 - 1 \otimes z).$$

PROOF. Since φ is skew, u satisfies

$$Q_E u = -u.$$

Thus Formula (8.6) yields

$$\tau_\wedge u = \pi_{E^*}(\iota \otimes \sigma)u \otimes 1 - 1 \otimes \pi_E(\sigma^{-1} \otimes \iota)u.$$

In view of the second diagram (8.1)

$$\pi_{E^*}(\iota \otimes \sigma)u = \pi_{E^*}(\iota \otimes \sigma)T_E^{-1}(\varphi) = \sigma_\wedge \Psi_E \pi(\varphi).$$

Since φ is skew, $\pi(\varphi) = 2\varphi$, and so we obtain

$$\pi_{E^*}(\iota \otimes \sigma)u = 2\sigma_\wedge \Psi_E(\varphi) = 2\sigma_\wedge z. \tag{8.7}$$

On the other hand, the first diagram (8.1) yields

$$\pi_E(\sigma^{-1} \otimes \iota)u = \pi_E(\sigma^{-1} \otimes \iota)T_E^{-1}(\varphi) = \Psi_E^{-1}\pi(\varphi) = 2\Psi_E^{-1}(\varphi) = 2z. \tag{8.8}$$

Relations (8.7), and (8.8) yield

$$\tau_\wedge u = 2(\sigma_\wedge z \otimes 1 - 1 \otimes z)$$

and so the lemma is proved. □

Theorem 8.3.1. *The characteristic coefficients of a skew linear transformation are given by*

$$C_{2p}(\varphi) = ((\Psi_E \varphi)^p, (\Psi_E \varphi)^p)$$

and

$$C_{2p+1}(\varphi) = 0 \qquad p \geqq 0.$$

PROOF. Set $\varphi = T_E(u)$, $u \in E^* \otimes E$. Then, according to (7.10),

$$C_k(\varphi) = \langle u^k, t_k \rangle = \langle u^k, t^k \rangle \qquad (k = 0, \ldots, n).$$

Next, observe that (see Section 6.17)

$$u^k = (-1)^{k(k-1)/2} u^{\hat{k}}$$

and

$$\bar{t}^k = (-1)^{k(k-1)/2} t^{\hat{k}}$$

(see Formula (6.33) of Section 6.17). Using Formula (6.34) in Section 6.18 we obtain

$$C_k(\varphi) = \langle u^{\hat{k}}, t^{\hat{k}} \rangle = (-1)^k \langle\!\langle u^{\hat{k}}, t^{\hat{k}} \rangle\!\rangle. \tag{8.9}$$

Now consider the automorphism τ. Since φ is skew, the lemma shows that

$$\tau_\wedge u = 2(z^* \otimes 1 - 1 \otimes z), \tag{8.10}$$

where

$$z = \Psi_E \varphi \quad \text{and} \quad z^* = \sigma_\wedge \Psi_E \varphi.$$

On the other hand, since $Q_E \bar{t} = \bar{t}$, Formula (8.4) yields

$$\tau_\wedge \bar{t} = 2\bar{t}.$$

Thus, since $\tau_\wedge$ preserves products in the algebra $\wedge E^* \hat{\otimes} \wedge E$,

$$\tau_\wedge(\bar{t}^{\hat{k}}) = 2^k \bar{t}^{\hat{k}}.$$

Inserting this into (8.9) and observing that $\tau_\wedge$ is selfadjoint with respect to the inner product $\langle\!\langle , \rangle\!\rangle$ we obtain

$$\begin{aligned} C_k(\varphi) &= (-1)^k 2^{-k} \langle\!\langle u^{\hat{k}}, \tau_\wedge(\bar{t}^{\hat{k}}) \rangle\!\rangle \\ &= (-1)^k 2^{-k} \langle\!\langle \tau_\wedge(u^{\hat{k}}), \bar{t}^{\hat{k}} \rangle\!\rangle \\ &= (-1)^k 2^{-k} \langle\!\langle (\tau_\wedge u)^{\hat{k}}, \bar{t}^{\hat{k}} \rangle\!\rangle. \end{aligned} \tag{8.11}$$

Raising (8.10) to the kth power (in the algebra $\wedge E^* \hat{\otimes} \wedge E$) and using the binomial formula yields

$$(\tau_\wedge u)^{\hat{k}} = 2^k (z^* \otimes 1 - 1 \otimes z)^{\hat{k}} = 2^k \sum_{i+j=k} (-1)^j z^{*i} \otimes z^j. \tag{8.12}$$

From (8.11) and (8.12) we obtain

$$C_k(\varphi) = (-1)^k \sum_{i+j=k} (-1)^j \langle\!\langle z^{*\hat{i}} \otimes z^{\hat{j}}, \bar{t}^{\hat{k}} \rangle\!\rangle. \tag{8.13}$$

Now observe that

$$z^{*\hat{i}} \otimes z^{\hat{j}} \in \wedge^{2i} E^* \otimes \wedge^{2j} E$$

while

$$\bar{t}^{\hat{k}} \in \wedge^k E^* \otimes \wedge^k E.$$

Thus Formula (8.13) shows that

$$C_k(\varphi) = 0 \quad \text{if } k \text{ is odd}$$

(see the definition of the inner product $\langle\!\langle\, , \rangle\!\rangle$ in Section 6.18).

On the other hand, if $k = 2p$, then, (8.13) yields

$$C_{2p}(\varphi) = (-1)^p \langle\!\langle z^{*\hat{p}} \otimes z^{\hat{p}}, \widehat{t^{2p}} \rangle\!\rangle$$

Since

$$\widehat{t^{2p}} = (-1)^{p(2p-1)} t^{2p} = (-1)^p t^{2p}$$

(see (6.33), Section 6.18) it follows that

$$C_{2p}(\varphi) = \langle\!\langle z^{*p} \otimes z^p, t^{2p} \rangle\!\rangle$$

and so, in view of (6.33)

$$C_{2p}(\varphi) = \langle z^{*p} \otimes z^p, t^{2p} \rangle.$$

Finally, applying (6.17), we obtain

$$C_{2p}(\varphi) = \langle z^{*p}, z^p \rangle = \langle (\sigma_\wedge z)^p, z^p \rangle = (z^p, z^p) = ((\Psi_E \varphi)^p, (\Psi_E \varphi)^p).$$

This completes the proof of the theorem. □

The Pfaffian of a Skew Linear Transformation

8.4. The Pfaffian

Let E be an inner product space of dimension $n = 2m$ and let $\varphi_1, \ldots, \varphi_m$ be skew linear transformations. Every φ_μ determines an element

$$\Psi_E(\varphi_\mu) \in \wedge^2 E.$$

Thus,

$$\Psi_E(\varphi_1) \wedge \cdots \wedge \Psi_E(\varphi_m) \in \wedge^n E.$$

Now fix a basis vector a of $\wedge^n E$ and set

$$\mathrm{Pf}_a(\varphi_1, \ldots, \varphi_m) = (\Psi_E(\varphi_1) \wedge \cdots \wedge \Psi_E(\varphi_m), a).$$

Equivalently,

$$\Psi_E(\varphi_1) \wedge \cdots \wedge \Psi_E(\varphi_m) = \frac{1}{(a, a)} \mathrm{Pf}_a(\varphi_1, \ldots, \varphi_m)\, a.$$

The scalar $\mathrm{Pf}_a(\varphi_1, \ldots, \varphi_m)$ is called the *Pfaffian* of $\varphi_1, \ldots, \varphi_m$. Since every two elements $\Psi_E(\varphi_\mu)$ commute, it follows that Pf_a is a *symmetric* m-linear function in $\mathrm{Sk}(E)$.

The Pfaffian of a single skew transformation φ is defined by

$$\mathrm{Pf}_a(\varphi) = \frac{1}{m!}\mathrm{Pf}_a(\underbrace{\varphi, \ldots, \varphi}_{m}).$$

This can be written in the form

$$\mathrm{Pf}_a(\varphi) = (\Psi_E(\varphi)^m, a).$$

Clearly,

$$\mathrm{Pf}_a(\lambda\varphi) = \lambda^m \mathrm{Pf}_a(\varphi) \qquad \lambda \in \Gamma.$$

Proposition 8.4.1. *If τ is an isometry of E, then*

$$\mathrm{Pf}_a(\tau \circ \varphi_1 \circ \tau^{-1}, \ldots, \tau \circ \varphi_m \circ \tau^{-1}) = \det \tau \cdot \mathrm{Pf}_a(\varphi_1, \ldots, \varphi_m).$$

PROOF. Since τ preserves the inner product we have

$$\Phi_E(\tau x \wedge \tau y) = \tau \circ \Phi_E(x \wedge y) \circ \tau^{-1} \qquad x, y \in E$$

whence

$$\tau_\wedge \Psi_E(\varphi) = \Psi_E(\tau \circ \varphi \circ \tau^{-1}) \qquad \varphi \in \mathrm{Sk}(E).$$

It follows that

$$\begin{aligned}&\Psi_E(\tau \circ \varphi_1 \circ \tau^{-1}) \wedge \cdots \wedge \Psi_E(\tau \circ \varphi_m \circ \tau^{-1})\\ &\qquad = \tau_\wedge \Psi_E(\varphi_1) \wedge \cdots \wedge \tau_\wedge \Psi_E(\varphi_m) = \tau_\wedge(\Psi_E(\varphi_1) \wedge \cdots \wedge \Psi_E(\varphi_m))\end{aligned}$$

whence

$$\begin{aligned}&\mathrm{Pf}_a(\tau \circ \varphi_1 \circ \tau^{-1}, \ldots, \tau \circ \varphi_m \circ \tau^{-1})\\ &\qquad = (\tau_\wedge(\Psi_E(\varphi_1) \wedge \cdots \wedge \Psi_E(\varphi_m)), a) = (\Psi_E(\varphi_1) \wedge \cdots \wedge \Psi_E(\varphi_m), \tau_\wedge^{-1} a).\end{aligned}$$

But since τ is an isometry

$$\tau_\wedge^{-1} a = \det \tau^{-1} \cdot a = \det \tau \cdot a$$

and so we obtain

$$\begin{aligned}\mathrm{Pf}_a(\tau \circ \varphi_1 \circ \tau^{-1}, \ldots, \tau \circ \varphi_m \circ \tau^{-1}) &= \det \tau(\Psi_E(\varphi_1) \wedge \cdots \wedge \Psi_E(\varphi_m), a)\\ &= \det \tau \cdot \mathrm{Pf}_a(\varphi_1, \ldots, \varphi_m).\end{aligned}$$ □

Proposition 8.4.2. *Let a and b be basis vectors of $\wedge^n E$. Then*

$$\mathrm{Pf}_a(\varphi) \cdot \mathrm{Pf}_b(\varphi) = (a, b)\det \varphi.$$

In particular,

$$\mathrm{Pf}_a(\varphi)^2 = (a, a)\det \varphi.$$

PROOF. In fact, since

$$(\Psi_E \varphi)^m = \frac{1}{(a, a)} \operatorname{Pf}_a(\varphi) a = \frac{1}{(b, b)} \operatorname{Pf}_b(\varphi) b,$$

it follows that

$$((\Psi_E \varphi)^m, (\Psi_E \varphi)^m) = \frac{\operatorname{Pf}_a(\varphi)\operatorname{Pf}_b(\varphi)}{(a, a)(b, b)} (a, b) = \frac{1}{(a, b)} \operatorname{Pf}_a(\varphi)\operatorname{Pf}_b(\varphi)$$

(since $b = \lambda a$, $\lambda \in \Gamma$). Now applying Theorem 8.3.1 with $p = m$ and observing that $C_n(\varphi) = \det \varphi$, we obtain

$$\det \varphi = \frac{1}{(a, b)} \operatorname{Pf}_a(\varphi) \cdot \operatorname{Pf}_b(\varphi).$$

8.5. Direct Sums

Let E and F be inner product spaces. Then an inner product is induced in the direct sum $E \oplus F$ by

$$(x_1 \oplus y_1, x_2 \oplus y_2) = (x_1, x_2) + (y_1, y_2) \qquad x_1, x_2 \in E, y_1, y_2 \in F.$$

Let

$$i_E : E \to E \oplus F, \qquad i_F : F \to E \oplus F$$

and

$$\pi_E : E \oplus F \to E, \qquad \pi_F : E \oplus F \to F$$

denote the natural inclusions and projections. Then the isomorphisms Φ_E, Φ_F, and $\Phi_{E \oplus F}$ (see Section 8.2) are connected by the relations

$$\Phi_{E \oplus F}(i_E)_\wedge(x \wedge y) = i_E \circ \Phi_E(x \wedge y) \circ \pi_E \qquad x \in E, y \in E$$

and

$$\Phi_{E \oplus F}(i_F)_\wedge(x \wedge y) = i_F \circ \Phi_F(x \wedge y) \circ \pi_F \qquad x \in F, y \in F,$$

as is easily checked. The relations above imply that

$$(i_E)_\wedge \Psi_E(\varphi) = \Psi_{E \oplus F}(i_E \circ \varphi \circ \pi_E) \qquad \varphi \in \operatorname{Sk}(E)$$

and

$$(i_F)_\wedge \Psi_F(\psi) = \Psi_{E \oplus F}(i_F \circ \psi \circ \pi_F) \qquad \psi \in \operatorname{Sk}(F).$$

Since

$$\varphi \oplus \psi = i_E \circ \varphi \circ \pi_E + i_F \circ \psi \circ \pi_F,$$

we obtain

$$(i_E)_\wedge \Psi_E(\varphi) + (i_F)_\wedge \Psi_F(\psi) = \Psi_{E \oplus F}(\varphi \oplus \psi) \qquad \varphi \in \operatorname{Sk}(E), \psi \in \operatorname{Sk}(F). \quad (8.14)$$

Proposition 8.5.1. *Let E and F be inner product spaces of dimensions 2p and 2q respectively. Choose basis vectors a and b of* $\wedge^{2p}E$ *and* $\wedge^{2q}F$*. Then*

$$\mathrm{Pf}_{a\otimes b}(\varphi \oplus \psi) = \mathrm{Pf}_a(\varphi)\cdot \mathrm{Pf}_b(\varphi) \qquad \varphi \in \mathrm{Sk}(E),\ \psi \in \mathrm{Sk}(F).$$

PROOF. Set $\Psi_E(\varphi) = u$, $\Psi_F(\psi) = v$ and $\Psi_{E\oplus F}(\varphi \oplus \psi) = w$, where

$$w = (i_E)_\wedge u + (i_F)_\wedge v.$$

It follows that

$$w^{p+q} = \sum_{k+l=p+q} ((i_E)_\wedge u)^k \wedge ((i_F)_\wedge v)^l = \sum_{k+l=p+q} u^k \otimes v^l.$$

Since $u^k = 0$ for $k > p$ and $v^l = 0$ for $l > q$, this formula reduces to

$$w^{p+q} = u^p \otimes v^q.$$

It follows that (see Section 5.15)

$$\begin{aligned}\mathrm{Pf}_{a\otimes b}(\varphi \oplus \psi) &= (w^{p+q}, a \otimes b)\\ &= (u^p \otimes v^q, a\otimes b) = (u^p, a)\cdot(v^q, b)\\ &= \mathrm{Pf}_a(\varphi)\cdot \mathrm{Pf}_b(\psi).\end{aligned}$$

□

8.6. Euclidean Spaces

Let E be an oriented Euclidean space of dimension $n = 2m$ and let e be the unique unit vector in $\wedge^n E$ which represents the orientation. Then we set

$$\mathrm{Pf}_E(\varphi_1, \ldots, \varphi_m) = (\Psi_E(\varphi_1) \wedge \cdots \wedge \Psi_E(\varphi_m), e) \qquad \varphi_\nu \in \mathrm{Sk}(E)$$

and

$$\mathrm{Pf}_E(\varphi) = \frac{1}{m!}\mathrm{Pf}_E(\varphi, \ldots, \varphi) \qquad \varphi \in \mathrm{Sk}(E).$$

Observe that if the orientation of E is reversed then Pf_E is changed to $-\mathrm{Pf}_E$.

Proposition 8.4.1 shows that the Pfaffian is invariant under proper rotations. Proposition 8.4.2 implies that

$$\mathrm{Pf}_E(\varphi)^2 = \det \varphi \qquad \varphi \in \mathrm{Sk}(E).$$

Finally, by Proposition 8.5.1,

$$\mathrm{Pf}_{E\oplus F}(\varphi \oplus \psi) = \mathrm{Pf}_E(\varphi)\cdot \mathrm{Pf}_E(\psi) \qquad \varphi, \psi \in \mathrm{Sk}(E).$$

EXAMPLE 1. Choose an orthonormal basis $\{e_1, \ldots, e_n\}$ of E and define φ by

$$\varphi e_{2\mu-1} = \lambda_\mu e_{2\mu} \qquad \lambda_\mu \in \Gamma$$

$$\varphi e_{2\mu} = -\lambda_\mu e_{2\mu-1} \qquad (\mu = 1, \ldots, m).$$

Then, when $n = 2$, $\Psi_E(\varphi) = \lambda_1(e_1 \wedge e_2)$, and so

$$\mathrm{Pf}_E(\varphi) = \lambda_1 \cdots \lambda_m$$

as follows from Proposition 8.5.1.

EXAMPLE 2. Let E be an n-dimensional complex vector space with a Hermitian inner product $(\,,\,)_{\mathrm{H}}$ and let $E_{\mathbb{R}}$ denote the underlying real vector space. Give $E_{\mathbb{R}}$ the natural orientation (see Section 7.9) and the positive definite inner product

$$(x, y) = \mathrm{Re}(x, y)_{\mathrm{H}} \qquad x, y \in E_{\mathbb{R}}.$$

Let $i_{\mathbb{R}}$ be the skew transformation of $E_{\mathbb{R}}$ given by

$$i_{\mathbb{R}}(x) = i \cdot x \qquad x \in E_{\mathbb{R}}.$$

Then

$$\mathrm{Pf}(i_{\mathbb{R}}) = 1$$

as follows from Example 1.

Skew-Hermitian Transformations

In Section 8.7 we shall derive an analogue of Theorem 8.3.1 for skew-Hermitian maps.

8.7. The Isomorphisms σ and τ

Let E be an n-dimensional complex vector space and let E^* be the complex dual space. Define an inner product in $E^* \oplus E$ by setting

$$\langle\!\langle x^* \oplus x, y^* \oplus y \rangle\!\rangle = \langle x^*, y\rangle + \langle y^*, x\rangle. \tag{8.15}$$

Extend this inner product to the (complex) exterior algebra $\wedge(E^* \oplus E)$. This inner product determines via the canonical isomorphism

$$\wedge(E^* \oplus E) \xrightarrow{\cong} \wedge E^* \mathbin{\hat{\otimes}} \wedge E$$

(see Section 5.15) an inner product in $\wedge E^* \mathbin{\hat{\otimes}} \wedge E$. A simple calculation shows that (see Formula (6.34))

$$\langle\!\langle u^* \otimes u, v^* \otimes v \rangle\!\rangle = (-1)^p \langle u^*, v\rangle \langle v^*, u\rangle \qquad u^*, v^* \in \wedge^p E^*, u, v \in \wedge^p E.$$

Next, assume that E is equipped with a Hermitian inner product $(\,,\,)_{\mathrm{H}}$. Then an inner product is defined in the underlying real vector space $E_{\mathbb{R}}$ by

$$(x, y) = \mathrm{Re}(x, y)_{\mathrm{H}} \qquad x, y \in E_{\mathbb{R}}.$$

The multiplication by i in E determines a skew linear transformation in $E_{\mathbb{R}}$ which will be denoted by $i_{\mathbb{R}}$ (see Example 2, Section 8.6)

$$i_{\mathbb{R}}(x) = i \cdot x.$$

Recall from Section 8.2 the inverse isomorphisms

$$\Phi_E\colon \wedge^2 E_{\mathbb{R}} \xrightarrow{\cong} \mathrm{Sk}E_{\mathbb{R}} \quad \text{and} \quad \Psi_E\colon \mathrm{Sk}(E_{\mathbb{R}}) \to \wedge^2 E_{\mathbb{R}}$$

and set

$$J = \Psi_E(i_{\mathbb{R}}).$$

Then $J \in \wedge^2 E_{\mathbb{R}}$.

Next define a (real) linear isomorphism $\sigma\colon E_{\mathbb{R}} \to E^*$ by

$$\langle \sigma x, y \rangle = (x, y) - i(x, i_{\mathbb{R}} y) \qquad x \in E_{\mathbb{R}},\ y \in E.$$

Since $i_{\mathbb{R}}$ is skew and an isometry it follows that

$$\sigma i_{\mathbb{R}}(x) = -i\sigma(x) \qquad x \in E_{\mathbb{R}}. \tag{8.16}$$

The map σ determines a (real) linear map $\tau\colon E_{\mathbb{R}} \to E^* \oplus E$ by

$$\tau(x) = i\sigma(x) \oplus x \qquad x \in E_{\mathbb{R}}.$$

Formula (8.15) implies that

$$\begin{aligned} \langle\!\langle \tau x, \tau y \rangle\!\rangle &= \langle i\sigma x, y \rangle + \langle i\sigma y, x \rangle \\ &= i(x, y) + (x, i_{\mathbb{R}} y) + i(y, x) + (y, i_{\mathbb{R}} x) \\ &= 2i(x, y) + (x, i_{\mathbb{R}} y) + (y, i_{\mathbb{R}} x) = 2i(x, y). \end{aligned}$$

Thus τ satisfies

$$\langle\!\langle \tau x, \tau y \rangle\!\rangle = 2i(x, y) \qquad x, y \in E_{\mathbb{R}}. \tag{8.17}$$

Next observe that $\tau x \wedge_{\mathbb{C}} \tau x = 0$, $x \in E_{\mathbb{R}}$, where the symbol $\wedge_{\mathbb{C}}$ indicates that the multiplication is taken in the *complex* algebra $\wedge(E^* \oplus E)$, and so τ extends to a homomorphism (see Section 5.5)

$$\tau_\wedge\colon \wedge E_{\mathbb{R}} \to \wedge(E^* \oplus E).$$

Relation (8.17) implies that

$$\langle\!\langle \tau_\wedge u, \tau_\wedge v \rangle\!\rangle = (2i)^p (u, v) \qquad u, v \in \wedge^p E_{\mathbb{R}}. \tag{8.18}$$

Lemma I. *Consider the direct decomposition*

$$\wedge^2(E^* \oplus E) = [\wedge^2 E^* \otimes 1] \oplus [E^* \otimes E] \oplus [1 \otimes \wedge^2 E]$$

and write

$$\tau_\wedge u = \tau_{\mathbf{L}} u \otimes 1 + \tau_{\mathbf{M}} u + 1 \otimes \tau_{\mathbf{R}} u \qquad u \in \wedge^2 E_{\mathbb{R}},$$

where $\tau_L u \in \wedge^2 E^*$, $\tau_M u \in E^* \otimes E$, $\tau_R u \in \wedge^2 E$. *Then the following relations hold*:

$$\tau_L \Psi_E[i_{\mathbb{R}}, \varphi] = -2i\tau_L \Psi_E(\varphi) \qquad \varphi \in \mathrm{Sk}(E_{\mathbb{R}})$$

$$\tau_R \Psi_E[i_{\mathbb{R}}, \varphi] = 2i\tau_R \Psi_E(\varphi) \qquad \varphi \in \mathrm{Sk}(E_{\mathbb{R}}).$$

Here $[\,,\,]$ *denotes the commutator and* $\Psi_E\colon \mathrm{Sk}(E_{\mathbb{R}}) \to \wedge^2 E_{\mathbb{R}}$.

PROOF. It follows from the definition of τ that

$$\tau_L(x \wedge y) = i\sigma x \wedge_{\mathbb{C}} i\sigma y = -\sigma x \wedge_{\mathbb{C}} \sigma y.$$

This equation yields, in view of (8.16),

$$\begin{aligned} \tau_L(i_{\mathbb{R}} x \wedge y + x \wedge i_{\mathbb{R}} y) &= -\sigma i_{\mathbb{R}}(x) \wedge_{\mathbb{C}} \sigma y - \sigma x \wedge_{\mathbb{C}} \sigma i_{\mathbb{R}} y \\ &= i\sigma x \wedge_{\mathbb{C}} \sigma y + \sigma x \wedge_{\mathbb{C}} i\sigma y \\ &= 2i\sigma x \wedge_{\mathbb{C}} \sigma y = -2i\tau_L(x \wedge y). \end{aligned} \tag{8.19}$$

Next observe that the isomorphism Φ_E satisfies

$$[\varphi, \Phi_E(x \wedge y)] = \Phi_E(\varphi x \wedge y + x \wedge \varphi y) \qquad \varphi \in \mathrm{Sk}(E_{\mathbb{R}}).$$

Applying this relation with $\varphi = i_{\mathbb{R}}$ we obtain

$$[i_{\mathbb{R}}, \Phi_E(x \wedge y)] = \Phi_E(i_{\mathbb{R}} x \wedge y + x \wedge i_{\mathbb{R}} y)$$

whence

$$\Psi_E[i_{\mathbb{R}}, \Phi_E(x \wedge y)] = i_{\mathbb{R}} x \wedge y + x \wedge i_{\mathbb{R}} y. \tag{8.20}$$

Formulas (8.19) and (8.20) imply that

$$\tau_L \Psi_E[i_{\mathbb{R}}, \Phi_E(x \wedge y)] = -2i\tau_L(x \wedge y)$$

whence

$$\tau_L \Psi_E[i_{\mathbb{R}}, \Phi_E u] = -2i\tau_L(u) \qquad u \in \wedge^2 E_{\mathbb{R}}.$$

Since $\Phi_E\colon \wedge^2 E_{\mathbb{R}} \to \mathrm{Sk}(E_{\mathbb{R}})$ is an isomorphism, the first formula in the lemma follows. The second formula is proved in the same way.

Corollary. *If* φ *is a skew-Hermitian transformation of E, then*

$$\tau_L \Psi_E(\varphi) = 0 \tag{8.21}$$

and

$$\tau_R \Psi_E(\varphi) = 0. \tag{8.22}$$

PROOF. Apply the lemma observing that $[i_{\mathbb{R}}, \varphi] = 0$.

Lemma II. *Let* φ *be a skew linear transformation of* $E_{\mathbb{R}}$ *and consider the complex linear transformation* $i_{\mathbb{R}} \circ \varphi + \varphi \circ i_{\mathbb{R}}$. *Then*

$$\tau_M \Psi_E(\varphi) = T_E^{-1}(i_{\mathbb{R}} \circ \varphi + \varphi \circ i_{\mathbb{R}}).$$

PROOF. It follows from the definition of τ that

$$\tau_M(x \wedge y) = i\sigma x \otimes_{\mathbb{C}} y - i\sigma y \otimes_{\mathbb{C}} x.$$

This yields, in view of (8.16),

$$\begin{aligned} T_E\tau_M(x \wedge y)(z) &= -\langle i\sigma y, z\rangle x + \langle i\sigma x, z\rangle y = -i\langle \sigma y, z\rangle x + i\langle \sigma x, z\rangle y \\ &= -i\{(y, z) - i(y, i_{\mathbb{R}} z)\}x + i\{(x, z) - i(x, i_{\mathbb{R}} z)\}y \\ &= i\{(x, z)y - (y, z)x\} + \{(x, i_{\mathbb{R}} z)y - (y, i_{\mathbb{R}} z)x\} \\ &= i_{\mathbb{R}}\Phi_E(x \wedge y)z + \Phi_E(x \wedge y)i_{\mathbb{R}} z. \end{aligned}$$

Thus we have the relation

$$T_E\tau_M(x \wedge y) = i_{\mathbb{R}}\Phi_E(x \wedge y) + \Phi_E(x \wedge y) \circ i_{\mathbb{R}}$$

whence

$$T_E\tau_M(u) = i_{\mathbb{R}} \circ \Phi_E(u) + \Phi_E(u) \circ i_{\mathbb{R}} \qquad u \in \wedge^2 E_{\mathbb{R}}.$$

Setting $\Phi_E(u) = \varphi$ we obtain

$$T_E\tau_M\Psi_E(\varphi) = i_{\mathbb{R}} \circ \varphi + \varphi \circ i_{\mathbb{R}} \qquad \varphi \in \mathrm{Sk}(E_{\mathbb{R}})$$

and so the lemma is proved. □

Corollary. A skew-Hermitian transformation φ of E satisfies the relation

$$T_E\tau_M\Psi_E(\varphi) = 2i\varphi. \tag{8.23}$$

Lemma III. *A skew-Hermitian transformation of E satisfies*

$$\tau_\wedge \Psi_E\varphi = 2iT_E^{-1}(\varphi).$$

In particular,

$$\tau_\wedge J = -2t.$$

PROOF. Apply formulas (8.21)–(8.23) observing that

$$\tau_\wedge u = \tau_L u \otimes 1 + \tau_M u + 1 \otimes \tau_R u \qquad u \in \wedge^2 E_{\mathbb{R}}.$$ □

Theorem 8.7.1. *The characteristic coefficients of a skew-Hermitian transformation are given by*

$$C_p(\varphi) = i^p((\Psi_E\varphi)^p, J^p) \qquad (p = 0, \ldots, n).$$

Thus, if p is even, then $C_p(\varphi)$ is real and if p is odd, then $C_p(\varphi)$ is imaginary.

PROOF. Set $T_E^{-1}(\varphi) = u$. Then we have, in view of formulas (7.10), (6.33), and (6.34)

$$C_p(\varphi) = \langle u^p, t^p\rangle = \langle u^{\hat{p}}, t^{\hat{p}}\rangle = (-1)^p\langle\!\langle u^{\hat{p}}, t^{\hat{p}}\rangle\!\rangle.$$

Now set $\Psi_E\varphi = z$. Then, by Lemma III,

$$\tau_\wedge z = 2iu$$

and

$$\tau_\wedge J = -2t.$$

It follows that

$$C_p(\varphi) = \frac{1}{2^p}\frac{1}{(2i)^p}\langle\!\langle \tau_\wedge z^{\hat{p}}, \tau_\wedge J^{\hat{p}}\rangle\!\rangle.$$

Now using relation (8.18) and (6.33), we obtain

$$C_p(\varphi) = \frac{1}{2^p}\frac{1}{(2i)^p}(2i)^{2p}(z^p, J^p) = i^p(z^p, J^p) = i^p((\Psi_E\varphi)^p, J^p). \qquad \square$$

Corollary. *Let $E_{\mathbb{R}}$ have the natural orientation (see Section 7.9). Then the Pfaffian of $\varphi_{\mathbb{R}}$ is given by*

$$\mathrm{Pf}\varphi_{\mathbb{R}} = \frac{1}{i^n}\det\varphi.$$

PROOF. Applying Theorem 8.7.1 with $p = n$ we obtain

$$\det\varphi = i^n(J^n, (\Psi_E\varphi)^n).$$

Now write

$$J^n = \lambda \cdot e,$$

where e is the positive unit vector in $\wedge^{2n}E_{\mathbb{R}}$. Since $J = \Psi_E(i_{\mathbb{R}})$, it follows that

$$\lambda = \mathrm{Pf}(i_{\mathbb{R}}) = 1$$

(see Example 2, Section 8.6). Thus,

$$\det\varphi = i^n(e, (\Psi_E\varphi)^n) = i^n\mathrm{Pf}(\varphi_{\mathbb{R}}). \qquad \square$$

Symmetric Tensor Algebra 9

In this chapter E denotes a vector space over a field of characteristic zero.

Symmetric Tensor Algebra

The results of this chapter are, in most cases, isomorphic to analogous results in Chapter 5. Consequently, most proofs are omitted or presented in a highly abbreviated form. Modulo occasional changes in terminology the reader should be able to read the proofs in Chapter 5 as substitutes for the proofs omitted here.

9.1. Symmetric Mappings

Let E and F be vector spaces and let

$$\varphi : \underbrace{E \times \cdots \times E}_{p} \to F$$

be a p-linear mapping. Then φ is called *symmetric* if $\varphi = \sigma\varphi$ for every permutation σ (see Section 5.1). Since every permutation is a product of transpositions, it follows, that a mapping φ is symmetric if and only if $\varphi = \tau\varphi$ for every transposition τ. Every p-linear mapping φ determines a symmetric p-linear mapping $S\varphi$ by

$$S\varphi = \frac{1}{p!} \sum_{\sigma} \sigma\varphi.$$

$S\varphi$ is called the *symmetric part* of φ and S is called the symmetry operator. If φ is symmetric, we have $S\varphi = \varphi$.

Proposition 9.1.1. *Let φ be a p-linear mapping of $E \times \cdots \times E$ into F and $f: \otimes^p E \to F$ be the induced linear map. Then φ is symmetric if and only if $M^p(E) \subset \ker f$ (cf. Section 4.9).*

PROOF. If φ is symmetric, we have, for each transposition τ,

$$f(x_1 \otimes \cdots \otimes x_p - \tau^{-1}(x_1 \otimes \cdots \otimes x_p)) = \varphi(x_1, \ldots, x_p) - \tau\varphi(x_1, \ldots, x_p) = 0.$$

Since the products $x_1 \otimes \cdots \otimes x_p$ generate $\otimes^p E$, it follows that f reduces to zero in $M^p(E)$. Conversely, if $M^p(E) \subset \ker f$, it follows that

$$(\varphi - \tau\varphi)(x_1, \ldots, x_p) = f[(x_1 \otimes \cdots \otimes x_p) - (\tau^{-1}(x_1 \otimes \cdots \otimes x_p)] = 0$$

for every transposition τ and hence φ is symmetric.

9.2. The Universal Property

Let

$$\vee^p: \underbrace{E \times \cdots \times E}_{p} \to S$$

be a symmetric p-linear mapping from E into a vector space S. We shall say that $\vee^p$ has the *universal property* (with respect to symmetric maps) if it satisfies the following conditions:

$\vee_1$: The vectors $\vee^p(x_1, \ldots, x_p)$, $x_\nu \in E$, generate S.
$\vee_2$: If

$$\varphi: \underbrace{E \times \cdots \times E}_{p} \to H$$

is any symmetric p-linear mapping, then there exists a linear map $f: S \to H$ such that the diagram

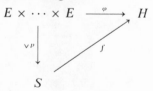

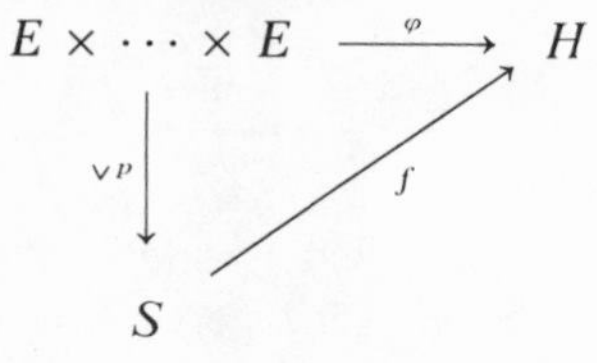

commutes.

Conditions $\vee_1$ and $\vee_2$ are equivalent to the following condition

$\vee$: If

$$\varphi: \underbrace{E \times \cdots \times E}_{p} \to H$$

is a symmetric p-linear mapping, then there is a *unique* linear map $f: S \to H$ which makes the diagram above commute.

Suppose now that

$$\vee^p:\underbrace{E\times\cdots\times E}_{p}\to S \quad\text{and}\quad \tilde{\vee}^p:\underbrace{E\times\cdots\times E}_{p}\to \tilde{S}$$

are symmetric maps with the universal property. Then there is a linear isomorphism $f:S\xrightarrow{\cong}\tilde{S}$ such that

$$f\circ\vee^p=\tilde{\vee}^p.$$

This is shown in exactly the same way as in the skew-symmetric case (cf. Section 5.3).

To establish existence, set

$$\vee^p E=\otimes^p E/M^p(E)$$

(where $M^p(E)$, the subspace of $\otimes^p E$, is defined as in Section 4.9) and define $\vee^p$ by

$$\vee^p(x_1,\ldots,x_p)=\pi(x_1\otimes\cdots\otimes x_p),$$

where π denotes the projection.

Definition. The pth *symmetric power of* E is a pair $(S,\ \vee^p)$, where

$$\vee^p:\underbrace{E\times\cdots\times E}_{p}\to S$$

is a symmetric p-linear mapping with the universal property. The space S is also called the pth *symmetric power of* E and is denoted by $\vee^p E$.

9.3. Symmetric Algebra

Consider the direct sum

$$\vee E=\bigoplus_{p=0}^{\infty}\vee^p E$$

(where $\vee^0 E=\Gamma$ and $\vee^1 E=E$) and identify each $\vee^p E$ with its image under the canonical injection i_p: $\vee^p E\to\vee E$. We thus obtain

$$\vee E=\sum_{p=0}^{\infty}\vee^p E.$$

Assigning to the elements of $\vee^p E$ the degree p, we make $\vee E$ into a graded vector space.

As in Section 5.4 on exterior algebra we construct the homogeneous linear isomorphism

$$f:\otimes E/M(E)\to\vee E$$

such that

$$f\pi(x_1 \otimes \cdots \otimes x_p) = x_1 \vee \cdots \vee x_p$$

and use it to induce a multiplication in $\vee E$. Then $\vee E$ becomes an associative commutative graded algebra with the scalar 1 as unit element. It follows from the definition of the product that

$$(x_1 \vee \cdots \vee x_p)(x_{p+1} \vee \cdots \vee x_{p+q}) = x_1 \vee \cdots \vee x_{p+q}.$$

Hence we shall denote the product of two elements u and v by $u \vee v$. Then the above formula reads

$$(x_1 \vee \cdots \vee x_p) \vee (x_{p+1} \vee \cdots \vee x_{p+q}) = x_1 \vee \cdots \vee x_{p+q}. \tag{9.1}$$

The graded algebra $\vee E$ is called the *symmetric algebra over E*. It is clear that the vectors of E together with the scalar 1 form a system of generators for the algebra $\vee E$.

As in Section 5.4 on exterior algebra, we define the kth power of an element $u \in \vee E$ by

$$u^k = \frac{1}{k!}\underbrace{u \vee \cdots \vee u}_{k} \qquad k \geq 1 \tag{9.2}$$

$$u^0 = 1.$$

Then we have the binomial formula

$$(u + v)^k = \sum_{i+j=k} u^i \vee v^j \qquad u, v \in \vee E. \tag{9.3}$$

The algebra $\vee E$ has the following *universal symmetric algebra property*: Let A be an associative algebra with unit element e and $\varphi : E \to A$ be a *linear* map such that

$$\varphi x \cdot \varphi y = \varphi y \cdot \varphi x.$$

Then there exists a unique homomorphism $h : \vee E \to A$ such that $h(1) = e$ and $h \circ i = \varphi$, where i denotes the injection $E \to \vee E$.

Moreover, if U is any associative algebra with unit element and $\varepsilon : E \to U$ is a linear map such that the pair (U, ε) satisfies the universal property above, then U is the symmetric algebra over E.

9.4. Symmetric Algebras Over Dual Spaces

Let $\vee E$, $\vee E^*$ be symmetric algebras over a pair of dual spaces E, E^*, and consider the induced isomorphisms

$$f : \otimes E/M(E) \xrightarrow{\cong} \vee E, \qquad g : \otimes E^*/M(E^*) \xrightarrow{\cong} \vee E^*.$$

It follows from Section 4.16 that f and g induce a scalar product $\langle\,,\rangle$ in $\vee E$ and $\vee E^*$ such that

$$\begin{aligned}\langle x^{*1} \vee \cdots \vee x^{*p}, x_1 \vee \cdots \vee x_p\rangle &= \operatorname{perm}(\langle x^{*i}, x_j\rangle) \qquad p \geq 1\\ \langle \lambda, \mu\rangle &= \lambda\mu \\ \langle \vee^p E^*, \vee^q E\rangle &= 0 \quad \text{if } p \neq q.\end{aligned} \tag{9.4}$$

From $\vee_1$ we obtain that $\langle\,,\rangle$ is uniquely determined by (9.4). It follows from (9.4) that the restriction of $\langle\,,\rangle$ to the pair $\vee^p E^*$, $\vee^p E$ is nondegenerate for each p, and so induces duality between these spaces. In particular, the restriction of $\langle\,,\rangle$ to $\vee^1 E^* = E^*$ and $\vee^1 E = E$ is just the original scalar product.

9.5. Homomorphisms and Derivations

Suppose that $\varphi: E \to F$ is a linear map. Then φ can be extended in a unique way to a homomorphism $\varphi_\vee: \vee E \to \vee F$ such that $\varphi_\vee(1) = 1$. The homomorphism $\varphi_\vee$ is given by

$$\varphi_\vee(x_1 \vee \cdots \vee x_p) = \varphi x_1 \vee \cdots \vee \varphi x_p \qquad x_i \in E$$

and is homogeneous of degree zero. If $\psi: F \to G$ is a linear map into a third vector space G, then we have

$$(\psi \circ \varphi)_\vee = \psi_\vee \circ \varphi_\vee \tag{9.5}$$

and the identity map of E induces the identity in $\vee E$,

$$\iota_\vee = \iota. \tag{9.6}$$

It follows from (9.5) and (9.6) that if φ is injective (surjective) then $\varphi_\vee$ is also injective (surjective). The fact that $\varphi_\vee$ preserves products can be expressed by the relation (see Section 9.6 for $\mu(a)$)

$$\varphi_\vee \circ \mu(a) = \mu(\varphi_\vee a) \circ \varphi_\vee \qquad a \in \vee E. \tag{9.7}$$

Suppose now that $\varphi^*: E^* \leftarrow F^*$ is a linear map dual to φ. Then the induced homomorphism $(\varphi^*)_\vee: \vee E^* \leftarrow \vee F^*$ is dual to $\varphi_\vee$,

$$(\varphi^*)_\vee = (\varphi_\vee)^*.$$

The homomorphism $(\varphi^*)_\vee$ will be denoted by $\varphi^\vee$.

If $\psi^*: F^* \leftarrow G^*$ is a linear map dual to ψ we have the composition formula

$$(\psi \circ \varphi)^\vee = \varphi^\vee \circ \psi^\vee.$$

Now let φ be a linear map of E into itself. Then φ extends in a unique way to a derivation $\theta_\vee(\varphi)$ in the algebra $\vee E$. The derivation $\theta_\vee(\varphi)$ is given by

$$\theta_\vee(\varphi)(x_1 \vee \cdots \vee x_p) = \sum_{j=1}^{p} x_1 \vee \cdots \vee \varphi x_j \vee \cdots \vee x_p$$

and is homogeneous of degree zero. The derivation property of $\theta_\vee(\varphi)$ can be expressed by the formula

$$\theta_\vee(\varphi) \circ \mu(a) = \mu(\theta_\vee(\varphi)a) + \mu(a) \circ \theta_\vee(\varphi) \qquad a \in \vee E. \tag{9.8}$$

If $\psi: E \to E$ is a second linear map, we have the composition formula

$$\theta_\vee(\varphi \circ \psi - \psi \circ \varphi) = \theta_\vee(\varphi) \circ \theta_\vee(\psi) - \theta_\vee(\psi) \circ \theta_\vee(\varphi).$$

If $\varphi^*: E^* \leftarrow E^*$ is dual to φ, then the induced derivation $\theta_\vee(\varphi^*)$ of the algebra $\vee E^*$ will be denoted by $\theta^\vee(\varphi)$. The linear maps $\theta_\vee(\varphi)$ and $\theta^\vee(\varphi)$ are dual,

$$\theta^\vee(\varphi) = \theta_\vee(\varphi)^*.$$

If $\psi^*: E^* \leftarrow E^*$ is dual to ψ, we have the composition formula

$$\theta^\vee(\varphi \circ \psi - \psi \circ \varphi) = \theta^\vee(\psi) \circ \theta^\vee(\varphi) - \theta^\vee(\varphi) \circ \theta^\vee(\psi).$$

9.6. The Operator $i(a)$

Fix $a \in \vee E$ and consider the linear map $\mu(a): \vee E \to \vee E$ given by

$$\mu(a)u = a \vee u \qquad u \in \vee E.$$

Clearly,

$$\mu(a \vee b) = \mu(a) \circ \mu(b) \qquad a, b \in \vee E. \tag{9.9}$$

Now let

$$i(a): \vee E^* \leftarrow \vee E^*$$

be the dual map. It is determined by the equation

$$\langle i(a)u^*, v\rangle = \langle u^*, a \wedge v\rangle \qquad u^* \in \vee E^*, v \in \vee E.$$

If $a \in \vee^p E$, then $i(a)$ is homogeneous of degree p. In particular,

$$i(a)u^* = \langle u^*, a\rangle \qquad u^* \in \vee^p E$$

and

$$i(a)u^* = 0 \qquad u^* \in \vee^r E, r < p.$$

Dualizing Formulas (9.9) and (9.7), we obtain

$$i(a \vee b) = i(b) \circ i(a) \qquad a, b \in \vee E,$$

and

$$i(a) \circ \varphi^{\vee} = \varphi^{\vee} \circ i(\varphi_{\vee} a) \qquad a \in \vee E, \varphi \in L(E; F).$$

Finally, if $\varphi: E \to E$, $\varphi^*: E^* \leftarrow E^*$ is a pair of dual linear transformations, then, dualizing (9.8), we have

$$i(a) \circ \theta^{\vee}(\varphi) = i(\theta_{\vee}(\varphi)a) + \theta^{\vee}(\varphi) \circ i(a) \qquad a \in \vee E.$$

Now consider the operator $i(h)$, where $h \in E$. Observe that $i(h)$ is homogeneous of degree -1. As in Proporition 5.14.1 one may show that $i(h)$ is a derivation in the algebra $\vee E^*$,

$$i(h)(u^* \vee v^*) = i(h)u^* \vee v^* + u^* \vee i(h)v^*.$$

This yields the *Leibniz formula*

$$i(h)^r(u^* \vee v^*) = \sum_{p+q=r} \binom{r}{p} i(h)^p u^* \vee i(h)^q v^*.$$

Finally, observe that if an element $u^* \in \vee^p E$ $(p \geq 1)$ satisfies the equation $i(h)u^* = 0$, for every $h \in E$, then $u^* = 0$ (cf. Proposition 5.14.2).

9.7. Zero Divisors

In this section it will be shown that the algebra $\vee E^*$ has no zero divisors (of course, the same holds for $\vee E$).

Consider, for each $p \geq 0$, the subspace $I_p \subset \vee^p E^*$ given by

$$I_p = \sum_{\mu \geq p} \vee^{\mu} E^*.$$

Clearly, I_p is an ideal in the algebra $\vee E^*$ and we have the sequence

$$\vee E^* = I_0 \supset I_1 \supset I_2 \supset \cdots.$$

The ideal I_1 is also denoted by $\vee^+ E^*$. Every two ideals

$$I_p = \sum_{\mu \geq p} \vee^{\mu} E^* \quad \text{and} \quad I^p = \sum_{\mu \geq p} \vee^{\mu} E$$

are dual. If $a \in \vee^p E$, then $i(a)$ restricts to an operator from I_q to I_{q-p} for $q \leqq p$.

Lemma. *Let $u^* \in I_p$ be an element such that*

$$i(h)^p u^* = 0 \quad \text{for every } h \in E.$$

Then $u^* = 0$.

PROOF. If $p = 1$, the lemma follows from the remark at the end of Section 9.6. Now assume that the lemma is true for $p - 1$ ($p \geq 2$) and let $u^* \in I_p$, be an element satisfying

$$i(h)^p u^* = 0 \qquad h \in E.$$

Replacing h by $\lambda h + k$ ($\lambda \in \Gamma$) we obtain

$$\sum_{\mu=0}^{p} \binom{p}{\mu} \lambda^\mu i(h)^\mu i(k)^{p-\mu} u^* = 0.$$

Since λ is arbitrary, this yields

$$i(h)^\mu i(k)^{p-\mu} u^* = 0 \qquad (\mu = 0, \ldots, p).$$

In particular,

$$i(h)^{p-1} i(k) u^* = 0 \qquad h, k \in E.$$

Thus, by induction,

$$i(k)u^* = 0 \qquad k \in E.$$

Now applying the lemma for $p = 1$ we obtain $u^* = 0$ and so the induction is closed. □

Proposition 9.7.1. *The algebra $\vee E^*$ has no zero divisors.*

PROOF. Let u^* and v^* be two nonzero elements in $\vee E^*$. Assume first that u^* and v^* are homogeneous of degree p and q respectively and that $p \geq 1$ and $q \geq 1$. In view of the lemma, there exists $h \in E$ such that $i(h)^p u^* \neq 0$. Now consider the elements $i(h)^\mu u^*$ ($\mu = 0, 1, \ldots$). Since $i(h)^0 v = v \neq 0$ and $i(h)^{q+1} v^* = 0$, there is an integer $r \geq 0$ such that

$$i(h)^r v^* \neq 0 \quad \text{while} \quad i(h)^{r+1} v^* = 0.$$

Now the Leibniz formula yields

$$i(h)^{p+r}(u^* \vee v^*) = \binom{p + r}{p} i(h)^p u^* \vee i(h)^r v^*.$$

Since $i(h)^p u^*$ is a nonzero scalar and $i(h)^r v^* \neq 0$, it follows that

$$i(h)^{p+r}(u^* \vee v^*) \neq 0$$

whence $u^* \vee v^* \neq 0$.

In the general case

$$u^* = \sum_{\lambda=0}^{p} u_\lambda^* \qquad u_\lambda^* \in \vee^\lambda E^*, u_p^* \neq 0$$

and

$$v^* = \sum_{\kappa=0}^{q} v_n^* \qquad v_\kappa^* \in \vee^\kappa E^*, v_q^* \neq 0.$$

Then

$$u^* \vee v^* = \sum_{\lambda+\kappa<p+q} u_\lambda^* \vee v_\kappa^* + u_p^* \vee v_q^*.$$

Since $u_p^* \vee v_q^* \neq 0$, it follows that $u^* \vee v^* \neq 0$. This completes the proof of the proposition. □

9.8. Symmetric Algebra Over a Direct Sum

Consider two vector spaces E and F and the direct sum $E \oplus F$. In this section we shall establish an isomorphism between $\vee(E \oplus F)$ and the canonical tensor product $\vee E \otimes \vee F$. Define a linear map

$$f: \vee E \otimes \vee F \to \vee(E \oplus F)$$

by

$$f(u \otimes v) = (i_1)_\vee u \vee (i_2)_\vee v,$$

where i_1 and i_2 are the inclusions. A straightforward calculation shows that f is an algebra homomorphism (cf. Section 5.15).

To show that f is an isomorphism, consider the linear map

$$\eta: E \oplus F \to \vee E \otimes \vee F$$

given by

$$\eta(z) = \pi_1 z \otimes 1 + 1 \otimes \pi_2 z \qquad z \in E \oplus F,$$

where π_1 and π_2 denote the projections. Since

$$\eta(z_1) \cdot \eta(z_2) = \eta(z_2) \cdot \eta(z_1) \qquad z_1, z_2 \in E \oplus F,$$

η extends to an algebra homomorphism

$$\eta: \vee(E \oplus F) \to \vee E \otimes \vee F$$

(see Section 9.3).

It is easy to check that

$$\eta f(x \otimes 1) = x \otimes 1, \qquad \eta f(1 \otimes y) = 1 \otimes y$$

and

$$f\eta(x \otimes y) = x \otimes y \qquad x \in E, y \in F.$$

These relations imply that

$$\eta \circ f = \iota \quad \text{and} \quad f \circ \eta = \iota.$$

Thus f is an isomorphism.

Next, let E^* and F^* be spaces dual to E and F respectively. Define a scalar product between $E^* \oplus F^*$ and $E \oplus F$ in the usual way and consider the induced scalar product between $\vee(E^* \oplus F^*)$ and $\vee(E \oplus F)$. Then we have the relation (cf. Formula (5.51))

$$\langle f(u^* \otimes v^*), f(u \otimes v)\rangle = \langle u^* \otimes v^*, u \otimes v\rangle = \langle u^*, u\rangle\langle v^*, v\rangle.$$

Finally, assume that $F = E$ and let $\Delta: E \to E \oplus E$ be the diagonal map. Then we have the relation (cf. Formula (5.52))

$$\Delta^\vee f(u^* \otimes v^*) = u^* \vee v^* \qquad u^*, v^* \in \vee E^*.$$

9.9. Symmetric Tensor Algebras Over a Graded Space

Let $E = \sum_{i=1} E_i$ be a graded vector space and let the vectors of E_i be homogeneous of degree k_i. Then there exists precisely one gradation in the algebra $\vee E$ such that the injection $i: E \to \vee E$ is homogeneous of degree zero. $\vee E$ together with this gradation is called the *graded symmetric algebra over the graded vector space E*. The subspace of homogeneous elements of degree k is given by

$$(\vee E)_k = \sum_{(p)} (\vee^{p_1} E_1) \otimes \cdots \otimes (\vee^{p_r} E_r),$$

where the sum is extended over all r-tuples $(p_1, \ldots, p_r)$ subject to

$$\sum_{i=1}^{r} p_i k_i = k.$$

9.10. Symmetric Algebra Over a Vector Space of Finite Dimension

Suppose now that E is a vector space of dimension n and that e_α $(\alpha = 1, \ldots, n)$ is a basis of E. Then the products

$$e_{\alpha_1} \vee \cdots \vee e_{\alpha_p} \qquad \alpha_1 \le \alpha_2 \le \cdots \le \alpha_p \tag{9.10}$$

form a basis of $\vee^p E$. In fact, it follows immediately from $\mathbf{V_1}$ and the commutativity of $\vee E$ that the products (9.10) generate $\vee E$. To prove linear independence let E^* be a dual space of E and $e^{*\alpha}$ $(\alpha = 1, \ldots, n)$ be a dual basis. Then Formula (9.4) yields

$$\langle e^{*\beta_1} \vee \cdots \vee e^{*\beta_p}, e_{\alpha_1} \vee \cdots \vee e_{\alpha_p}\rangle = \operatorname{perm}(\langle e^{*\beta_j}, e_{\alpha_i}\rangle) = \operatorname{perm}(\delta_{\alpha_i}^{\beta_j})$$

and thus the products (9.10) are linearly independent.

The above result shows in particular that

$$\dim \vee^p E = \binom{n+p-1}{p} \qquad p \ge 0. \tag{9.11}$$

The basis vectors of $\vee^p E$ can be written in the form

$$\prod_{i=1}^{n} k_i!\, e_1^{k_1} \vee \cdots \vee e_n^{k_n} \qquad k_\nu \geq 0, \sum_{\nu=1}^{n} k_\nu = p.$$

9.11. Poincaré Series

For the Poincaré series of the graded algebra $\vee E$ we obtain, from (9.11)

$$P(t) = \sum_{p=0}^{\infty} \binom{n+p-1}{p} t^p = \sum_{p=0}^{\infty} (-1)^p \binom{-n}{p} t^p = \frac{1}{(1-t)^n},$$

whence

$$P(t) = (1-t)^{-n}.$$

(Here E has dimension n.)

Now suppose that $E = \sum_{i=1}^{r} E_i$ is a positively graded vector space of finite dimension and that the vectors of E_i are homogeneous of degree k_i. Then the Poincaré series of $\vee E_i$ is given by

$$P_i(t) = (1 - t^{k_i})^{-n_i} \qquad n_i = \dim E_i.$$

Hence, the Poincaré series of E is

$$P(t) = (1 - t^{k_1})^{-n_1} \cdots (1 - t^{k_r})^{-n_r}.$$

9.12. Homogeneous Functions

A homogeneous function of degree p in E is a map $h: E \to \Gamma$ which satisfies

$$h(\lambda x) = \lambda^p h(x) \qquad \lambda \in \Gamma.$$

The homogeneous functions of degree p form a vector space $H^p(E)$. The *product* of two homogeneous functions h and k of degree p and q respectively is the homogeneous function $h \cdot k$ of degree $p + q$ given by

$$(h \cdot k)(x) = h(x)k(x) \qquad x \in E.$$

This multiplication makes the direct sum

$$H(E) = \sum_p H^p(E)$$

into a commutative associative algebra. Its unit element is the homogeneous function h_0 of degree zero given by

$$h_0(x) = 1 \qquad x \in E.$$

Now let Φ be a symmetric p-linear function in E. Then a homogeneous function h_Φ of degree p is given by

$$h_\Phi(x) = \Phi(x, \ldots, x).$$

In terms of the substitution operator we can write

$$h_\Phi(x) = i(x)^p\Phi.$$

It follows from the definition (see Section 9.15 for $S^p(E)$) that

$$h_{\lambda\Phi+\mu\Psi} = \lambda h_\Phi + \mu h_\Psi \qquad \Phi, \Psi \in S^p(E), \lambda, \mu \in \Gamma$$

and that

$$h_{\Phi \vee \Psi} = h_\Phi \cdot h_\Psi \qquad \Phi, \Psi \in S^p(E).$$

Thus, the correspondence $\Phi \to h_\Phi$ determines an algebra homomorphism

$$\tau: S^\bullet(E) \to H(E).$$

The map τ is injective. In fact, assume that $\tau\Phi = 0$ for some $\Phi \in S^p(E)$. Then

$$i(x)^p\Phi = 0$$

for every $x \in E$ and so the lemma in Section 9.7 implies that $\Phi = 0$.

On the other hand, τ is not in general surjective. As an example let E be a Euclidean space and define $h \in H^1(E)$ by

$$h(x) = \varepsilon(x)(x, x)^{1/2},$$

where ε is a function in E satisfying

$$\varepsilon(\lambda x) = \begin{cases} \varepsilon(x) & \lambda > 0, \\ -\varepsilon(x) & \lambda < 0. \end{cases}$$

Then h is homogeneous of degree 1 but it is not additive and hence not a linear function. Thus it is not contained in Im τ.

Problems

1. Consider the problems of Chapter 5, and carry them over to symmetric algebra whenever possible.

2. Let $F \subset E$ be a subspace, and define I_F to be the ideal in $\vee E$ generated by the vectors of F. If F_1 is a complementary subspace, prove that

$$\vee E = I_F \oplus \vee F_1$$

and

$$I_F = \vee^+ F \otimes \vee F_1.$$

3. If $\varphi: E \to F$ is a linear map, prove that

$$\operatorname{Im} \varphi_\vee = \vee \operatorname{Im} \varphi \quad \text{and} \quad \ker \varphi_\vee = I_{\ker \varphi}.$$

4. If

$$\varphi: \underbrace{E \times \cdots \times E}_{p} \to F$$

is a simultaneously symmetric and skew-symmetric p-linear mapping ($p \geq 2$), prove that $\varphi = 0$.

Polynomial Algebras

9.13. Polynomial Algebras

A *monomial of degree p* in n variables in a field Γ is a function

$$P: \underbrace{\Gamma \times \cdots \times \Gamma}_{p} \to \Gamma$$

which satisfies

$$P(t_1, \ldots, t_n) = P(1, \ldots, 1)t_1^{k_1} \cdots t_n^{k_n},$$

where $k_1 + \cdots + k_n = p$.

Thus every monomial of degree p can be written in the form

$$P(t_1, \ldots, t_n) = at_1^{k_1} \cdots t_n^{k_n} \qquad \sum k_\nu = p.$$

In particular, a monomial of degree zero is an element of Γ.

The monomial of degree 1 given by

$$P_i(t_1, \ldots, t_n) = t_i \qquad (i = 1, \ldots, n)$$

will be denoted by t_i.

The monomials of degree p generate a vector space with respect to the usual operators. It will be denoted by Γ_n^p.

The product of a monomial P of degree p and a monomial Q of degree q is the monomial of degree $p + q$ defined by

$$(P \cdot Q)(t_1, \ldots, t_n) = P(t_1, \ldots, t_n)Q(t_1, \ldots, t_n).$$

This multiplication makes the direct sum

$$\Gamma_n = \sum_p \Gamma_n^p$$

into a commutative associative algebra called the *polynomial algebra in n variables over* Γ with the scalar 1 as unit element. It is generated by 1 and the monomials t_i ($i = 1, \ldots, n$). The elements of Γ_n^p are called *homogeneous polynomials of degree p.*

9.14.

Now we shall establish an isomorphism between $\vee E$ ($\dim E = n$) and the polynomial algebra in n variables. In fact, fix a basis $\{e_1, \ldots, e_n\}$ of E. Then every element $u \in \vee^p E$ can be uniquely written in the form

$$u = \sum_{(\nu)} c_{\nu_1 \cdots \nu_n} e_1^{\nu_1} \cdots e_n^{\nu_n} \qquad \sum_i \nu_i = p, \; c_{\nu_1 \cdots \nu_p} \in \Gamma$$

(cf. Section 9.10). Thus it determines a homogeneous polynomial P_u of degree p given by

$$P_u(t_1, \ldots, t_n) = \sum_{(\nu)} c_{\nu_1 \cdots \nu_n} t_1^{\nu_1} \cdots t_n^{\nu_n} \qquad \sum_i \nu_i = p.$$

Since,

$$P_{\lambda u + \mu v} = \lambda P_u + \mu P_v \qquad \lambda, \mu \in \Gamma \qquad u, v \in \vee^p E,$$

this defines a linear map

$$\varphi : \vee^p E \to \Gamma_n^p.$$

Conversely, every polynomial

$$P(t_1, \ldots, t_n) = \sum_{(\nu)} c_{\nu_1, \ldots, \nu_n} t_1^{\nu_1} \cdots t_n^{\nu_n},$$

homogeneous of degree p, determines an element $u_P \in \vee^p E$ given by

$$u_P = \sum_{(\nu)} c_{\nu_1, \ldots, \nu_n} e_1^{\nu_1} \cdots e_n^{\nu_n} \qquad \sum_i \nu_i = p$$

and so we have a linear map

$$\psi : \Gamma_n^p \to \vee^p E.$$

It follows from the definition that

$$\psi \circ \varphi = \iota, \qquad \varphi \circ \psi = \iota$$

and so φ and ψ are inverse isomorphisms. Finally, observe that

$$P_{u \vee v} = P_u \cdot P_v \qquad u, v \in \vee E$$

and thus φ is an algebra isomorphism,

$$\varphi : \vee E \xrightarrow{\cong} \Gamma_n.$$

In particular, we have

$$\varphi(e_i) = t_i \qquad (i = 1, \ldots, n).$$

It follows that the polynomial algebra has no zero divisors since $\vee E$ has no zero divisors (see Proposition 9.7.1).

PROBLEMS

1. Prove that $\vee E$ is a principal ideal domain if and only if $\dim E = 1$. *Hint*: See Chapter XII of *Linear Algebra*.
2. Let E be a pseudo-Euclidean space of dimension n (see Section 9.17 of *Linear Algebra* for the definition) and index r. Consider the symmetric tensor algebra $\vee E$ and choose for each $p \geq 1$ a subspace $T^p \subset \vee^p E$ of maximal dimension such that the restriction of the scalar product to T^p is negative definite. Prove that the Poincaré polynomial of the graded space $T = \sum_{p \geq 1} T^p$ is given by

$$P_T(t) = \frac{1}{2(1-t)}\left[\frac{1}{(1-t)^s} - \frac{1}{(1+t)^s}\right] \qquad s = n - r.$$

   Prove an analogous formula for the exterior algebra:

$$P_T(t) = \tfrac{1}{2}(1-t)^r[(1+t)^s - (1-t)^s] \qquad s = n - r.$$

The Algebra of Symmetric Functions

9.15. Symmetric Functions

A p-linear function Φ in E is called *symmetric*, if

$$\sigma\Phi = \Phi \qquad \sigma \in S_p.$$

The symmetric functions form a subspace of $T^p(E)$ (cf. Section 3.18) denoted by $S^p(E)$.

Every p-linear function Φ determines a symmetric function $S\Phi$ by

$$S\Phi = \frac{1}{p!}\sum_\sigma \sigma\Phi$$

called the *symmetric part* of Φ.

If $\Phi \in S^p(E)$, then $S\Phi = \Phi$ and so S is a projection operator. Moreover, S satisfies (cf. Formula (5.81))

$$S(\Phi \otimes \Psi) = S(S\Phi \otimes \Psi) = S(\Phi \otimes S\Psi) \qquad \Phi, \Psi \in T^{\cdot}(E)$$

whence

$$S(\Phi \otimes \Psi) = S(S\Phi \otimes S\Psi).$$

The *symmetric product* of $\Phi \in S^p(E)$ and $\Psi \in S^q(E)$ is defined by

$$\Phi \vee \Psi = \frac{(p+q)!}{p!q!} S(\Phi \otimes \Psi).$$

Explicitly,

$$(\Phi \vee \Psi)(x_1, \ldots, x_{p+q}) = \frac{1}{p!q!} \sum_{\sigma} \Phi(x_{\sigma(1)}, \ldots, x_{\sigma(p)}) \Psi(x_{\sigma(p+1)}, \ldots, x_{\sigma(p+q)}).$$

The symmetric product satisfies the relations

$$\Phi \vee \Psi = \Psi \vee \Phi$$

and

$$(\Phi \vee \Psi) \vee X = \Phi \vee (\Psi \vee X).$$

This multiplication makes the direct sum

$$S^{\cdot}(E) = \sum_{p=0}^{\infty} S^p(E)$$

into a commutative associative algebra, called the algebra of symmetric functions in E.

A linear map $\varphi : E \to F$ induces a homomorphism $\varphi^* : S^{\cdot}(E) \leftarrow S^{\cdot}(F)$ given by

$$(\varphi^*\Psi)(x_1, \ldots, x_p) = \Psi(\varphi x_1, \ldots, \varphi x_p) \qquad \Psi \in S^p(F)$$

and a linear transformation φ of E determines a derivation $\theta^S(\varphi)$ in the algebra $S^{\cdot}(E)$. It is defined by

$$(\theta^S(\varphi)\Phi)(x_1, \ldots, x_p) = \sum_{\nu=1}^{p} \Phi(x_1, \ldots, \varphi x_\nu, \ldots, x_p).$$

This is shown in the same way as for the algebra $A^{\cdot}(E)$ in Section 5.31.

9.16. The Operator $i_S(h)$

Let $i_S(h) : S^{\cdot}(E) \to S^{\cdot}(E)$ also denote the restriction of the operator $i_S(h)$ defined in Section 3.19 to $S^{\cdot}(E)$. Thus,

$$(i_S(h)\Phi)(x_1, \ldots, x_{p-1}) = \Phi(h, x_1, \ldots, x_{p-1}) \qquad \Phi \in S^p(E).$$

$i_S(h)$ is called the *substitution operator* in the algebra $S^{\cdot}(E)$.

In exactly the same way as in Section 5.32 it is shown that

$$i_S(h) \circ S = S \circ i_S(h)$$

(see the lemma in Section 5.32) and consequently,

$$i_S(h)(\Phi \vee \Psi) = i_S(h)\Phi \vee \Psi + \Phi \vee i_S(h)\Psi$$

(see Proposition 5.32.1). Thus, $i_S(h)$ is a derivation in the algebra $S^{\cdot}(E)$.

Note that the operator $i_S(h)$ is dual to the multiplication operator $\mu_S(h)$ in $S^{\cdot}(E)$.

Finally, we have the commutative diagram (see Section 3.20)

$$\begin{array}{ccc} \otimes^p E^* & \xrightarrow[\cong]{\alpha} & T^p(E) \\ \Big\downarrow{\scriptstyle \pi_S} & & \Big\downarrow{\scriptstyle S} \\ \otimes^p E^* & \xrightarrow[\cong]{\alpha} & T^p(E) \end{array}$$

where π_S denotes the symmetrizer (see Section 4.15). It implies that α restricts to a linear isomorphism $\vee^p E^* \xrightarrow{\cong} S^p(E)$ for each p and in fact to an algebra isomorphism $Y(E^*) \xrightarrow{\cong} S(E)$.

Since $\vee^p E^* \cong Y^p(E)$ (see Section 4.15), it follows that $\vee^p E^* \cong S^p(E)$. In particular,

$$\dim S^p(E) = \binom{n+p-1}{p} \qquad (p = 0, 1, \ldots).$$

9.17. The Algebra $S_\bullet(E)$

In exactly the same way we obtain the algebra

$$S_\bullet(E) = \sum_{p=0}^{\infty} S_p(E),$$

where $S_p(E)$ denotes the space of symmetric p-linear functions in E^*. The scalar product between $T^p(E)$ and $T_p(E)$ determines a scalar product between $S^p(E)$ and $S_p(E)$ given by

$$\langle \Phi, \Psi \rangle_S = \frac{1}{p!}\langle \Phi, \Psi \rangle \qquad \Phi \in S^p(E), \Psi \in S_p(E).$$

It follows from the definition that (cf. Section 5.34)

$$\langle \Phi, x_1 \vee \cdots \vee x_p \rangle_S = \Phi(x_1, \ldots, x_p) \qquad x_\nu \in E$$

and

$$\langle f_1 \vee \cdots \vee f_p, \Psi \rangle_S = \Psi(f_1, \ldots, f_p) \qquad f_\nu \in E^*.$$

In particular,

$$\langle f_1 \vee \cdots \vee f_p, x_1 \vee \cdots \vee x_p \rangle = \operatorname{perm}(f_i(x_j)).$$

9.18. Homogeneous Functions and Homogeneous Polynomials

Let

$$P(t_1, \ldots, t_n) = \sum_{(k)} c_{k_1,\ldots,k_n} t_1^{k_1} \cdots t_n^{k_n} \qquad \sum_{\nu} k_\nu = p$$

be a polynomial of degree p. It determines a homogeneous function h_P of degree p by

$$h_P(x) = \sum_{(k)} c_{k_1,\ldots,k_n} (\xi^1)^{k_1} \cdots (\xi^n)^{k_n}$$

(here the k_ν are *exponents*!), where $x = \sum_\nu \xi^\nu e_\nu$. This yields a homomorphism

$$\sigma: \Gamma_n \to H(E).$$

On the other hand, the polynomial P determines an element $u_P \in S^p(E)$ given by (cf. Section 9.14)

$$u_P = \sum_{(k)} c_{k_1,\ldots,k_n} (f^1)^{k_1} \cdots (f^n)^{k_n} \qquad \sum_{\nu} k_\nu = p,$$

where $\{f^1, \ldots, f^n\}$ is the dual basis of $\{e_1, \ldots, e_n\}$ (the k_ν are again exponents). Moreover, the correspondence $P \mapsto u_P$ defines an isomorphism

$$\psi: \Gamma_n \xrightarrow{\cong} S^{\bullet}(E)$$

(see Section 9.16).

We shall show that the diagram (see Section 9.12 for τ).

$$\begin{array}{ccc} \Gamma_n & \xrightarrow[\cong]{\psi} & S^{\bullet}(E) \\ & \sigma \searrow \quad \swarrow \tau & \\ & H(E) & \end{array} \qquad (9.12)$$

commutes. Since all maps are algebra homomorphisms, it is sufficient to show that

$$\tau\psi(t_i) = \sigma(t_i) \qquad (i = 1, \ldots, n).$$

But, since $\psi t_i = f^i$, we have $\tau\psi(t_i) = f^i$. On the other hand, $\sigma f^i = \xi^i = f^i$ whence (9.12).

Clifford Algebras 10

In this chapter E denotes a vector space over a field with characteristic zero and $(\ ,\ )$ denotes a (possibly degenerate) symmetric bilinear function in E.

Basic Properties

10.1. The Universal Property

Let A be an associative algebra with unit element e_A. A *Clifford map* from E to A is a linear map φ which satisfies

$$(\varphi x)^2 = (x, x)e_A \qquad x \in E$$

or equivalently,

$$\varphi(x)\varphi(y) + \varphi(y)\varphi(x) = 2(x, y)e_A \qquad x, y \in E.$$

A Clifford algebra over E is an associative algebra C_E with unit element e together with a Clifford map $i_E: E \to C_E$ subject to the following conditions:

C_1: The subspace Im i_E generates the algebra C_E.
C_2: To every Clifford map $\varphi: E \to A$ there exists a homomorphism $f: C_E \to A$ which makes the diagram

$$\begin{array}{ccc} E & \xrightarrow{\varphi} & A \\ {\scriptstyle i_E}\downarrow & \nearrow_{f} & \\ C_E & & \end{array} \qquad (10.1)$$

commutative.

Conditions **C_1** and **C_2** are equivalent to the following condition

C: To every Clifford map $\varphi: E \to A$ there exists a *unique* homomorphism $f: C_E \to A$ such that Diagram (10.1) commutes.

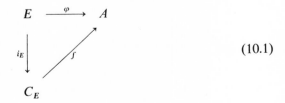

In fact, it is easy to check that $\mathbf{C}_1$ and $\mathbf{C}_2$ imply $\mathbf{C}$. Conversely, assume that i_E satisfies $\mathbf{C}$. Then $\mathbf{C}_2$ follows immediately. To establish $\mathbf{C}_1$ denote by A the subalgebra of C_E generated by Im i_E and by $\tilde{i}_E$ the map i_E considered as a map into A. Then, clearly, $\tilde{i}_E$ is a Clifford map. Hence there is a unique homomorphism $f: C_E \to A$ *such that*

$$f \circ i_E = \tilde{i}_E.$$

On the other hand, if $j: A \to C_E$ is the inclusion map, we have

$$j \circ \tilde{i}_E = i_E.$$

Now consider the map $j \circ f: C_E \to C_E$. Then the relations above imply that

$$(j \circ f) \circ i_E = j \circ (f \circ i_E) = j \circ \tilde{i}_E = i_E.$$

On the other hand,

$$\iota \circ i_E = i_E.$$

Thus the uniqueness part of Condition $\mathbf{C}$ implies that

$$j \circ f = \iota.$$

Hence j is surjective and so $\mathbf{C}_1$ follows.

10.2. Examples

(1) Let $\mathbb{R}$ be the real axis with the negative definite inner product given by

$$(x, y) = -xy \qquad x, y \in \mathbb{R}.$$

We show that $\mathbb{C}$ is a Clifford algebra over $\mathbb{R}$. Let $i_E: \mathbb{R} \to \mathbb{C}$ be the linear map given by

$$i_E(x) = i \cdot x \qquad x \in \mathbb{R}.$$

Then i_E satisfies $\mathbf{C}_1$. To establish $\mathbf{C}_2$, let

$$\varphi: \mathbb{R} \to A$$

be any Clifford map. Since φ is linear,

$$\varphi(x) = x \cdot \varphi(1) \qquad x \in \mathbb{R}.$$

Set $\varphi(1) = a$. Since φ is a Clifford map, it follows that $a^2 = -e_A$. Thus φ is of the form

$$\varphi(x) = x \cdot a, \qquad a^2 = -e_A.$$

Now define $f: \mathbb{C} \to A$ by setting

$$f(x + iy) = x \cdot e_A + y \cdot a \qquad x, y \in \mathbb{R}.$$

Then f is an algebra homomorphism as is easily checked. Moreover,

$$fi_E(y) = f(iy) = y \cdot a = \varphi(y) \qquad y \in \mathbb{R}$$

and so i_E satisfies $\mathbf{C_2}$.

(2) Let $\mathbb{R}^2$ be the plane with a negative definite inner product. We show that the algebra of quaternions (see *Linear Algebra* Section 7.23) is a Clifford algebra over $\mathbb{R}^2$.

Choose an orthonormal basis $\{x_1, x_2\}$ of $\mathbb{R}^2$ and let $\{e, e_1, e_2, e_3\}$ be an orthonormal basis of $\mathbb{H}$. Define a linear map $i_E: \mathbb{R}^2 \to \mathbb{H}$ by setting

$$i_E(x_1) = e_1, \qquad i_E(x_2) = e_2.$$

Then

$$(i_E(x))^2 = -(x, x) \cdot e = (x, x)^- \cdot e \qquad x \in \mathbb{R}^2$$

and so i_E is a Clifford map. It is easily checked that i_E satisfies $\mathbf{C_1}$. To establish $\mathbf{C_2}$, let $\varphi: \mathbb{R}^2 \to A$ be a Clifford map and set $\varphi(x_1) = a_1$, $\varphi(x_2) = a_2$. Then φ is of the form

$$\varphi x = \lambda^1 a_1 + \lambda^2 a_2,$$

where

$$x = \lambda^1 x_1 + \lambda^2 x_2$$

and the vectors a_i $(i = 1, 2)$ satisfy the relations

$$a_1^2 = a_2^2 = -e$$

$$a_1 a_2 + a_2 a_1 = 0.$$

Now define $f: \mathbb{H} \to A$ by

$$f(\lambda e + \lambda^1 e_1 + \lambda^2 e_2 + \lambda^3 e_3) = \lambda e_A + \lambda^1 a_1 + \lambda^2 a_2 + \lambda^3 a_1 a_2.$$

Then f is a homomorphism and satisfies

$$f i_E(x) = f i_E(\lambda^1 x_1 + \lambda^2 x_2) = \lambda^1 a_1 + \lambda^2 a_2 = \varphi x \qquad x \in \mathbb{R}^2;$$

i.e.,

$$f \circ i_E = \varphi.$$

10.3. Uniqueness and Existence

In this section we shall show that there is a Clifford algebra over every inner product space, unique up to an isomorphism.

First suppose that C_E, $\tilde{C}_E$ are Clifford algebras over E and let $i_E: E \to C_E$ and $\tilde{i}_E: E \to \tilde{C}_E$ denote the corresponding Clifford maps. Then, by $\mathbf{C_2}$, there are homomorphisms $f: C_E \to \tilde{C}_E$ and $g: \tilde{C}_E \to C_E$ such that

$$\tilde{i}_E = f \circ i_E \quad \text{and} \quad i_E = g \circ \tilde{i}_E.$$

It follows that

$$\tilde{i}_E = (f \circ g) \circ \tilde{i}_E \quad \text{and} \quad i_E = (g \circ f) \circ i_E.$$

Now Condition $\mathbf{C}_1$ implies that

$$f \circ g = \tilde{\iota} \quad \text{and} \quad g \circ f = \iota.$$

Thus f and g are inverse isomorphisms. This shows that a Clifford algebra over E, if it exists, is uniquely determined up to an isomorphism.

To prove existence consider the tensor algebra $\otimes E$ (see Section 3.2) and let J denote the two-sided ideal in $\otimes E$ generated by the elements

$$x \otimes x - (x, x) \cdot 1 \qquad x \in E$$

where 1 is the 1-element of $\otimes^0 E = \Gamma$. Define C_E to be the quotient algebra

$$C_E = \otimes E/J$$

and let

$$\pi : \otimes E \to C_E$$

be the canonical projection. Let $i_E : E \to C_E$ be the linear map

$$i_E = \pi \circ j_E,$$

where $j_E : E \to \otimes E$ denotes the inclusion map. We shall show that (C_E, i_E) is a Clifford algebra over E.

First observe that, for $x \in E$,

$$(i_E x)^2 = (\pi j_E x)^2 = \pi (j_E x)^2 = \pi(x \otimes x) = (x, x) \cdot 1$$

and so i_E is a Clifford map. Since the algebra $\otimes E$ is generated by E and since π is surjective, it follows that i_E satisfies $\mathbf{C}_1$. To establish $\mathbf{C}_2$, let $\varphi : E \to A$ be any Clifford map. In view of Section 3.3, φ extends to a homomorphism

$$h : \otimes E \to A.$$

This homomorphism satisfies

$$h(x \otimes x - (x, x) \cdot 1) = (\varphi x)^2 - (x, x) e_A = 0.$$

It follows that $J \subset \ker h$ and so h factors over π to induce a homomorphism

$$f : C_E \to A.$$

Clearly,

$$f i_E(x) = f \pi j_E(x) = f \pi(x) = h(x) = \varphi(x) \qquad x \in E$$

and so $\mathbf{C}_2$ follows.

Thus to every inner product space E there exists a unique Clifford algebra C_E (up to an isomorphism). It is called *the* Clifford algebra over E.

EXAMPLE. If $(\,,\,) = 0$, then J is generated by the products $x \otimes x$ and so $C_E = \wedge E$. Thus the Clifford algebra is a generalization of the exterior algebra.

10.4. The Injectivity of i_E

Proposition 10.4.1. *The map $i_E : E \to C_E$ is injective.*

PROOF. If $(\,,\,) = 0$, then $C_E = \wedge E$ and the injectivity of i_E follows from Section 5.8. Now assume that $(\,,\,)$ is nondegenerate. Then, if $x_0 \in \ker i_E$, we have for $y \in E$

$$2(x_0, y)e = i_E(x_0) \cdot i_E y + i_E y \cdot i_E(x_0) = 0$$

whence $x_0 = 0$.

In general write $E = E_0 \oplus E_1$, where E_0 is the nullspace of $(\,,\,)$ and E_1 is a subspace such that the restriction $(\,,\,)_1$ of $(\,,\,)$ to E_1 is nondegenerate. Then the Clifford map $i_1 : E_1 \to C_{E_1}$ is injective.

Let $\pi_0 : E \to E_0$ and $\pi_1 : E \to E_1$ be the projections and consider the map

$$\varphi_1 : E \xrightarrow{\pi_1} E_1 \xrightarrow{i_1} C_{E_1}.$$

Then

$$(\varphi_1 x)^2 = (i_1 \pi_1 x)^2 = (\pi_1 x, \pi_1 x) \cdot e_1,$$

where e_1 denotes the unit element of C_{E_1}. Now,

$$(\pi_1 x, \pi_1 x) = (x, x) - 2(x, \pi_0 x) + (\pi_0 x, \pi_0 x) = (x, x) \qquad x \in E$$

and so we obtain

$$(\varphi_1 x)^2 = (x, x)e_1 \qquad x \in E. \tag{10.2}$$

Next consider the linear map $\varphi : E \to \wedge E_0 \hat{\otimes} C_{E_1}$ given by

$$\varphi x = \pi_0 x \otimes e_1 + 1 \otimes \varphi_1 x \qquad x \in E.$$

Then we have, in view of (10.2),

$$\begin{aligned}(\varphi x)^2 &= (\pi_0 x)^2 \otimes e_1 + \pi_0 x \otimes \varphi_1 x - \pi_0 x \otimes \varphi_1 x + 1 \otimes (\varphi_1 x)^2 \\ &= (x, x) \cdot (1 \otimes e_1)\end{aligned}$$

and so φ is a Clifford map. Thus, by $\mathbf{C_2}$, there is a homomorphism $f : C_E \to \wedge E_0 \hat{\otimes} C_{E_1}$ such that $\varphi = f \circ i_E$. Since φ is injective, it follows that i_E is injective. □

Henceforth we shall identify E with its image under i_E. Then E becomes a subspace of C_E and we have the relation

$$x \cdot y + y \cdot x = 2(x, y) \cdot e \qquad x, y \in E.$$

In particular,

$$x^2 = (x, x) \cdot e \qquad x \in E.$$

Now properties $\mathbf{C_1}$ and $\mathbf{C_2}$ can be rephrased as follows:

$\mathbf{C_1}$: The algebra C_E is generated by the subspace E.
$\mathbf{C_2}$: Every Clifford map $\varphi: E \to A$ extends to a homomorphism $f: C_E \to A$.

10.5. Homomorphisms

Let F be a second vector space and let $(\ ,\ )$ be a symmetric bilinear function in F. A linear map $\varphi: E \to F$ is called an *isometry*, if

$$(\varphi x, \varphi y) = (x, y) \qquad x, y \in E.$$

Let $\varphi: E \to F$ be an isometry. Then there is a unique homomorphism $\varphi_C: C_E \to C_F$ which makes the diagram

$$\begin{array}{ccc} E & \xrightarrow{\varphi} & F \\ {\scriptstyle i_E}\downarrow & & \downarrow{\scriptstyle i_F} \\ C_E & \xrightarrow[\varphi_C]{} & C_F \end{array} \tag{10.3}$$

commutative. In fact, consider the linear map

$$\varphi_F: E \overset{\varphi}{\to} F \overset{i_F}{\to} C_F.$$

Then

$$(\varphi_F x)^2 = (i_F \varphi x)^2 = (\varphi x, \varphi x) e_F = (x, x) e_F \qquad x \in E,$$

and so φ_F is a Clifford map. Thus it extends to a homomorphism $\varphi_C: C_E \to C_F$.

To prove uniqueness, suppose that $\tilde{\varphi}_C: C_E \to C_F$ is a second homomorphism making Diagram (10.3) commute. Then

$$\varphi_C i_E(x) = \tilde{\varphi}_C i_E(x) \qquad x \in E$$

and so $\mathbf{C_1}$ implies that $\tilde{\varphi}_C = \varphi_C$.

If $\psi: F \to H$ is a second isometry, then clearly,

$$(\psi \circ \varphi)_C = \psi_C \circ \varphi_C.$$

Moreover, the identity map of E induces the identity in C_E,

$$\iota_C = \iota.$$

These properties imply that if φ is an isometric isomorphism, then φ_C is an algebra isomorphism.

10.6. The $\mathbb{Z}_2$-Gradation of C_E

Consider the linear automorphism ω of E given by

$$\omega(x) = -x \qquad x \in E.$$

Since ω is an isometry, it induces a homomorphism

$$\omega_E : C_E \to C_E.$$

Moreover, since $\omega^2 = \iota$, it follows that

$$\omega_E^2 = \iota.$$

Thus ω_E is an involution of the algebra C_E.

Next, consider the subspaces C_E^0 and C_E^1 of C_E consisting of the elements

$$C_E^0 = \ker(\omega_E - \iota)$$

and

$$C_E^1 = \ker(\omega_E + \iota).$$

Since ω_E is an involution, it follows that

$$C_E = C_E^0 \oplus C_E^1.$$

Moreover,

$$C_E^0 \cdot C_E^0 \subset C_E^0$$

$$C_E^0 \cdot C_E^1 \subset C_E^1, \qquad C_E^1 \cdot C_E^0 \subset C_E^1$$

$$C_E^1 \cdot C_E^1 \subset C_E^0.$$

In particular, C_E^0 is a subalgebra of C_E.

The elements of C_E^0 (respectively C_E^1) are called *homogeneous* of even (respectively odd) degree. This defines a $\mathbb{Z}_2$-gradation of C_E. The map ω_E is called the *degree involution of* C_E.

It follows from the definitions that the subspaces C_E^0 and C_E^1 are linearly generated by the products

$$C_E^0: \quad x_1 \cdots x_p \qquad x_i \in E,\ p \text{ even}$$

and

$$C_E^1: \quad x_1 \cdots x_p \qquad x_i \in E,\ p \text{ odd}.$$

A subspace U of C_E which is stable under the degree involution is a *graded* subspace. To prove this, set

$$U_0 = U \cap C_E^0, \qquad U_1 = U \cap C_E^1.$$

Then

$$U = U_0 \oplus U_1.$$

In fact, let $u \in U$ and set

$$u_0 = \tfrac{1}{2}(u + \omega_E u), \qquad u_1 = \tfrac{1}{2}(u - \omega_E u)$$

Then $u_0 \in U_0$, $u_1 \in U_1$, and $u = u_0 + u_1$.

10.7. Direct Decompositions

Let E and F be inner product spaces. Define an inner product in the direct sum $E \oplus F$ by setting

$$(x_1 \oplus y_1, x_2 \oplus y_2) = (x_1, x_2) + (y_1, y_2) \qquad x_1, x_2 \in E, y_1, y_2 \in F.$$

Now consider the Clifford algebras C_E, C_F, and $C_{E \oplus F}$.

Theorem 10.7.1. *The $\mathbb{Z}_2$-graded algebra $C_{E\oplus F}$ is isomorphic to the skew tensor product of the $\mathbb{Z}_2$-graded algebras C_E and C_F,*

$$C_{E\oplus F} \cong C_E \mathbin{\hat{\otimes}} C_F.$$

PROOF. To simplify notation we write $E \oplus F = H$. Let $i: E \to H$ and $j: F \to H$ denote the inclusion maps. Since these maps are isometries they induce homomorphisms

$$i_C: C_E \to C_H \quad \text{and} \quad j_C: C_F \to C_H.$$

Now define a linear map $f: C_E \mathbin{\hat{\otimes}} C_F \to C_H$ by

$$f(a \otimes b) = i_C(a) \cdot j_C(b) \qquad a \in C_E, b \in C_F.$$

We show that f is an algebra homomorphism. Since i_C and j_C are homomorphisms it is sufficient to show that

$$i_C(a) \cdot j_C(b) = (-1)^{\alpha\beta} j_C(b) \cdot i_C(a) \qquad \alpha = \deg a, \beta = \deg b.$$

Moreover, in view of $\mathbf{C}_1$, we may assume that

$$a = x_1 \cdots x_p \qquad x_i \in E$$

and

$$b = y_1 \cdots y_q \qquad y_j \in F.$$

Then

$$i_C(a) \cdot j_C(b) = i_C(x_1) \cdots i_C(x_p) \cdot i_C(y_1) \cdots i_C(y_q).$$

Since every two vectors $i_C(x_i)$ and $j_C(y_q)$ are orthogonal with respect to the inner product in H, it follows that

$$i_C(x_i) \cdot j_C(y_j) + j_C(y_j) \cdot i_C(x_i) = 0.$$

Thus we obtain

$$\begin{aligned} i_C(a) \cdot j_C(b) &= (-1)^{pq} j_C(y_1) \cdots j_C(y_q) i_C(x_1) \cdots i_C(x_p) \\ &= (-1)^{pq} j_C(b) \cdot i_C(a) = (-1)^{\alpha\beta} j_C(b) \cdot i_C(a). \end{aligned}$$

To show that f is an isomorphism we construct an inverse homomorphism. Consider the linear map

$$\eta: E \oplus F \to C_E \hat{\otimes} C_F.$$

given by

$$\eta(x \oplus y) = x \otimes e_F + e_E \otimes y \qquad x \in E,\ y \in F.$$

It satisfies

$$\begin{aligned} (\eta(x \oplus y))^2 &= x^2 \otimes e_F + x \otimes y - x \otimes y + e_E \otimes y^2 \\ &= x^2 \otimes e_F + e_E \otimes y^2 = [(x, x) + (y, y)] e_E \otimes e_F \\ &= (x \oplus y, x \oplus y) e_H. \end{aligned}$$

Thus η is a Clifford map and so it extends to a homomorphism

$$g: C_H \to C_E \hat{\otimes} C_F.$$

It follows from the definitions of f and g that

$$gf(x \otimes e_F) = gi_C(x) = g(x \oplus 0) = \eta(x \oplus 0) = x \otimes e_F \qquad x \in E.$$

Similarly,

$$gf(e_E \oplus y) = e_E \otimes y \qquad y \in F.$$

Finally,

$$gf(e_E \otimes e_F) = e_E \otimes e_F.$$

Since the algebra $C_E \hat{\otimes} C_F$ is generated by the elements $x \otimes e_F$, $e_E \otimes y$, and $e_E \otimes e_F$; this implies that

$$g \circ f = \iota.$$

On the other hand,

$$\begin{aligned} fg(x \oplus y) &= f\eta(x \oplus y) = f(x \otimes e_F) + f(e_E \otimes y) \\ &= i(x) + j(y) = x \oplus y \qquad x \in E,\ y \in F \end{aligned}$$

and

$$fg(e_H) = e_H.$$

These relations imply that

$$f \circ g = \iota.$$

Thus f is an isomorphism. □

EXAMPLE. Suppose that the inner product in E is degenerate and let E_0 denote the null-space. Choose a second subspace F of E such that

$$E = E_0 \oplus F.$$

Then we have, in view of Theorem 10.7.1 and the example in Section 10.3,

$$C_E \cong \wedge E_0 \hat{\otimes} C_F.$$

Proposition 10.7.2. *Let $\varphi: E \to F$ be an isometry. Then*

1. If φ *is injective, so is* φ_C;
2. *If* φ *is surjective, so is* φ_C.

PROOF OF 1. Set $\operatorname{Im} \varphi = F_1$. Then φ determines an isometric isomorphism $\varphi_1: E \xrightarrow{\cong} F_1$. Now write $F = F_1 \oplus F_2$ where F_2 is orthogonal to F_1. Then the diagram

$$\begin{array}{ccccc} C_E & \xrightarrow[\cong]{(\varphi_1)_C} & C_{F_1} & \xrightarrow{i_1} & C_{F_1} \hat{\otimes} C_{F_2} \\ & \searrow^{\varphi_C} & & \swarrow_{f}^{\cong} & \\ & & C_F & & \end{array}$$

commutes as is easily checked (i_1 denotes the obvious inclusion map). The diagram shows that φ_C is injective.

PROOF OF 2. Set $\ker \varphi = E_1$ and write $E = E_1 \oplus E_2$ where E_2 is orthogonal to E_1. Then φ induces an isometric isomorphism $\varphi_2: E_2 \xrightarrow{\cong} F$ and the diagram

$$\begin{array}{ccccc} C_E & \xrightarrow[\cong]{f^{-1}} & C_{E_1} \hat{\otimes} C_{E_2} & \xrightarrow{\pi_2} & C_{E_2} \\ & \searrow^{\varphi_C} & & \swarrow_{(\varphi_2)_C}^{\cong} & \\ & & C_F & & \end{array}$$

commutes, where π_2 is the obvious projection. Since π_2 is surjective, it follows that φ_C is surjective. □

10.8. The Involution S_E

Given a Clifford algebra C_E consider the opposite algebra C_E^{opp} (cf. *Linear Algebra* Section 5.1) and let $\cdot$ denote the multiplication in C_E^{opp}. Then the inclusion

$$j: E \to C_E^{\text{opp}}$$

is a Clifford map and so it extends to a homomorphism

$$S_E: C_E \to C_E^{\text{opp}}.$$

For $x_i \in E$ we have

$$S_E(x_1 \cdots x_p) = x_1 \cdot x_2 \cdot \cdots \cdot x_p = x_p \cdots x_1.$$

Clearly,

$$S_E^2 = \iota$$

and so S_E is an involution. Moreover,

$$S_E x = x \qquad x \in E.$$

Finally, S_E commutes with the degree involution,

$$S_E \circ \omega_E = \omega_E \circ S_E.$$

Next, let $u \in C_E$ and set

$$\bar{u} = S_E \omega_E(u).$$

In particular,

$$\bar{x} = -x \qquad x \in E.$$

The correspondence $u \mapsto \bar{u}$ defines a linear involution of C_E. It satisfies

$$\overline{u \cdot v} = \bar{v} \cdot \bar{u} \qquad u, v \in C_E.$$

In fact,

$$\begin{aligned}\overline{u \cdot v} &= S_E \omega_E(u \cdot v) = S_E(\omega_E(u) \cdot \omega_E(v)) \\ &= S_E(\omega_E(v)) \cdot S_E(\omega_E(u)) = \bar{v} \cdot \bar{u}.\end{aligned}$$

Problem

Show that the map $a \to \bar{a}$ (see Section 10.8) in the cases $C_E = \mathbb{C}$ and $C_E = \mathbb{H}$ coincides with the usual conjugation.

Clifford Algebras Over a Finite-Dimensional Space

In this paragraph E denotes an n-dimensional vector space.

10.9

Proposition 10.9.1. *Let* $\dim E = n$. *Then*

$$\dim C_E = 2^n.$$

Proof. First consider the case $n = 1$. Fix a nonzero vector a in E and let A denote the vector space generated by e and a. Then

$$e \cdot a = a \cdot e = a \quad \text{and} \quad a^2 = (a, a)e.$$

Thus A is an algebra. It is easy to check that the inclusion map $E \to A$ extends to an isomorphism $C_E \xrightarrow{\cong} A$. Thus, $\dim A = 2$.

In the general case choose an orthogonal basis $\{e_i\}$ $(i = 1, \ldots, n)$ of E and denote by E_i the 1-dimensional subspace of E generated by e_i $(i = 1,,,, n)$. Then we have the orthogonal decomposition

$$E = E_1 \oplus \cdots \oplus E_n.$$

Thus, by Theorem 10.7.1 (cf. Section 5.20)

$$C_E = C_{E_1} \hat{\otimes} \cdots \hat{\otimes} C_{E_n}$$

and so

$$\dim C_E = 2^n. \qquad \square$$

Corollary. *If $\{x_i\}$ $(i = 1, \ldots, n)$ is any basis of E, then the 2^n vectors*

$$e, x_i, x_i \cdot x_j \; (i < j), \ldots, x_1 \cdots x_n$$

form a basis of C_E.

PROOF. It follows from the relation $x_i x_j + x_j x_i = 2(x_i, x_j)e$ that the vectors above generate the space C_E. Since, by the proposition, $\dim C_E = 2^n$, they must form a basis of C_E. $\square$

10.10. The Canonical Element e_Δ

Consider the linear map

$$\xi_E \colon \wedge E \to C_E$$

given by

$$\xi_E(x_1 \wedge \cdots \wedge x_p) = \frac{1}{p!} \sum_\sigma \varepsilon_\sigma x_{\sigma(1)} \cdots x_{\sigma(p)} \qquad (1 \le p \le n).$$

We show that ξ_E is a linear isomorphism. In fact, choose an orthogonal basis $\{e_1, \ldots, e_n\}$ of E. Then $e_i \cdot e_j = -e_j \cdot e_i$ $(i \neq j)$ and so

$$\xi_E(e_{i_1} \wedge \cdots \wedge e_{i_p}) = e_{i_1} \cdot \ldots \cdot e_{i_p} \qquad (i_1 < i_2 < \cdots < i_p).$$

Thus ξ_E takes a basis of $\wedge E$ into a basis of C_E and so it is a linear isomorphism.

Now choose a nonzero determinant function Δ in E. Then ξ_E determines an element $e_\Delta \in C_E$ by the equation

$$\xi_E(x_1 \wedge \cdots \wedge x_n) = \Delta(x_1, \ldots, x_n) \cdot e_\Delta \qquad x_\nu \in E. \tag{10.4}$$

e_Δ is called the *canonical element* in C_E (with respect to the determinant function Δ).

Now choose a basis $\{e_1, \ldots, e_n\}$ of E such that $(e_i, e_j) = 0$ $(i \neq j)$ and $\Delta(e_1, \ldots, e_n) = 1$. Then

$$e_\Delta = e_1 \cdots e_n. \tag{10.5}$$

Next observe that the determinant function Δ determines a scalar λ_Δ such that the Lagrange identity

$$\det((x_i, y_j)) = \lambda_\Delta \Delta(x_1, \ldots, x_n)\Delta(y_1, \ldots, y_n) \qquad x_\nu \in E,\ y_\nu \in E, \tag{10.6}$$

holds. Setting $x_\nu = y_\nu = e_\nu$ we obtain

$$(e_1, e_1) \cdots (e_n, e_n) = \lambda_\Delta. \tag{10.7}$$

Relations (10.5) and (10.7) imply that

$$\begin{aligned} e_\Delta^2 &= (-1)^{n(n-1)/2} e_1^2 \cdots e_n^2 \\ &= (-1)^{n(n-1)/2} (e_1, e_1) \cdots (e_n, e_n) \cdot e \\ &= (-1)^{n(n-1)/2} \lambda_\Delta \cdot e. \end{aligned}$$

Thus the square of the canonical element is given by

$$e_\Delta^2 = (-1)^{n(n-1)/2} \lambda_\Delta \cdot e. \tag{10.8}$$

Since $\lambda_\Delta \neq 0$ if and only if the inner product in E is nondegenerate it follows that

1. If the inner product is nondegenerate, then e_Δ is invertible in C_E,
2. If the inner product is degenerate, then $e_\Delta^2 = 0$.

Proposition 10.10.1. *The canonical element e_Δ satisfies the relation*

$$e_\Delta \cdot x = (-1)^{n-1} x \cdot e_\Delta \qquad x \in E.$$

Thus,

$$e_\Delta \cdot u = \omega_E^{n-1}(u) \cdot e_\Delta \qquad u \in C_E.$$

In particular, if n is odd,

$$e_\Delta \cdot u = u \cdot e_\Delta \qquad u \in C_E$$

and if n is even,

$$e_\Delta \cdot u = \omega_E(u) \cdot e_\Delta \qquad u \in C_E.$$

PROOF. Choose an orthogonal basis $\{e_1, \ldots, e_n\}$ of E and write

$$e_\Delta = \lambda \cdot e_1 \cdots e_n \qquad \lambda \in \Gamma.$$

(see Formula (10.5)). Then we have

$$e_\Delta \cdot e_i = \lambda e_1 \cdots e_n \cdot e_i = (-1)^{n-i} \lambda (e_i, e_i) e_1 \cdots \hat{e}_i \cdots e_n$$

and

$$e_i \cdot e_\Delta = \lambda e_i \cdot e_1 \cdots e_n = (-1)^{i-1} \lambda (e_i, e_i) e_1 \cdots \hat{e}_i \cdots e_n$$

whence

$$e_\Delta \cdot e_i = (-1)^{n-1} e_i \cdot e_\Delta.$$

Thus, by linearity,

$$e_\Delta \cdot x = (-1)^{n-1} x \cdot e_\Delta \qquad x \in E. \qquad \square$$

10.11. Center and Anticenter

The *center* of a Clifford algebra C_E, denoted by Z_E, consists of the elements a which satisfy

$$a \cdot u = u \cdot a \qquad u \in C_E.$$

Clearly, Z_E is a subalgebra of C_E. Since C_E is generated by E, it follows that an element $a \in C_E$ is in the center if and only if

$$a \cdot x = x \cdot a \qquad x \in E.$$

Next observe that Z_E is stable under the degree involution. In fact, if $a \in Z_E$, then

$$\begin{aligned}\omega_E(a) \cdot x &= -\omega_E(a) \cdot \omega_E(x) = -\omega_E(a \cdot x) = -\omega_E(x \cdot a) \\ &= -\omega_E(x) \cdot \omega_E(a) = x \cdot \omega_E(a)\end{aligned}$$

and so $\omega_E(a) \in Z_E$. Thus Z_E is a graded subspace of C_E and hence a *graded* subalgebra,

$$Z_E = Z_E^0 \oplus Z_E^1, \qquad Z_E^0 = Z_E \cap C_E^0, \qquad Z_E^1 = C_E^1 \cap Z_E.$$

The *anticenter* of C_E, denoted by AZ_E, consists of the elements a which satisfy

$$a \cdot u = \omega_E(u) \cdot a \qquad u \in C_E.$$

Since C_E is generated by E, an element $a \in C_E$ is in AZ_E if and only if

$$a \cdot x = -x \cdot a \qquad x \in E.$$

As above it follows that the anticenter is stable under the degree involution and hence it is a *graded* subspace of C_E.

Proposition 10.10.1 shows that

1. If n is odd, then $e_\Delta \in Z_E$,
2. If n is even, then $e_\Delta \in AZ_E$.

Next, assume that the inner product in E is nondegenerate and consider the linear map $\varphi_E: C_E \to C_E$ given by

$$\varphi_E(u) = e_\Delta \cdot u \qquad u \in C_E.$$

Since e_Δ is invertible, φ_E is a linear isomorphism.

Now let $u \in Z_E$. Then we have, in view of Proposition 10.10.1,

$$\varphi_E(u) \cdot x = e_\Delta \cdot u \cdot x = e_\Delta \cdot x \cdot u = (-1)^{n-1} x \cdot e_\Delta \cdot u$$
$$= (-1)^{n-1} x \cdot \varphi_E(u) \qquad x \in E.$$

Similarly, if $a \in AZ_E$, then

$$\varphi_E(a) \cdot x = (-1)^n x \cdot \varphi_E(a) \qquad x \in E.$$

These relations show that

1. If n is odd, then φ_E restricts to linear automorphisms of Z_E and AZ_E,
2. If n is even, then φ_E interchanges Z_E and AZ_E.

Lemma I. *Assume the inner product in E is nondegenerate. Then* $(AZ_E)^1 = 0$.

Proof. Let $a \in (AZ_E)^1$. Then

$$a \cdot x = -x \cdot a \qquad x \in E$$

and so

$$a \cdot e_\Delta = (-1)^n e_\Delta \cdot a.$$

On the other hand, Proposition 10.10.1 yields, since $a \in C_E^1$,

$$e_\Delta \cdot a = \omega_E^{n-1}(a) \cdot e_\Delta = (-1)^{n-1} a \cdot e_\Delta.$$

These relations imply that

$$a \cdot e_\Delta = 0.$$

Since the inner product is nondegenerate, e_Δ is invertible and we obtain $a = 0$. □

Lemma II. *Assume that the inner product in E is nondegenerate. Then* $Z_E^0 = (e)$.

Proof. The lemma is trivial for $n = 1$. Assume that it holds for $n - 1$ and let E be an n-dimensional inner product space. Choose a vector $a \in E$ such that $(a, a) \neq 0$ and write

$$E = (a) \oplus F,$$

where F denotes the orthogonal complement of a. Thus we can write

$$u = 1 \otimes v + a \otimes w \qquad u \in C_E^0 \qquad v \in C_F^0, w \in C_F^1$$

and

$$x = 1 \otimes y + a \otimes e_F \qquad x \in E, y \in F.$$

These relations imply that

$$u \cdot x = a \otimes v - a^2 \otimes w + 1 \otimes vy + a \otimes wy$$
$$= a \otimes v - (a, a)1 \otimes w + 1 \otimes vy + a \otimes wy.$$

Similarly,

$$x \cdot u = a \otimes v + (a, a)1 \otimes w + 1 \otimes yv - a \otimes yw.$$

Now assume that u is in the center of E. Then $u \cdot x = x \cdot u$ and we obtain

$$a \otimes (wy + yw) = 0 \qquad y \in F \tag{10.9}$$

and

$$1 \otimes (vy - yv) - 2(a, a)1 \otimes w = 0. \tag{10.10}$$

Thus, $w \in AZ_E$ and so $w \in (AZ_E)^1$. Hence Lemma I implies that $w = 0$. Now Formula (10.10) yields

$$vy = yv \qquad y \in F.$$

This shows that $v \in Z_F$ and so $v \in Z_F^0$. Hence, by induction, $v = \lambda \cdot e_F$. It follows that

$$u = 1 \otimes v = \lambda(1 \otimes e_F) = \lambda \cdot e$$

and so the induction is closed. □

Proposition 10.11.1. *Assume that the inner product in E is nondegenerate. Then*

1. *If n is odd,* $Z_E = (e) + (e_\Delta)$, $AZ_E = 0$,
2. *If n is even,* $Z_E = (e)$, $AZ_E = (e_\Delta)$.

PROOF OF 1. *Let n be odd.* Assume that $a \in AZ_E$. Then $ax = -xa$, $x \in E$, and so

$$a \cdot e_\Delta = -e_\Delta \cdot a.$$

It follows from Proposition 10.10.1 that $a \cdot e_\Delta = 0$ whence $a = 0$.

Next observe that, by Lemma II, $Z_E^0 = (e)$. Since n is odd, the map φ_E restricts to an isomorphism $Z_E^0 \xrightarrow{\cong} Z_E^1$. Since $\varphi_E(e) = e_\Delta$ we obtain $Z_E^1 = (e_\Delta)$ whence $Z_E = (e) + (e_\Delta)$.

PROOF OF 2. Let n be even. Then we have, for $a \in Z_E$, $ax = xa$, $x \in E$, and so

$$a \cdot e_\Delta = e_\Delta \cdot a.$$

On the other hand, by Proposition 10.10.1,

$$a \cdot e_\Delta = \omega_E(a) \cdot e_\Delta.$$

It follows that $\omega_E(a) = a$ whence $a \in Z_E^0$. Now Lemma II implies that $a = \lambda \cdot e$ whence $Z_E = (e)$.

Since the map φ_E for even n interchanges center and anticenter, it follows that $AZ_E = (e_\Delta)$. This completes the proof of the proposition. □

10.12. The Algebra C_{-E}

Given an inner product space E denote by $-E$ the space E with the inner product

$$(x, y)_- = -(x, y) \qquad x, y \in E$$

and let C_{-E} be the Clifford algebra over $-E$. Denote the multiplication in C_{-E} by $\circ$. Thus,

$$x \circ x = -x \cdot x \qquad x \in E.$$

Now fix a nonzero determinant function Δ in E. Then it is easy to check that

$$\lambda_\Delta^- = (-1)^n \lambda_\Delta \tag{10.11}$$

(cf. Formula (10.6)).

Now define a linear map $\varphi: E \to C_{-E}$ by

$$\varphi x = e_\Delta^- \circ x \qquad x \in E,$$

where e_Δ^- denotes the canonical element of C_{-E}. Then we have, in view of Proposition 10.10.1,

$$\varphi x \circ \varphi x = e_\Delta^- \circ x \circ e_\Delta^- \circ x = (-1)^{n-1} e_\Delta^- \circ e_\Delta^- \circ x \circ x.$$

Since, by (10.8),

$$e_\Delta^- \circ e_\Delta^- = (-1)^{n(n-1)/2} \lambda_\Delta^- e = (-1)^{n(n-1)/2} (-1)^n \lambda_\Delta e$$

and

$$x \circ x = (x, x)_- \cdot e = -(x, x) \cdot e,$$

it follows that

$$\varphi x \circ \varphi x = (-1)^{n(n-1)/2} (x, x) \lambda_\Delta \cdot e.$$

Now assume that Δ can be chosen such that

$$\lambda_\Delta = (-1)^{n(n-1)/2}. \tag{10.12}$$

Then the relation above reads

$$\varphi x \circ \varphi x = (x, x) \cdot e \qquad x \in E.$$

Thus φ extends to a homomorphism

$$\varphi: C_E \to C_{-E}.$$

We show that

$$\varphi(e_\Delta) = (-1)^{n(n-1)/2} (e_\Delta^-)^{n+1}. \tag{10.13}$$

In fact, choose an orthogonal basis $\{e_1, \ldots, e_n\}$ of E such that $\Delta(e_1, \ldots, e_n) = 1$. Then $e_\Delta = e_1 \cdots e_n$ and thus

$$\begin{aligned}\varphi(e_\Delta) &= \varphi(e_1) \circ \cdots \circ \varphi(e_n) \\ &= (e_\Delta^- \circ e_1) \circ \cdots \circ (e_\Delta^- \circ e_n) \\ &= (-1)^{n(n-1)/2}(e_\Delta^-)^{n+1}.\end{aligned}$$

Similarly, consider the linear map $\psi: -E \to C_E$ given by

$$\psi x = e_\Delta \cdot x \qquad x \in -E.$$

Suppose that Δ can be chosen such that

$$\lambda_\Delta = (-1)^{n(n+1)/2}.$$

Then

$$\lambda_\Delta^- = (-1)^{n(n-1)/2}$$

and hence, by the result above (applied to $-E$), ψ extends to a homomorphism

$$\psi: C_{-E} \to C_E.$$

Proposition 10.12.1. *Assume that n is even, $n = 2m$, and that Δ can be chosen such that $\lambda_\Delta = (-1)^m$. Then the algebras C_E and C_{-E} are isomorphic.*

PROOF. Since $n = 2m$ and $\lambda_\Delta = (-1)^m$,

$$\lambda_\Delta = (-1)^{n(n-1)/2} = (-1)^{n(n+1)/2}.$$

Thus we have the homomorphisms

$$\varphi: C_E \to C_{-E} \quad \text{and} \quad \psi: C_{-E} \to C_E.$$

To show that φ and ψ are isomorphisms we establish the relations

$$\psi \circ \varphi = \omega_E^m \quad \text{and} \quad \varphi \circ \psi = \omega_{-E}^m.$$

In fact, let $x \in E$. Then

$$\begin{aligned}\psi\varphi(x) &= \psi(e_\Delta^- \circ x) = \psi(e_\Delta^-) \cdot \psi x \\ &= (-1)^m e_\Delta^{n+1} \cdot \psi x = (-1)^m e_\Delta^{n+1} \cdot e_\Delta \cdot x \\ &= (-1)^m e_\Delta^{n+2} \cdot x = (-1)^m (e_\Delta^2)^{m+1} \cdot x.\end{aligned}$$

Since $e_\Delta^2 = e$, (by (10.8)), it follows that

$$\psi\varphi(x) = (-1)^m x = \omega_E^m(x) \qquad x \in E.$$

But φ and ψ are homomorphisms and so the equation above implies that

$$\psi \circ \varphi = \omega_E^m.$$

In the same way it is shown that $\varphi \circ \psi = \omega_{-E}^m$. □

10.13. The Canonical Tensor Product of Clifford Algebras

If E is an inner product space and $\varepsilon = \pm 1$ we shall set $E_\varepsilon = E$ if $\varepsilon = 1$ and $E_\varepsilon = -E$ if $\varepsilon = -1$.

Theorem 10.13.1. *Let E be an even-dimensional inner product space which admits a determinant function Δ such that $e_\Delta^2 = \varepsilon \cdot e$ ($\varepsilon = \pm 1$). Then, for any inner product space F,*

$$C_{E \oplus \varepsilon F} \cong C_E \otimes C_F.$$

PROOF. Write $E \oplus \varepsilon F = H$ and let $i: E \to H$ and $j: F \to H$ denote the inclusion maps. Since E has even dimension, Proposition 10.10.1 implies that

$$x \cdot e_\Delta + e_\Delta \cdot x = 0 \qquad x \in E. \tag{10.14}$$

Moreover, if $i_C: C_E \to C_H$ and $j_C: C_{\varepsilon F} \to C_H$ are the induced homomorphisms, then (cf. the proof of Theorem 10.7.1)

$$i_C(e_\Delta) \cdot j_C(b) = j_C(b) \cdot i_C(e_\Delta) \qquad b \in C_{\varepsilon F}. \tag{10.15}$$

Now let $\varphi: F \to C_H$ be the linear map given by

$$\varphi(y) = i_C(e_\Delta) \cdot j_C(y) \qquad y \in F.$$

Then, in view of (10.15),

$$\begin{aligned}\varphi(y)^2 &= i_C(e_\Delta^2) j_C(y^2) = \varepsilon(y, y)_\varepsilon i_C(e_E) \cdot j_C(e_F)\\ &= \varepsilon^2(y, y) \cdot e_H = (y, y) \cdot e_H.\end{aligned}$$

Thus, φ extends to a homomorphism

$$\varphi: C_F \to C_H.$$

Next note that, in view of (10.14),

$$\begin{aligned}i_C(x) \cdot \varphi(y) &= i_C(x) i_C(e_\Delta) j_C(y) = -i_C(e_\Delta) i_C(x) j_C(y)\\ &= \varphi(y) \cdot i_C(x) \qquad x \in E,\ y \in F\end{aligned}$$

whence

$$i_C(a) \cdot \varphi(b) = \varphi(b) \cdot i_C(a) \qquad a \in C_E,\ b \in C_F.$$

Next, define

$$\Phi: C_E \otimes C_F \to C_H$$

by setting

$$\Phi(a \otimes b) = i_C(a) \cdot \varphi(b) \qquad a \in C_E,\ b \in C_F.$$

Then Φ is a homomorphism. To show that it is an isomorphism we construct an inverse map.

First define a linear map $\psi: H \to C_E \otimes C_F$ by

$$\psi(x \oplus y) = x \otimes e_F + \varepsilon e_\Delta \otimes y.$$

Then, in view of 10.14,

$$\begin{aligned}(\psi(x \oplus y))^2 &= x^2 \otimes e_F + \varepsilon[(xe_\Delta + e_\Delta x) \otimes y] + \varepsilon^2(e_\Delta^2 \otimes y^2)\\ &= [(x,x) + \varepsilon(y,y)]e_E \otimes e_F = (x \oplus y, x \oplus y)_H e_E \otimes e_F.\end{aligned}$$

Thus ψ extends to a homomorphism

$$\Psi: C_H \to C_E \otimes C_F.$$

Note that

$$\Psi i_C(a) = a \otimes e_F \qquad a \in C_E.$$

It follows that

$$\begin{aligned}(\Psi \circ \Phi)(x \otimes e_F + e_E \otimes y) &= \Psi[i(x) + i_C(e_\Delta)j(y)]\\ &= x \otimes e_F + \varepsilon(e_\Delta \otimes e_F)(e_\Delta \otimes y)\\ &= x \otimes e_F + \varepsilon^2(e_E \otimes y)\\ &= x \otimes e_F + e_E \otimes y \qquad x \in E, y \in F.\end{aligned}$$

This shows that $\Psi \circ \Phi = \iota$. On the other hand,

$$\begin{aligned}(\Phi \circ \Psi)(x \oplus y) &= \Phi[x \otimes e_F + \varepsilon(e_\Delta \otimes y)]\\ &= i(x) + \varepsilon i_C(e_\Delta) i_C(e_\Delta) j_C(y)\\ &= i(x) + \varepsilon^2 i_C(e_E) j(y)\\ &= i(x) + j(y) = x \oplus y \qquad x \in E, y \in F\end{aligned}$$

and so $\Phi \circ \Psi = \iota$. It follows that Φ is an isomorphism. □

10.14. The Direct Sum of Dual Spaces

Let E_1 and E_2 be dual n-dimensional spaces and consider the direct sum

$$E = E_1 \oplus E_2.$$

Then a nondegenerate inner product is defined in E by

$$(x_1 \oplus x_2, y_1 \oplus y_2) = \tfrac{1}{2}[\langle x_1, y_2\rangle + \langle y_1, x_2\rangle] \qquad x_1, y_1 \in E_1, x_2, y_2 \in E_2$$

(note that this is not the usual inner product in the direct sum!). In particular, the restriction of twice the inner product in E to $E_1 \times E_2$ coincides with the scalar product between E_1 and E_2.

Proposition 10.14.1. *There is an isomorphism* $C_E \xrightarrow{\cong} L(\wedge E_1)$.

PROOF. First recall the definition of the multiplication and the substitution operators in $\wedge E_1$. Identifying E_1 with E_2^* we have the relations

$$\mu(x_1)^2 = 0 \qquad x_1 \in E_1$$

$$i(x_2)^2 = 0 \qquad x_2 \in E_2$$

and (see Corollary I to Proposition 5.14.1)

$$i(x_2) \circ \mu(x_1) + \mu(x_1) \circ i(x_2) = \langle x_1, x_2 \rangle \iota \qquad x_1 \in E_1, x_2 \in E_2.$$

Now define a linear map $\varphi: E \to L(\wedge E_1)$ by setting

$$\varphi(x) = \mu(x_1) + i(x_2) \qquad x \in E,$$

where $x = x_1 \oplus x_2$, $x_1 \in E_1$, $x_2 \in E_2$. Then the relations above yield for all $x \in E$

$$\begin{aligned}\varphi(x)^2 &= \mu(x_1) \circ \mu(x_1) + \mu(x_1) \circ i(x_2) + i(x_2) \circ \mu(x_1) + i(x_2) \circ i(x_2) \\ &= \langle x_1, x_2 \rangle \cdot \iota = (x, x) \cdot e.\end{aligned}$$

Thus φ extends to a homomorphism

$$\varphi: C_H \to L(\wedge E_1).$$

We show that φ is an isomorphism. Since

$$\dim L(\wedge E_1) = (2^n)^2 = 2^{2n} = \dim C_E,$$

it is sufficient to show that φ is surjective. This is a consequence of the following.

Lemma. *Let E^*, E be a pair of finite-dimensional dual spaces. Then the algebra $L(\wedge E)$ is generated by the operators $\mu(x)$ and $i(x^*)$, $x \in E$, $x^* \in E^*$.*

PROOF. Recall from Section 6.2 the linear isomorphism

$$T_E: \wedge^p E^* \otimes \wedge E \xrightarrow{\cong} L(\wedge^p E; \wedge E)$$

given by

$$T_E(a^* \otimes b)u = \langle a^*, u \rangle b.$$

This can be written in the form

$$T_E(a^* \otimes b) = \mu(b) \circ i(a^*).$$

Since every vector $a^* \in \wedge^p E^*$ is generated by products $x_1^* \wedge \cdots \wedge x_p^*$, $x_i^* \in E^*$ and every vector $b \in \wedge^q E$ is generated by products $y_1 \wedge \cdots \wedge y_q$, $y_j \in E$, and since

$$\mu(y_1 \wedge \cdots \wedge y_q) = \mu(y_1) \circ \cdots \circ \mu(y_q)$$

and

$$i(x_1^* \wedge \cdots \wedge x_p^*) = i(x_p^*) \circ \cdots \circ i(x_1^*)$$

the lemma follows because $L(\wedge E) = \oplus_p L(\wedge^p E; \wedge E)$. □

Proposition 10.14.2. *Let E be a $2n$-dimensional vector space with a non-degenerate inner product. Assume that there exists an involution ω of E such that $\tilde{\omega} = -\omega$. Then the Clifford algebra C_E is isomorphic to the algebra of linear transformations of $\wedge E_1$, where $E_1 = \ker(\omega - \iota)$.*

PROOF. Consider the subspaces E_1 and E_2 consisting of the vectors x which satisfy, respectively, $\omega x = x$ and $\omega x = -x$. Then $E = E_1 \oplus E_2$. In fact, let $x \in E$ and set

$$x_1 = \tfrac{1}{2}(x + \omega x)$$
$$x_2 = \tfrac{1}{2}(x - \omega x).$$

For $x_1 \in E_1$, $y_1 \in E_1$, we have

$$(x_1, y_1) = (\omega x_1, \omega y_1) = -(\omega^2 x_1, y_1) = -(x_1, y_1)$$

whence $(x_1, y_1) = 0$. Similarly,

$$(x_2, y_2) = 0 \qquad x_2, y_2 \in E_2.$$

Thus the restrictions of the inner product to $E_1 \times E_1$ and $E_2 \times E_2$ are zero. On the other hand, the restriction of the inner product to $E_1 \times E_2$ is non-degenerate. In fact, fix $x_1 \in E_1$ and assume that $(x_1, y_2) = 0$ for every $y_2 \in E_2$. Then we have for $y \in E$

$$(x_1, y) = (x_1, y_1) + (x_1, y_2) = 0$$

whence $x_1 = 0$. Thus a scalar product between E_1 and E_2 is defined by

$$\langle x_1, x_2 \rangle = 2(x_1, x_2) \qquad x_1 \in E_1, x_2 \in E_2.$$

It satisfies the relation

$$\begin{aligned}(x_1 \oplus x_2, y_1 \oplus y_2) &= (x_1, y_2) + (y_1, x_2)\\ &= \tfrac{1}{2}[\langle x_1, y_2 \rangle + \langle y_1, x_2 \rangle].\end{aligned}$$

Thus Proposition 10.14.1 implies that $C_E \cong L(\wedge E_1)$. □

PROBLEMS

1. Let C_E be the Clifford algebra over an n-dimensional inner product space and denote the left multiplication by an element $a \in C_E$ by $\mu(a)$. Show that

$$\det \mu(x) = (x, x)^{2^{n-1}} \qquad x \in E$$

and

$$\operatorname{tr} \mu(a) = 0 \quad \text{if } a \in C_E^1.$$

2. *The isomorphisms ξ_E and η_E.* Let E be an n-dimensional vector space with a non-degenerate inner product. Identify E with the dual space and denote by $i(x)$ ($x \in E$) the substitution operator in $\wedge E$.

i. Show that the isomorphism ξ_E defined in Section 10.10 satisfies the relations for all $x \in E$ and all $a \in \wedge^p E$

$$\xi_E(x \wedge a) = x \cdot \xi_E(a) - \xi_E i(x)a$$

and

$$\xi_E(a \wedge x) = \xi_E(a) \cdot x + (-1)^p \xi_E i(x)a.$$

ii. Let η_E denote the inverse isomorphism. Show that for all $x \in E$ and all $u \in C_E$

$$\eta_E(x \cdot u) = x \wedge \eta_E(u) + i(x)\eta_E(u)$$

and

$$\eta_E(u \cdot x) = \eta_E(u) \wedge x - i(x)\eta_E(\omega_E u).$$

iii. Let $\pi_E: C_E \to \Gamma$ be the linear map given by

$$\pi_E u = \pi_0 \eta_E(u) \qquad u \in C_E$$

where $\pi_0: \wedge E \to \Gamma$ is the obvious projection. Prove the following relations for all $x \in E$ and all $u, v \in C_E$:

$$\pi_E(x \cdot u) = \pi_0 i(x)\eta_E(u)$$
$$\pi_E(u \cdot x) = -\pi_0 i(x)\eta_E(\omega_E u)$$
$$\pi_E \circ \omega_E = \pi_E$$
$$\pi_E(u \cdot v) = \pi_E(v \cdot u).$$

3. Use the linear map π_E (see Problem 2 (iii),) to define a bilinear function Q_E in C_E by setting

$$Q_E(u, v) = \pi_E(u \cdot v).$$

i. Show that Q_E is symmetric and nondegenerate.
ii. Show that $Q_E(x, y) = (x, y)$, $x, y \in E$.
iii. Prove the relation

$$Q_E(u \cdot w, v \cdot w) = Q_E(w \cdot u, w \cdot v) \qquad u, v, w \in C_E.$$

Clifford Algebras over a Complex Vector Space

10.15. Clifford Algebras Over Complex Vector Spaces

Let $\mathbb{C}^n$ be an n-dimensional complex vector space and let $(\, , \,)$ be a nondegenerate symmetric bilinear function in $\mathbb{C}^n$. Note that if $(\, , \,)_1$ and $(\, , \,)_2$ are two such bilinear functions, then there is a linear isomorphism φ of $\mathbb{C}^n$ such that

$$(\varphi x, \varphi y)_1 = (x, y)_2 \qquad x, y \in \mathbb{C}^n.$$

Thus the corresponding Clifford algebras are isomorphic. We shall denote the Clifford algebra over $\mathbb{C}^n$ by C_n.

Example : $n = 1$. The Clifford algebra C_1 is isomorphic to the algebra $\mathbb{C} \oplus \mathbb{C}$. In fact, choose a vector $e_1 \in \mathbb{C}$ such that $(e_1, e_1) = 1$. Then the vectors $\{e, e_1\}$ form a basis of C_1. Now define a linear map $\Phi: C_1 \to \mathbb{C} \oplus \mathbb{C}$ by setting

$$\Phi(\alpha e + \beta e_1) = (\alpha + \beta, \alpha - \beta) \qquad \alpha, \beta \in \mathbb{C}.$$

Then Φ is an algebra isomorphism as is easily checked. Thus,

$$C_1 \cong \mathbb{C} \oplus \mathbb{C}.$$

The Element e_Δ. A *normed determinant function* in $\mathbb{C}^n$ is a determinant function Δ which satisfies

$$\Delta(x_1, \ldots, x_n) \cdot \Delta(y_1, \ldots, y_n) = \det((x_i, y_j)) \qquad x_i, y_j \in \mathbb{C}^n.$$

It is easy to see that a normed determinant function always exists and that it is determined up to sign. Thus the canonical element e_Δ corresponding to a normed determinant function satisfies (see (10.8))

$$e_\Delta^2 = (-1)^{n(n-1)/2} e.$$

Hence Theorem 10.13.1 yields an isomorphism

$$C_{2m+k} \cong C_{2m} \otimes C_k. \tag{10.16}$$

Proposition 10.15.11. *Let n be even, n = 2m, and assume that the inner product is nondegenerate. Then*

$$C_{2m} \cong L(\wedge \mathbb{C}^m). \tag{10.17}$$

Proof. Write $\mathbb{C}^n$ as an orthogonal sum

$$\mathbb{C}^n = A \oplus B$$

where $\dim A = m$ and $\dim B = m$. Choose orthonormal bases $\{a_\mu\}$ and $\{b_\mu\}$, $\mu = 1, \ldots, m$ in A and B respectively and define an involution ω of $\mathbb{C}^n$ by setting

$$\omega(a_\mu) = ib_\mu \qquad \mu = 1, \ldots, m$$

$$\omega(b_\mu) = -ia_\mu \qquad \mu = 1, \ldots, m.$$

Then for $\mu, \nu = 1, \ldots, m$ we have the relations

$$(\omega a_\mu, a_\nu) = 0, \qquad (\omega b_\mu, b_\nu) = 0$$

and

$$(\omega a_\mu, b_\nu) = i(b_\mu, b_\nu) = i\delta_{\mu\nu} = -(a_\mu, \omega b_\nu).$$

Thus ω is a skew transformation. Now Proposition 10.14.2 shows that

$$C_n \cong L(\wedge E_1),$$

where E_1 is the subspace of E determined by the equation $\omega x = x$. □

Combining Formulas (10.16) and (10.17) we obtain for all $m \geq 1$ the relations

$$C_{2m} \cong L(\wedge \mathbb{C}^m)$$

and (using the example above)

$$C_{2m+1} \cong L(\wedge \mathbb{C}^m) \oplus L(\wedge \mathbb{C}^m).$$

10.16. Complexification of Real Clifford Algebras

Let F be a real inner product space (not necessarily of finite dimension) and consider the complexification $E = \mathbb{C} \otimes F$. Define an inner product in E by

$$(\lambda \otimes x, \mu \otimes y)_E = \lambda\mu(x, y) \qquad \lambda, \mu \in \mathbb{C}, x, y \in F.$$

Then the inclusion map $j: F \to E$ is an isometry and so it extends to a (real) homomorphism $j_C: C_F \to C_E$. Now consider the complex linear map

$$\varphi: \mathbb{C} \otimes C_F \to C_E$$

given by

$$\varphi(\lambda \otimes a) = \lambda \cdot j_C(a) \qquad \lambda \in \mathbb{C}, a \in C_F.$$

To show that φ is an isomorphism we construct an inverse homomorphism. Consider the linear map $\psi: E \to \mathbb{C} \otimes C_F$ given by

$$\psi(x + iy) = 1 \otimes x + i \otimes y.$$

Then

$$\begin{aligned}(\psi(x + iy))^2 &= 1 \otimes x^2 - 1 \otimes y^2 + i(x \cdot y + y \cdot x)\\ &= [(x, x) - (y, y)](1 \otimes e) + 2i(x, y)(1 \otimes e)\\ &= (x + iy, x + iy)_E \cdot (1 \otimes e) \qquad x, y \in F.\end{aligned}$$

Thus ψ extends to a homomorphism $\psi: C_E \to \mathbb{C} \otimes C_F$ It follows from the definitions that $\psi \circ \varphi = \iota$ and $\varphi \circ \psi = \iota$. Thus φ is an isomorphism,

$$\varphi: \mathbb{C} \otimes C_F \xrightarrow{\cong} C_E.$$

Clifford Algebras Over a Real Vector Space

10.17

Let E be a real n-dimensional vector space with a nondegenerate inner product. Recall that E can be decomposed in the form

$$E = E^+ \oplus E^-,$$

where the restriction of the inner product to E^+ (respectively E^-) is positive (respectively negative) definite. If $\dim E^+ = p$ and $\dim E^- = q$, we shall say that the inner product is of type (p, q). Clearly, $p + q = n$. The difference $s = p - q$ is called the *signature* of the inner product.

The Clifford algebra over an inner product space of type (p, q) will be denoted by $C(p, q)$. We shall write

$$C(n, 0) = C_n(+)$$

and

$$C(0, n) = C_n(-).$$

Thus, by Examples 1 and 2 of Section 10.2,

$$C_1(-) \cong \mathbb{C} \quad \text{and} \quad C_2(-) \cong \mathbb{H}.$$

Next recall from Section 9.19 of *Linear Algebra* that a normed *determinant function* in E is a determinant function Δ which satisfies

$$\det((x_i, y_j)) = (-1)^q \Delta(x_1, \ldots, x_n) \cdot \Delta(y_1, \ldots, y_n) \qquad x_\nu \in E,\ y_\nu \in E.$$

Every inner product space admits a normed determinant function Δ and Δ is uniquely determined up to sign. Thus the scalar λ_Δ (see Section 10.10) corresponding to a normed determinant function is given by

$$\lambda_\Delta = (-1)^q.$$

This implies that the corresponding canonical element e_Δ of the Clifford algebra $C(p, q)$ satisfies

$$e_\Delta^2 = (-1)^{(1/2)n(n-1)+q} e.$$

In particular,

1. If $(\ ,\ )$ is positive definite, $e_\Delta^2 = (-1)^{(1/2)n(n-1)} e$
2. If $(\ ,\ )$ is negative definite, $e_\Delta^2 = (-1)^{(1/2)n(n+1)} e$,
3. If $n = 2m$ and $p = q = m$, $e_\Delta^2 = e$.

Theorem 10.17.1. *There are algebra isomorphisms*

$$C(p, q) \otimes C_2(+) \cong C(q + 2, p) \qquad p, q \geq 0$$

and

$$C(p, q) \otimes C_2(-) \cong C(q, p + 2) \qquad p, q \geq 0.$$

In particular, for $n \geq 0$,

$$C_n(-) \otimes C_2(+) \cong C_{n+2}(+)$$

and

$$C_n(+) \otimes C_2(-) \cong C_{n+2}(-).$$

PROOF. Observe that the canonical elements of $C_2(+)$ and $C_2(-)$ satisfy $e_\Delta^2 = -e$ and apply Theorem 10.13.1. □

Theorem 10.17.2. *Assume that* $s \equiv 0 \pmod 4$. *Then*

$$C(p, q) \cong C(q, p).$$

In particular,

$$C_n(+) \cong C_n(-) \quad \text{if } n \equiv 0 \pmod 4.$$

PROOF. Set $s = 4k$. Then

$$n = 2q + s = 2q + 4k = 2m, \qquad \text{where } m = q + 2k$$

and

$$m - q = 2k.$$

Thus, if Δ is the normed determinant function, then

$$\lambda_\Delta = (-1)^q = (-1)^m.$$

Now apply Proposition 10.12.1. □

10.18. Inner Product Spaces With Signature Zero

Suppose that E is an inner product space with signature zero. Then $p = q$ and so $\dim E = 2p$. We shall show that

$$C_E \cong L(\wedge E_1),$$

where $\dim E_1 = p$.

Choose an orthogonal decomposition $E = E^+ \oplus E^-$ such that the restriction of the inner product to E^+ (respectively E^-) is positive (respectively negative) definite. Thus $\dim E^+ = \dim E^- = p$. Let $\{a_\nu\}$ and $\{b_\nu\}$ ($\nu = 1, \ldots, p$) be orthonormal bases of E^+ and E^- respectively,

$$(a_\nu, a_\mu) = \delta_{\nu\mu} \qquad \nu, \mu = 1, \ldots, p$$

$$(b_\nu, b_\mu) = -\delta_{\nu\mu} \qquad \nu, \mu = 1, \ldots, p.$$

Define an involution ω of E by setting

$$\omega a_\nu = b_\nu \quad \text{and} \quad \omega b_\nu = a_\nu \qquad \nu = 1, \ldots, p.$$

Then ω is a skew transformation as is easily checked. Thus by Proposition 10.14.2,

$$C_E \cong L(\wedge E_1),$$

where E_1 is the kernel of $\omega - \iota$.

EXAMPLE. Consider the algebra $C(2, 2)$. Then, by the result above,

$$C(2, 2) \cong L(\mathbb{R}^4).$$

On the other hand, the second formula in Theorem 10.17.1 applied with $p = 0$ and $q = 2$ yields

$$C(2, 2) \cong C_2(-) \otimes C_2(-) \cong \mathbb{H} \otimes \mathbb{H}.$$

Thus,

$$\mathbb{H} \otimes \mathbb{H} \cong L(\mathbb{R}^4).$$

The following proposition establishes an explicit isomorphism between these algebras.

Proposition 10.18.1. *There is a canonical algebra isomorphism*

$$\Phi: \mathbb{H} \otimes \mathbb{H} \to L(\mathbb{R}^4).$$

PROOF. Identify $\mathbb{R}^4$ with $\mathbb{H}$ and define a linear map

$$\Phi: \mathbb{H} \otimes \mathbb{H} \to L(\mathbb{R}^4)$$

by setting

$$\Phi(p \otimes q)x = p \cdot x \cdot \bar{q} \qquad p, q \in \mathbb{H}, x \in \mathbb{R}^4,$$

where $\bar{q}$ denotes the conjugate of q. It is easily checked that Φ is an algebra homomorphism. To show that it is an isomorphism define positive definite inner products in $\mathbb{H} \otimes \mathbb{H}$ and $L(\mathbb{R}^4)$ by

$$(p \otimes q, p' \otimes q') = (p, p')(q, q')$$

and

$$(\varphi, \psi) = \tfrac{1}{4} \operatorname{tr}(\tilde{\varphi} \circ \psi) \qquad \varphi, \psi \in L(\mathbb{R}^4)$$

respectively. Observe that the adjoint transformation of $\Phi(p \otimes q)$ is given by

$$\widetilde{\Phi(p \otimes q)} = \Phi(\bar{p} \otimes \bar{q}).$$

Thus we have

$$\begin{aligned}(\Phi(p \otimes q), \Phi(p' \otimes q')) &= \tfrac{1}{4} \operatorname{tr}(\Phi(\bar{p} \otimes \bar{q}) \circ \Phi(p' \otimes q')) \\ &= \tfrac{1}{4} \operatorname{tr} \Phi(\bar{p}p' \otimes \bar{q}q').\end{aligned}$$

Now it is easy to check that for $a \in \mathbb{H}$ and $b \in \mathbb{H}$

$$\operatorname{tr} \Phi(a \otimes b) = 4(a, e)(b, e).$$

It follows that

$$\begin{aligned}(\Phi(p \otimes q), \Phi(p' \otimes q')) &= (\bar{p}p', e)(\bar{q}q', e) = (p', p)(q', q) \\ &= (p \otimes q, p' \otimes q').\end{aligned}$$

Thus Φ is an isometry and hence a linear isomorphism. □

10.19. Clifford Algebras of Low Dimensions

In this and the following section we shall determine the structure of $C_n(+)$ and $C_n(-)$ for $1 \le n \le 8$.

(1) $n = 1$. We show that

$$C_1(+) \cong \mathbb{R} \oplus \mathbb{R}.$$

In fact, let e_1 be a unit vector in $\mathbb{R}$. Then the vectors e, e_1 form a basis of $C_1(+)$. Now define a linear map $\Phi: C_1(+) \to \mathbb{R} \oplus \mathbb{R}$ by setting

$$\Phi(\alpha e + \beta e_1) = (\alpha + \beta, \alpha - \beta) \qquad \alpha, \beta \in \mathbb{R}.$$

Then Φ is a homomorphism as is easily checked. Moreover, Φ is a linear isomorphism and so an isomorphism of algebras. This proves that $C_1(+) \cong \mathbb{R} \oplus \mathbb{R}$.

On the other hand, it has been shown in Example 1, Section 10.2, that

$$C_1(-) \cong \mathbb{C}.$$

(2) $n = 2$. We show that

$$C_2(+) \cong L(\mathbb{R}^2).$$

Fix an orthonormal basis $\{e_1, e_2\}$ of $\mathbb{R}^2$ and define a linear map $\varphi: \mathbb{R}^2 \to L(\mathbb{R}^2)$ by setting

$$\varphi(x)e_1 = \alpha e_1 + \beta e_2$$

$$\varphi(x)e_2 = \beta e_1 - \alpha e_2,$$

where

$$x = \alpha e_1 + \beta e_2.$$

Then it is easy to check that

$$\varphi(x)^2 = (\alpha^2 + \beta^2)\iota = (x, x) \cdot \iota.$$

Thus φ extends to a homomorphism $\Phi: C_2(+) \to L(\mathbb{R}^2)$. To show that Φ is an isomorphism note that the transformations $\Phi(x)$, $x \in E$, are selfadjoint and have trace zero. On the other hand, $\Phi(e) = \iota$. Thus, if A denotes the subspace of C_E spanned by e, e_1, and e_2, Φ determines an isomorphism from A to the space of selfadjoint transformations of $\mathbb{R}^2$. Finally, it is easy to check that $\Phi(e_2 \cdot e_1)$ is the transformation $e_1 \to e_2$, $e_2 \to -e_1$ and so it is skew. Since every transformation of $\mathbb{R}^2$ is the sum of a selfadjoint and a skew transformation, it follows that Φ is an isomorphism.

10.20

Using the results of Section 10.19 and Theorem 10.17.1 we shall now determine the Clifford algebras $C_k(+)$ and $C_k(-)$ for $k \le 8$ explicitly. Recall that the direct sum of two algebras A and B is the algebra $A \oplus B$ with multiplication defined by

$$(a_1 \oplus b_1) \cdot (a_2 \oplus b_2) = (a_1 a_2 \oplus b_1 b_2).$$

We have the following isomorphisms:

$$n = 3:\quad C_3(+) \cong C_1(-) \otimes C_2(+) \cong \mathbb{C} \otimes L(\mathbb{R}^2)$$
$$C_3(-) \cong C_1(+) \otimes C_2(-) \cong (\mathbb{R} \oplus \mathbb{R}) \otimes \mathbb{H} \cong \mathbb{H} \oplus \mathbb{H}$$
$$n = 4:\quad C_4(\pm) \cong C_2(-) \otimes C_2(+) \cong \mathbb{H} \otimes L(\mathbb{R}^2)$$
$$n = 5:\quad C_5(+) \cong C_3(-) \otimes C_2(+) \cong (\mathbb{H} \oplus \mathbb{H}) \otimes L(\mathbb{R}^2)$$
$$\cong \mathbb{H} \otimes L(\mathbb{R}^2) \oplus \mathbb{H} \otimes L(\mathbb{R}^2)$$
$$C_5(-) \cong C_3(+) \otimes C_2(-) \cong \mathbb{C} \otimes L(\mathbb{R}^2) \otimes \mathbb{H}$$
$$n = 6:\quad C_6(+) \cong C_4(-) \otimes C_2(+) \cong \mathbb{H} \otimes L(\mathbb{R}^2) \otimes L(\mathbb{R}^2) \cong \mathbb{H} \otimes L(\mathbb{R}^4)$$
$$C_6(-) \cong C_4(+) \otimes C_2(-) = \mathbb{H} \otimes L(\mathbb{R}^2) \otimes \mathbb{H} \cong \mathbb{H} \otimes \mathbb{H} \otimes L(\mathbb{R}^2)$$
$$\cong L(\mathbb{R}^4) \otimes L(\mathbb{R}^2) \cong L(\mathbb{R}^8)$$
$$n = 7:\quad C_7(+) \cong C_5(-) \otimes C_2(+) \cong \mathbb{C} \otimes L(\mathbb{R}^2) \otimes \mathbb{H} \otimes L(\mathbb{R}^2)$$
$$\cong \mathbb{C} \otimes \mathbb{H} \otimes L(\mathbb{R}^4)$$
$$C_7(-) \cong C_5(+) \otimes C_2(-) \cong [\mathbb{H} \otimes L(\mathbb{R}^2) \oplus \mathbb{H} \otimes L(\mathbb{R}^2)] \otimes \mathbb{H}$$
$$\cong \mathbb{H} \otimes \mathbb{H} \otimes L(\mathbb{R}^2) \oplus \mathbb{H} \otimes \mathbb{H} \otimes L(\mathbb{R}^2)$$
$$\cong L(\mathbb{R}^4) \otimes L(\mathbb{R}^2) \oplus L(\mathbb{R}^4) \otimes L(\mathbb{R}^2) \cong L(\mathbb{R}^8) \oplus L(\mathbb{R}^8)$$
$$n = 8:\quad C_8(\pm) \cong C_4(-) \otimes C_4(+) \cong \mathbb{H} \otimes L(\mathbb{R}^2) \otimes \mathbb{H} \otimes L(\mathbb{R}^2)$$
$$\cong \mathbb{H} \otimes \mathbb{H} \otimes L(\mathbb{R}^2) \otimes L(\mathbb{R}^2)$$
$$\cong L(\mathbb{R}^4) \otimes L(\mathbb{R}^2) \otimes L(\mathbb{R}^2) \cong L(\mathbb{R}^{16}).$$

These results are combined in the following table:

n	$C(+)$	$C(-)$
1	$\mathbb{R} \otimes \mathbb{R}$	$\mathbb{C}$
2	$L(\mathbb{R}^2)$	$\mathbb{H}$
3	$\mathbb{C} \otimes L(\mathbb{R}^2)$	$\mathbb{H} \oplus \mathbb{H}$
4	$\mathbb{H} \otimes L(\mathbb{R}^2)$	$\mathbb{H} \otimes L(\mathbb{R}^2)$
5	$(\mathbb{H} \otimes L(\mathbb{R}^2) \oplus (\mathbb{H} \otimes L(\mathbb{R}^2))$	$\mathbb{C} \otimes L(\mathbb{R}^2) \otimes \mathbb{H}$
6	$\mathbb{H} \otimes L(\mathbb{R}^4)$	$L(\mathbb{R}^8)$
7	$\mathbb{C} \otimes \mathbb{H} \otimes L(\mathbb{R}^4)$	$L(\mathbb{R}^8) \oplus L(\mathbb{R}^8)$
8	$L(\mathbb{R}^{16})$	$L(\mathbb{R}^{16})$

10.21. The Algebras $C_n(+)$ and $C_n(-)$

Since the canonical element of $C_8(+) \cong C_8(-)$ satisfies $e_\Delta^2 = 1$, Theorem 10.13.1 yields isomorphisms

$$C_{n+8}(+) \cong C_n(+) \otimes L(\mathbb{R}^{16})$$

and

$$C_{n+8}(-) \cong C_n(-) \otimes L(\mathbb{R}^{16}).$$

These isomorphisms, together with the isomorphisms in the table, determine the structure of the Clifford algebras $C_n(+)$ and $C_n(-)$ for all n.

10.22. The Algebras $C(p, q)$

Finally, we determine the algebra $C(p, q)$ for an indefinite inner product of type (p, q).

Case 1: $p = q$. Then it was shown in Section 10.18 that

$$C(p, p) \cong L(\wedge \mathbb{R}^p). \tag{10.18}$$

Case 2: $p > q$. Write $p = q + s$. Then we have an orthogonal decomposition

$$\mathbb{R}^n = \mathbb{R}^p_+ \oplus \mathbb{R}^q_- = (\mathbb{R}^s_+ \oplus \mathbb{R}^q_+) \oplus \mathbb{R}^q_- = \mathbb{R}^s_+ \oplus (\mathbb{R}^q_+ \oplus \mathbb{R}^q_-).$$

Since $\mathbb{R}^q_+ \oplus \mathbb{R}^q_-$ has even dimension and since the canonical element of $C(q, q)$ satisfies $e_\Delta^2 = e$, Theorem 10.13.1 gives

$$C(p, q) \cong C(q, q) \otimes C_s(+).$$

Now using Formula (10.18) we obtain

$$C(p, q) \cong L(\wedge \mathbb{R}^q) \otimes C_s(+) \qquad p > q, s = p - q.$$

Case 3: $p < q$. Set $q - p = r$. Then we obtain

$$C(p, q) = C(p, p + r) \cong C(p, p) \otimes C_r(-).$$

Thus

$$C(p, q) \cong L(\wedge \mathbb{R}^p) \otimes C_r(-) \qquad p < q, r = q - p.$$

Now the results of Section 10.21 give the structure of $C(p, q)$ for all p, q.

EXAMPLE. The Clifford algebra over the Minkowski space ($p = 3, q = 1$) is given by

$$C(3, 1) = C_2(+) \otimes L(\wedge \mathbb{R}^1) \cong L(\mathbb{R}^2) \otimes L(\mathbb{R}^2) \cong L(\mathbb{R}^4).$$

PROBLEMS

1. Establish explicit isomorphisms

$$C_3(+) \cong L(\mathbb{C}^2)$$

$$C_3(-) \cong \mathbb{H} \oplus \mathbb{H}$$

$$C_7(+) \cong L(\mathbb{C}^8)$$

$$C_7(-) \cong L(\mathbb{R}^8) \oplus L(\mathbb{R}^8).$$

2. Let $\mathbb{C}^2$ be a complex vector space of dimension 2 with a positive definite Hermitian inner product

 i. Consider the linear transformations φ which satisfy

$$\varphi + \tilde{\varphi} = \operatorname{tr} \varphi \cdot \iota.$$

 Show that these transformations form an algebra (over $\mathbb{R}$) and that this algebra is isomorphic to $\mathbb{H}$.

 ii. Use this algebra to establish a canonical isomorphism

$$\mathbb{C} \otimes \mathbb{H} \cong L(\mathbb{C}^2).$$

3. Let E be an n-dimensional Euclidean space and consider the symmetric bilinear function Q_E in C_E defined in Problem 3 after Section 10.14.

 i. Show that the signature s_n of Q_E is given by

$$s_n = 2^{n/2}\left(\cos\frac{n\pi}{4} + \sin\frac{n\pi}{4}\right) \qquad n \geq 1.$$

 ii. Conclude that

$$s_{n+4} = -4s_n, \qquad s_{n+8} = 16s_n \qquad n \geq 1.$$

 iii. Show that Q_E is positive definite only if $n = 1$.

4 Derivations. Let E be a vector space over a field Γ.

 i. Let θ be a derivation in C_E which restricts to a linear transformation φ of E. Show that φ is skew.

 ii. Show that every skew transformation φ of E extends uniquely to a derivation in C_E.

5. Antiderivations. Recall that an antiderivation in C_E is a linear transformation Ω which satisfies

$$\Omega(ab) = \Omega(a)b + \omega_E(a)b \qquad a, b \in C_E$$

(cf. Section 5.8 of *Linear Algebra*).

 i. Show that the map $\Omega = \iota - \omega_E$ is an antiderivation in C_E.

 ii. Assume that the inner product in E is nondegenerate. Show that every antiderivation Ω in C_E which restricts to a linear transformation of E is of the form

$$\Omega = \lambda(\iota - \omega_E) \qquad \lambda \in \Gamma.$$

6. Clifford algebras over even dimensional spaces. Let C_E be the Clifford algebra over an even-dimensional real vector space with a nondegenerate inner product. Show that C_E is simple (that is, the only two-sided ideals in C_E are C_E and (0)). Conclude that a homomorphism φ from C_E to any associative algebra A which satisfies $\varphi(e) = e_A$ is injective. *Hint*: Consider the complexification of C_E and apply Proposition 10.15.1.

7. Clifford algebras over spaces of odd dimension. Let C_E be the Clifford algebra over an odd-dimensional space and assume that the canonical element satisfies $e_\Delta^2 = e$. Consider the linear transformations

$$\varphi a = \tfrac{1}{2}(e + e_\Delta)a \qquad a \in C_E$$

and

$$\psi a = \tfrac{1}{2}(e - e_\Delta)a \qquad a \in C_E.$$

Set $\mathscr{T}^+ = \operatorname{Im} \varphi$ and $\mathscr{T}^- = \operatorname{Im} \psi$. Show that $\mathscr{T}^+$ and $\mathscr{T}^-$ are two-sided ideals in C_E and that

$$C_E = \mathscr{T}^+ \oplus \mathscr{T}^-.$$

Apply the result to complex and real Clifford algebras.

8. Show that the center of a Clifford algebra over an infinite-dimensional vector space with a nondegenerate inner product is the subspace spanned by e.

11 Representations of Clifford Algebras

Basic Concepts

11.1. Representations of an Algebra

Let A be an associative algebra with unit element e. A *representation* of A in a vector space V is a homomorphism $R: A \to L(V)$ where $L(V)$ denotes the algebra of linear transformations of V such that $R(e) = \iota$. A representation is called *faithful*, if the map R is injective.

A subspace $W \subset V$ is called *stable* under R, if it is stable under every transformation $R(a)$, $a \in A$. A representation is called *irreducible*, if the only stable subspaces are $W = V$ and $W = 0$. In particular, if R is surjective, then the representation is irreducible. In fact, assume that W is a stable subspace. Then R is stable under every linear transformation of V. This is only possible if $W = 0$ or $W = V$.

Two representations R_1 and R_2 of A in V_1 and V_2 respectively are called *equivalent*, if there is a linear isomorphism $\Phi: V_1 \xrightarrow{\cong} V_2$ such that

$$\Phi \circ R_1(a) = R_2(a) \circ \Phi \qquad a \in A.$$

In this case we shall write $R_1 \sim R_2$.

Next let P and Q be representations of A in U and V respectively. Then a representation of A in $U \oplus V$, denoted by $P \oplus Q$, is given by

$$(P \oplus Q)(a) = P(a) \oplus Q(a) \qquad a \in A.$$

It is called the *direct sum* of P and Q. Similarly, the *tensor product* of P and Q, denoted by $P \otimes Q$, is the representation in $U \otimes V$ defined by

$$(P \otimes Q)(a) = P(a) \otimes Q(a) \qquad a \in A.$$

Clearly, if $P_1 \sim P_2$ and $Q_1 \sim Q_2$, then

$$P_1 \oplus Q_1 \sim P_2 \oplus Q_2 \quad \text{and} \quad P_1 \otimes Q_1 \sim P_2 \otimes Q_2.$$

11.2. Representations of a Clifford Algebra

Let C_E be the Clifford algebra over an inner product space E and let R be a representation of C_E in an n-dimensional vector space V. Then R restricts to a linear map $R_E: E \to L(V)$. We show that this map is injective if the inner product in V is nondegenerate. In fact, assume that $R_E(x_0) = 0$ for some $x_0 \in E$. Then

$$R(x_0 y + y x_0) = R(x_0) \circ R(y) + R(y) \circ R(x_0) = 0 \qquad \text{for every } y \in E.$$

Since

$$x_0 y + y x_0 = 2(x_0, y)e,$$

we obtain

$$(x_0, y) = 0 \qquad y \in E$$

whence $x_0 = 0$. Thus R_E is injective.

11.3. Orthogonal Representations

A representation of a Clifford algebra C_E in a Euclidean space V is called *orthogonal*, if

$$(R(x)u, R(x)v) = \varepsilon(x, x) \cdot (u, v) \qquad x \in E, u, v \in V,$$

where $\varepsilon = \pm 1$. It is called *positive orthogonal*, if $\varepsilon = +1$ and *negative orthogonal*, if $\varepsilon = -1$.

Thus, if R is positive orthogonal, then

$$(R(x)u, R(x)v) = (x, x) \cdot (u, v).$$

It follows from this equation that $\tilde{R}(x) \circ R(x) = (x, x) \cdot \iota$, $x \in E$.

On the other hand, $R(x) \circ R(x) = R(x^2) = (x, x) \cdot \iota$, $x \in E$. These relations imply that $\tilde{R}(x) = R(x)$. Similarly, if R is negative orthogonal, then the transformations $R(x)$ are skew.

Proposition 11.3.1. *Assume that the inner product in E is positive (respectively negative) definite. Then every representation of C_E in a Euclidean space is equivalent to a positive (respectively negative) orthogonal representation.*

PROOF. Let $\dim E = k$ and choose a basis $\{e_1, \ldots, e_k\}$ of E such that

$$(e_i, e_j) = \varepsilon \cdot \delta_{ij} \qquad (i, j = 1, \ldots, k).$$

Then

$$e_i e_j + e_j e_i = 2\varepsilon\delta_{ij} \cdot e \qquad (i, j = 1, \ldots, k).$$

In particular, $e_i^2 = \varepsilon \cdot e$, and so the elements e_i are invertible. Thus, if C_E^* denotes the multiplicative group of invertible elements in C_E, then $e_i \in C_E^*$. Let G denote the subgroup of C_E^* generated by the elements e_i $(i = 1, \ldots, k)$ and e. The relations above imply that G is a *finite* group.

Now introduce a new definite inner product in V by setting

$$(u, v)_0 = \sum_{a \in G} (R(a)u, R(a)v).$$

Then we have for $g \in G$

$$\begin{aligned}(R(g)u, R(g)v)_0 &= \sum_{a \in G} (R(a)R(g)u, R(a)R(g)v) \\ &= \sum_{a \in G} (R(ag)u, R(ag)v) = \sum_{a \in G} (R(a)u, R(a)v) \\ &= (u, v)_0 \qquad u, v \in V.\end{aligned}$$

Thus,

$$(R(g)u, R(g)v)_0 = (u, v)_0 \qquad g \in G. \tag{11.1}$$

Next observe that, since both inner products of V are definite of the same type, there is a linear automorphism Φ of V such that

$$(\Phi(u), \Phi(v)) = (u, v)_0 \qquad u, v \in V.$$

Now set

$$P(a) = \Phi \circ R(a) \circ \Phi^{-1} \qquad a \in C_E.$$

Then P is a representation of C_E equivalent to R. Moreover, Relation (11.1) yields

$$\begin{aligned}(P(g)u, P(g)v) &= (\Phi R(g)\Phi^{-1}(u), \Phi R(g)\Phi^{-1}(v)) \\ &= (R(g)\Phi^{-1}(u), R(g)\Phi^{-1}(v))_0 \\ &= (\Phi^{-1}(u), \Phi^{-1}(v))_0 = (u, v) \qquad g \in G, u, v \in V.\end{aligned}$$

Thus,

$$(P(g)u, P(g)v) = (u, v).$$

In particular, if we set $P(e_i) = \sigma_i$ $(i = 1, \ldots, k)$, then

$$(\sigma_i u, \sigma_i v) = (u, v) \qquad u, v \in V,$$

and so

$$\tilde{\sigma}_i \circ \sigma_i = \iota \qquad (i = 1, \ldots, k).$$

On the other hand, we have

$$\sigma_i \circ \sigma_i = \sigma_i^2 = P(e_i^2) = (e_i, e_i) \cdot \iota = \varepsilon \cdot \iota \qquad (i = 1, \ldots, k).$$

These relations yield

$$\tilde{\sigma}_i = \varepsilon \cdot \sigma_i \qquad (i = 1, \ldots, k)$$

and so, by linearity,

$$\widetilde{P(x)} = \varepsilon \cdot P(x) \qquad x \in E.$$

It follows that

$$\begin{aligned}(P(x)u, P(x)v) &= (\tilde{P}(x)P(x)u, v) = \varepsilon \cdot (P(x)^2 u, v) \\ &= \varepsilon \cdot (P(x^2)u, v) = \varepsilon \cdot (x, x)(u, v) \qquad x \in E.\end{aligned}$$

This relation shows that P is an orthogonal representation. In particular, if the inner product in E is positive (respectively negative) definite, P is a positive (respectively negative) orthogonal representation. □

The Twisted Adjoint Representation

11.4

Definition. Let E be an n-dimensional vector space with a nondegenerate inner product. Denote by C_E^* the multiplicative group of invertible elements in C_E. Then a representation of the group C_E^* in C_E is defined by

$$\mathrm{ad}(a)u = \omega_E(a)ua^{-1} \qquad a \in C_E^*, u \in C_E,$$

where ω_E denotes the degree involution. ad is called the *twisted adjoint representation* of C_E^*.

It follows from the definition that

$$\mathrm{ad}\, \omega_E(a) = \omega_E \circ \mathrm{ad}(a) \circ \omega_E^{-1}. \tag{11.2}$$

We show that the kernel K of ad consists of the elements λe, $\lambda \neq 0$. Clearly,

$$\mathrm{ad}(\lambda e)u = \lambda u \lambda^{-1} = u \qquad u \in C_E$$

and so $\lambda e \in K$. Conversely, assume that $\mathrm{ad}(a) = \iota$. Then

$$\omega_E(a)u = ua \qquad u \in C_E. \tag{11.3}$$

Setting $u = e$ we obtain $\omega_E(a) = a$. Now Equation (11.3) yields

$$au = ua \qquad u \in C_E$$

whence $a \in Z_E$ and so $a \in Z_E^0$. Now Lemma II, Section 10.11, implies that $a = \lambda \cdot e$, $\lambda \in \Gamma$. Since $a \in C_E^*$, it follows that $\lambda \neq 0$.

11.5. The Clifford Group

Let Γ_E denote the subgroup of C_E^* consisting of those elements a for which E is stable under $\mathrm{ad}(a)$. Γ_E is called the *Clifford group* of E. Every element $h \in E$ which satisfies $(h, h) \neq 0$ is contained in Γ_E. In fact, since for $x \in E$

$$\mathrm{ad}(h)x = -hxh^{-1} = x - 2\frac{(h, x)}{(h, h)}h,$$

it follows that $\mathrm{ad}(h)x \in E$, $x \in E$, and so $h \in \Gamma_E$.

Proposition 11.5.1. *The Clifford group is stable under the degree involution and under the antiautomorphism S_E (see Section* 10.8).

PROOF. (1) Let $a \in \Gamma_E$. Then Formula (11.2) yields

$$\mathrm{ad}(\omega_E a)x = \omega_E\,\mathrm{ad}(a)\omega_E^{-1}(x) = -\omega_E\,\mathrm{ad}(a)x = \mathrm{ad}(a)x \in E \qquad x \in E.$$

Thus, $\omega_E(a) \in \Gamma_E$.

(2) Let $a \in \Gamma_E$. Then $a^{-1} \in \Gamma_E$ and so we have

$$\omega_E(a^{-1})xa \in E \qquad x \in E.$$

Applying S_E yields

$$S_E(a)xS_E\,\omega_E(a^{-1}) \in E \qquad x \in E$$

whence, since ω_E commutes with S_E

$$\omega_E(S_E(a))x(S_E(a))^{-1} \in E \qquad x \in E.$$

Thus, $S_E(a) \in \Gamma_E$. □

11.6. The Map λ_E

Recall from Section 10.8 the antiautomorphism $a \to \bar{a}$ defined by $\bar{a} = S_E\,\omega_E(a)$. Proposition 11.5.1 implies that the Clifford group is stable under this map.

Now consider the (nonlinear) map $\theta: C_E \to C_E$ given by

$$\theta(a) = a\bar{a} \qquad a \in C_E.$$

Proposition 11.6.1. *If $a \in \Gamma_E$, then*

$$\theta(a) = \lambda_a \cdot e \qquad \lambda_a \in \Gamma^*$$

Moreover, the map $\lambda_E: \Gamma_E \to \Gamma^*$ given by $\lambda_E(a) = \lambda_a$ is a homomorphism from Γ_E to Γ^* (the multiplicative group of Γ).

PROOF. Since Γ_E is stable under conjugation, $\bar{a} \in \Gamma_E$. Now let $x \in E$ and set $y = \omega_E(\bar{a})x\bar{a}^{-1}$. Then $y \in E$ and so $S_E(y) = y$. It follows that

$$S_E(\bar{a})^{-1}xS_E(\omega_E\bar{a}) = \omega_E(\bar{a})x\bar{a}^{-1}$$

whence

$$xS_E(\omega_E\bar{a})\bar{a} = S_E(\bar{a})\omega_E(\bar{a})x \qquad x \in E.$$

Since

$$S_E(\omega_E\bar{a}) = a \quad \text{and} \quad S_E(\bar{a}) = \omega_E(a),$$

we obtain

$$x(a\bar{a}) = \omega_E(a\bar{a})x \qquad x \in E.$$

Thus, setting $a\bar{a} = b$, we have

$$xb = \omega_E(b)x \qquad x \in E.$$

Now write $b = b_0 + b_1$ with $b_0 \in C_E^0$ and $b_1 \in C_E^1$. Then the equation above yields

$$xb_0 = b_0x \quad \text{and} \quad xb_1 = -b_1x \qquad x \in E.$$

Thus, $b_0 \in Z_E^0$ and $b_1 \in (AZ_E)^1$. Now Lemmas I and II, Section 10.11, show that $b_0 = \lambda_a e$ and $b_1 = 0$ whence $b = \lambda_a \cdot e$ and so $\Phi_E(a) = \lambda_a \cdot e$. Finally, since a is invertible, it follows that $\lambda_a \neq 0$ and so $\lambda_a \in \Gamma^*$.

Now consider the map $\lambda_E : \Gamma_E \to \Gamma^*$ defined by

$$\Phi_E(a) = \lambda_E(a)e \qquad a \in \Gamma_E.$$

To show that λ_E is a homomorphism, let $a \in \Gamma_E$ and $b \in \Gamma_E$. Then

$$\Phi_E(ab) = ab \cdot \overline{ab} = ab\bar{b}\bar{a}.$$

By the first part of the proposition,

$$b\bar{b} = \lambda_E(b) \cdot e.$$

It follows that

$$\Phi_E(a \cdot b) = \lambda_E(b)a\bar{a} = \lambda_E(b)\lambda_E(a)e = \lambda_E(a)\lambda_E(b)e$$

whence

$$\lambda_E(ab) = \lambda_E(a)\lambda_E(b). \qquad \square$$

Corollary I. *The map θ satisfies*

$$\theta(\omega_E(a)) = \theta(a) \qquad a \in \Gamma_E.$$

PROOF. In fact,

$$\theta(\omega_E(a)) = \omega_E(a)\overline{\omega_E(a)} = \omega_E(a)\omega_E(\bar{a}) = \omega_E(a\bar{a}) = \lambda_E(a) \cdot e = \theta(a). \qquad \square$$

Corollary II. *The homomorphism λ_E satisfies the following two relations*:

1. $\lambda_E(\omega_E(a)) = \lambda_E(a) \qquad a \in \Gamma_E.$
2. $\lambda_E(\mathrm{ad}(b)a) = \lambda_E(a) \qquad a, b \in \Gamma_E.$

PROOF. (1) follows from Corollary I.

(2) Let $a \in \Gamma_E$ and $b \in \Gamma_E$. Then, since λ_E is a homomorphism, the first relation yields

$$\begin{aligned}\lambda_E(\mathrm{ad}(b)a) = \lambda_E(\omega(b)ab^{-1}) &= \lambda_E(\omega(b))\lambda_E(a)\lambda_E(b)^{-1}\\ &= \lambda_E(b)\lambda_E(a)\lambda_E(b)^{-1} = \lambda_E(a).\end{aligned}$$ □

Proposition 11.6.2. *Fix $a \in \Gamma_E$ and let τ_a denote the restriction of* ad(a) *to E. Then τ_a is an isometry.*

PROOF. Since for $x \in E$

$$\theta(x) = -(x, x) \cdot e,$$

it follows that

$$\lambda_E(x) = -(x, x) \qquad x \in E.$$

Now Part (2) of Corollary II to Proposition 11.6.1 yields

$$(\mathrm{ad}(a)x, \mathrm{ad}(a)x) = -\lambda_E(\mathrm{ad}(a)x) = -\lambda_E(x) = (x, x)$$

whence

$$(\tau_a x, \tau_a x) = (x, x) \qquad x \in E.$$ □

In particular, consider a vector $h \in E$ with $(h, h) \neq 0$ (recall that then $h \in \Gamma_E$). Since

$$hx + xh = 2(x, h)e,$$

we obtain

$$\tau_h(x) = x - 2\frac{(h, x)}{(h, h)}x \qquad x \in E.$$

This equation shows that

$$\tau_h(h) = -h,$$

while

$$\tau_h(y) = y \qquad \text{if } (h, y) = 0.$$

Thus τ_h is the reflection in the plane orthogonal to h.

11.7. The Homomorphism $\Phi_E: \Gamma_E \to O(E)$

Let $O(E)$ denote the group of isometries of E. Then a homomorphism

$$\Phi_E: \Gamma_E \to O(E)$$

is defined by

$$\Phi_E(a) = \tau_a \qquad a \in \Gamma_E.$$

Proposition 11.7.1. *The homomorphism Φ_E is surjective.*

PROOF. Let $h \in E$ be a vector such that $(h, h) \neq 0$ and let ρ_h denote the reflection in the plane perpendicular to h. Then $\Phi_E(h) = \rho_h$.

By Lemma I below, every isometry of E is generated by reflections and so the proposition follows. □

Lemma I. *Let E be an n-dimensional vector space with a nondegenerate inner product. Then every isometry τ of E is the product of at most $n + 1$ reflections.*

PROOF. We show first that if a and b are any two vectors such that $(a, a) = (b, b) \neq 0$, then there is a reflection ρ such that $\rho(a) = \pm b$.

In fact, since

$$(a + b, a + b) + (a - b, a - b) = 4(a, a) \neq 0,$$

it follows that $(a + b, a + b) \neq 0$ or $(a - b, a - b) \neq 0$. Replacing b by $-b$ if necessary, we may assume that $(a - b, a - b) \neq 0$. Now set $h = a - b$ and consider the reflection

$$\rho(x) = x - \frac{2(h, x)}{(h, h)} h.$$

Then $\rho(a) = a - (a - b) = b$.

Now we prove the lemma by induction on n. If $n = 1$, then $\tau x = \varepsilon \cdot x$, $\varepsilon = \pm 1$, and so τ is a product of at most two reflections. Suppose now that the lemma holds for a space with dimension $n - 1$ and let τ be an isometry of E ($\dim E = n$). Choose a vector $a \in E$ such that $(a, a) \neq 0$ and set $b = \tau(a)$. Then $(b, b) = (a, a)$ and so, by the remark above, there is a reflection ρ such that $\rho(a) = \pm b$. Now set

$$\tau_1 = \rho^{-1} \circ \tau.$$

Then $\tau_1(a) = \pm a$ and so τ_1 restricts to an isometry of the orthogonal complement, E_1, of a. Thus, by induction, τ_1 is the product of at most n reflections of E_1. Every such reflection extends to a reflection of E. Thus, τ is the product of at most $n + 1$ reflections and the induction is closed. □

Corollary I. *The Clifford group Γ_E is generated by the elements $h \in E$ which satisfy $(h, h) \neq 0$.*

PROOF. Let $a \in \Gamma_E$ and set $\Phi_E(a) = \tau$. By Lemma I, τ is of the form

$$\tau = \rho_{h_1} \circ \cdots \circ \rho_{h_r} \qquad h_i \in E$$

where $(h_i, h_i) \neq 0$. Since $\Phi_E(h_i) = \rho_{h_i}$, it follows that

$$\Phi_E(a^{-1}h_1 \cdots h_r) = \tau^{-1}\tau = \iota.$$

Thus,

$$a^{-1}h_1 \cdots h_r = \lambda \cdot e \qquad \lambda^* \in \Gamma^*$$

and so

$$a = \lambda^{-1}h_1 \cdots h_r = (\lambda^{-1}h_1)h_2 \cdots h_r.$$

□

Corollary II. *The homomorphism Φ_E satisfies*

$$\det \Phi_E(a) \cdot a = \omega_E(a) \qquad a \in \Gamma_E. \tag{11.4}$$

PROOF. Consider the homomorphism $\varphi: \Gamma_E \to \Gamma_E$ given by

$$\varphi(a) = \det \Phi_E(a) \cdot a.$$

Then we have, for $x \in E$, $(x, x) \neq 0$,

$$\varphi(x) = -x = \omega_E(x)$$

and so (11.4) holds in this case. Now apply Corollary I.

11.8. The Group *Pin*

From now on E will be a real vector space with a (nondegenerate) inner product of type (p, q). We shall write $C_E = C(p, q)$ and $\Gamma_E = \Gamma(p, q)$. Recall from Section 11.6, the homomorphism $\lambda_E: \Gamma(p, q) \to \mathbb{R}^*$. The group $Pin(p, q)$ is the subgroup of $\Gamma(p, q)$ consisting of the elements a which satisfy

$$\lambda_E(a) = \pm 1.$$

Since, for $x \in E$, $\lambda_E(x) = -(x, x)$, it follows that all the vectors $x \in E$ for which $(x, x) = \pm 1$ are contained in $Pin(p, q)$. In particular, $-e \in Pin(p, q)$. Moreover, Corollary I to Proposition 11.7.1 shows that $Pin(p, q)$ is generated by these vectors.

Next, denote by $O(p, q)$ the group of isometries of $\mathbb{R}^n$ and consider the homomorphism $\Phi_E: \Gamma(p, q) \to O(p, q)$ defined in Section 11.7. We shall show that the restriction Φ of Φ_E to $Pin(p, q)$ is still surjective. In fact, let $\alpha \in O(p, q)$. Then, by Proposition 11.7.1, there is an element $b \in \Gamma(p, q)$ such that $\Phi_E(b) = \alpha$. Now set

$$a = \frac{b}{|\lambda_E(b)|^{1/2}}.$$

Then

$$\lambda_E(a) = \frac{\lambda_E(b)}{|\lambda_E(b)|} = \pm 1$$

and so $a \in Pin(p, q)$. Moreover, $\Phi_E(a) = \Phi_E(b) = \alpha$. Thus Φ_E restricts to a surjective homomorphism

$$\Phi: Pin(p, q) \to O(p, q).$$

Now we show that $\ker \Phi = S^0$, where S^0 is the subgroup of $Pin(p, q)$ consisting of the elements e and $-e$. In fact, if $a \in \ker \Phi$, then $a \in \ker \mathrm{ad}$, and so $a = \lambda \cdot e$ (see Section 11.4). It follows that $\lambda_E(a) = \lambda^2$. On the other hand, since $a \in Pin(p, q)$, $|\lambda_E(a)| = 1$. These relations yield $\lambda^2 = 1$ whence $\lambda = \pm 1$ and so $a = \pm e$.

In view of the results above we have the exact sequence of groups

$$1 \to S^0 \xrightarrow{i} Pin(p, q) \xrightarrow{\Phi} O(p, q) \to 1$$

where i denotes the inclusion map.

11.9. The Group *Spin*

Let $Spin(p, q)$ denote the subgroup of $Pin(p, q)$ consisting of the elements which satisfy $\omega_E(a) = a$. Thus,

$$Spin(p, q) = Pin(p, q) \cap C^0(p, q).$$

Formula (11.4) shows that an element $a \in Pin(p, q)$ is contained in $Spin(p, q)$ if and only if $\det \Phi(a) = +1$; that is, if and only if $\Phi(a)$ is a *proper* isometry. Thus the homomorphism Φ restricts to a surjective homomorphism Ψ from $Spin(p, q)$ to the group $SO(p, q)$ of proper isometries. Since $S^0 \subset Spin(p, q)$ it follows that the kernel of this homomorphism is again S^0. Thus we have the exact sequence

$$1 \to S^0 \xrightarrow{i} Spin(p, q) \xrightarrow{\Psi} SO(p, q) \to 1.$$

Problems

1. Show that the homomorphism λ_E and the bilinear function Q_E (see Section 10.14, Problem 3) are connected by the relation

$$\lambda_E(a) = Q_E(a, \bar{a}) \qquad a \in \Gamma_E.$$

2. Let $\mathbb{R}^n$ be an n-dimensional Euclidean space and write $Pin(n, 0) = Pin(n)$ and $Spin(n, 0) = Spin(n)$. Determine the groups $Pin(n)$ and $Spin(n)$ ($n = 1, 2, 3$) explicitly. In particular, show that the group $Spin(3)$ is isomorphic to the group of unit quaternions.

The Spin Representation

11.10

Let F be a $2n$-dimensional Euclidean space. Recall that a complex structure in F is a linear transformation J which satisfies

$$J^2 = -\iota \quad \text{and} \quad (Jx, Jy) = (x, y) \qquad x, y \in E.$$

These relations imply that

$$\tilde{J} = -J$$

and so J is a skew transformation.

Next, consider the $2n$-dimensional complex vector space $E = \mathbb{C} \otimes F$ and define an inner product in E by

$$(\lambda \otimes x, \mu \otimes y) = \lambda\mu(x, y) \qquad \lambda, \mu \in \mathbb{C},\ x, y \in F.$$

Let ω denote the linear transformation of E given by

$$\omega(\lambda \otimes x) = i\lambda \otimes Jx \qquad \lambda \in \mathbb{C},\ x \in F.$$

Then we have

$$\omega^2(\lambda \otimes x) = (-\lambda) \otimes (-x) = \lambda \otimes x$$

whence

$$\omega^2 = \iota.$$

Thus ω is an involution. Moreover,

$$(\omega(\lambda \otimes x), \lambda \otimes x) = (i\lambda \otimes Jx, \lambda \otimes x) = i\lambda^2(Jx, x) = 0 \qquad \lambda \in \mathbb{C},\ x \in E$$

and so ω is skew.

Let E_1 and E_2 denote the following subspaces of E:

$$E_1 = \{x \mid \omega x = x\} \quad \text{and} \quad E_2 = \{x \mid \omega x = -x\}.$$

Then Proposition 10.14.2 shows that

$$C_E \cong L(\wedge E_1).$$

Moreover, the isomorphism $C_E \xrightarrow{\cong} L(\wedge E_1)$ is obtained as follows: Let $\varphi: E \to L(\wedge E_1)$ be the linear map given by

$$\varphi(x)u = x_1 \wedge u + i(x_2)u \qquad x \in E$$

$$x = x_1 \oplus x_2 \qquad x_1 \in E_1,\ x_2 \in E_2$$

and extend it to a homomorphism $R_E: C_E \to L(\wedge E_1)$. Then this homomorphism is an isomorphism,

$$R_E: C_E \xrightarrow{\cong} L(\wedge E_1).$$

In particular, the representation R_E is irreducible.

11.11. The Spin Representation

Recall that the inclusion map $j: F \to E$ induces a homomorphism $j_C: C_F \to C_E$. Thus the representation R_E determines a representation R_F of C_F by complex linear transformations of $\wedge E_1$ given by

$$R_F(a) = R_E(1 \otimes a) \qquad a \in C_F.$$

R_F is called the *spin representation* of C_F. A representation of a real algebra in a complex vector space V is called *irreducible* if the only stable (complex) subspaces are $W = V$ and $W = (0)$. We show that the spin representation is irreducible. In fact, assume that W is a stable subspace of $\wedge E_1$. Let $b \in C_E$ and write

$$b = \lambda \otimes a \qquad \lambda \in \mathbb{C},\, a \in C_F$$

(see Section 10.16). Then we have for $w \in W$

$$R_E(b)w = R_E(\lambda \otimes a)w = \lambda R_E(1 \otimes a)w = \lambda R_F(a).$$

Since W is a complex subspace of $\wedge E_1$, it follows that $R_E(b)w \in W$ and so W is stable under R_E. But R_E is irreducible and so it follows that $W = \wedge E_1$, or $W = (0)$.

11.12. The Hermitian Inner Product in $\wedge E_1$

Since $E = \mathbb{C} \otimes F$, we have the complex conjugation $z \mapsto \bar{z}$ in E given by $\lambda \otimes x \mapsto \bar{\lambda} \otimes x$. Now introduce a positive definite Hermitian inner product in E by setting

$$(z_1, z_2)_{\mathrm{H}} = (z_1, \bar{z}_2) \qquad z_1, z_2 \in E.$$

Then we have an induced Hermitian inner product in $\wedge E$. It is given by

$$(z_1 \wedge \cdots \wedge z_p, w_1 \wedge \cdots \wedge w_p)_{\mathrm{H}}$$
$$= (z_1 \wedge \cdots \wedge z_p, \overline{w}_1 \wedge \cdots \wedge \overline{w}_p) \qquad z_i \in E,\, w_i \in E.$$

Proposition 11.12.1. *Let $x \in F$. Then the transformation $R_F(x)$ is Hermitian symmetric.*

PROOF. Let $i_{\mathrm{H}}(z)$ denote the substitution operator in $\wedge E_1$ corresponding to the Hermitian inner product. We show that

$$i_{\mathrm{H}}(z) = i(\bar{z}) \qquad z \in E. \tag{11.5}$$

In fact, let $z, z_2, \ldots, z_p \in E_1$ and $w_1, \ldots, w_p \in E_1$. Then

$$\begin{aligned}(i_{\mathrm{H}}(z)(w_1 \wedge \cdots \wedge w_p), z_2 \wedge \cdots \wedge z_p)_{\mathrm{H}} \\ &= (w_1 \wedge \cdots \wedge w_p, z \wedge z_2 \wedge \cdots \wedge z_p)_{\mathrm{H}} \\ &= (w_1 \wedge \cdots \wedge w_p, \bar{z} \wedge \bar{z}_2 \wedge \cdots \wedge \bar{z}_p) \\ &= (i(\bar{z})(w_1 \wedge \cdots \wedge w_p), \bar{z}_2 \wedge \cdots \wedge \bar{z}_p) \\ &= (i(\bar{z})(w_1 \wedge \cdots \wedge w_p), z_2 \wedge \cdots \wedge z_p)_{\mathrm{H}},\end{aligned}$$

and so Formula (11.5) follows.

Next, let $x \in F$ and set

$$x_1 = \tfrac{1}{2}(x + \omega x) = \tfrac{1}{2}(1 \otimes x + i \otimes Jx)$$

and

$$x_2 = \tfrac{1}{2}(x - \omega x) = \tfrac{1}{2}(1 \otimes x - i \otimes Jx).$$

These relations show that $x_2 = \bar{x}_1$.

Now consider the linear transformation $R_F(x)$ of $\wedge E_1$. Then, since $i(x_2) = i(\bar{x}_1) = i_{\mathrm{H}}(x_1)$,

$$\begin{aligned}(R_F(x)u, v)_{\mathrm{H}} &= (x_1 \wedge u, v)_{\mathrm{H}} + (i(x_2)u, v)_{\mathrm{H}} \\ &= (u, i_{\mathrm{H}}(x_1)v)_{\mathrm{H}} + (i_{\mathrm{H}}(x_1)u, v)_{\mathrm{H}} \\ &= (u, i(x_2)v)_{\mathrm{H}} + (u, x_1 \wedge v)_{\mathrm{H}} = (u, R_F(x)v)_{\mathrm{H}}\end{aligned}$$

and so the proposition is proved. □

Corollary. *If $x \in F$, then*

$$(R_F(x)u, R_F(x)v)_{\mathrm{H}} = (x, x)(u, v)_{\mathrm{H}} \qquad u, v \in \wedge E_1.$$

In particular, if x is a unit vector, then $R_F(x)$ is a unitary transformation.

PROOF. In fact, by the proposition,

$$(R_F(x)u, R_F(x)v)_{\mathrm{H}} = (R_F(x)^2 u, v)_{\mathrm{H}} = (x, x)(u, v)_{\mathrm{H}}.$$ □

11.13. The Half Spin Representations

The spin representation restricts to a representation R_F^0 of the subalgebra C_F^0 in $\wedge E_1$. Now write

$$(\wedge E_1)^+ = \sum_{p \text{ even}} \wedge^p E_1 \quad \text{and} \quad (\wedge E_1)^- = \sum_{p \text{ odd}} \wedge^p E_1.$$

Then the spaces $(\wedge E_1)^+$ and $(\wedge E_1)^-$ are stable under the transformations $R_F(a)$, $a \in C_F^0$. Thus we have induced representations

$$R_F^+ : C_F^0 \to L(\wedge E_1)^+$$

and

$$R_F^- : C_F^0 \to L(\wedge E_1)^-$$

called the *half spin representations.*

Proposition 11.13.1. *The homomorphisms R_F^+ and R_F^- are isomorphisms. In particular, the half spin representations are irreducible.*

Proof. First observe that the maps R_F^+ and R_F^- are injective. Since $\dim_{\mathbb{C}}(\wedge E_1)^+ = 2^{n-1}$, we have $\dim_{\mathbb{C}} L(\wedge E_1)^+ = 2^{2n-2}$ and so $\dim_{\mathbb{R}} L(\wedge E_1)^+ = 2^{2n-1}$. On the other hand, $\dim C_F^0 = 2^{2n-1}$. Thus R_F^+ is an isomorphism. The same argument applies to R_F^-. □

The Wedderburn Theorems

In this paragraph we establish an algebraic structure theorem which will be used later to study the representations of $C_8(-)$.

11.14. Invariant Linear Maps

Let E and F be finite-dimensional vector spaces and let R be a representation of the algebra $L(E)$ in F. Thus R is a homomorphism

$$R: L(E) \to L(F).$$

A linear map $\alpha: E \to F$ will be called *R-invariant*, if it satisfies

$$\alpha \circ \varphi = R(\varphi) \circ \alpha \qquad \varphi \in L(E). \tag{11.6}$$

The R-invariant linear maps form a subspace of $L(E; F)$ denoted by $L_R(E; F)$.

A linear transformation ψ of F is called *R-invariant*, if

$$R(\varphi) \circ \psi = \psi \circ R(\varphi) \qquad \varphi \in L(E). \tag{11.7}$$

These transformations form a subspace of $L(F)$ denoted by $L_R(F)$.

11.15. The Isomorphism Φ_R

Let

$$\Phi_R: L_R(E; F) \otimes E \to F$$

be the linear map given by

$$\Phi_R(\alpha \otimes x) = \alpha(x) \qquad \alpha \in L_R(E; F), x \in E.$$

We show that the diagram

$$\begin{array}{ccc} L_R(E;F) \otimes E & \xrightarrow{\Phi_R} & F \\ \Big\downarrow{\scriptstyle \iota \otimes \varphi} & & \Big\downarrow{\scriptstyle R(\varphi)} \\ L_R(E;F) \otimes E & \xrightarrow{\Phi_R} & F \end{array}$$

commutes. In fact, let $\varphi \in L(E)$ and $\alpha \in L_R(E;F)$. Then we have, in view of (11.6),

$$\Phi_R[(1 \otimes \varphi)(\alpha \otimes x)] = R(\varphi)(\alpha x) = R(\varphi)\Phi_R(\alpha \otimes x).$$

Thus,

$$\Phi_R \circ (\iota \otimes \varphi) = R(\varphi) \circ \Phi_R \qquad \varphi \in L(E). \tag{11.8}$$

11.16

Theorem 11.16.1 (Wedderburn). *Φ_R is a linear isomorphism,*

$$\Phi_R : L_R(E; F) \otimes E \xrightarrow{\cong} F.$$

Proof. We shall construct an inverse map

$$\Psi : F \to L_R(E;F) \otimes E.$$

Choose a pair of dual bases $\{e^{*\nu}\}$, $\{e_\nu\}$ $(\nu = 1, \ldots, n)$ of E^* and E and set

$$\varphi_\nu^\mu = T_E(e^{*\mu} \otimes e_\nu),$$

where $T_E : E^* \otimes E \xrightarrow{\cong} L(E)$ is the canonical isomorphism defined by (6.5). Observe that T_E satisfies

$$\varphi \circ T_E(x^* \otimes x) = T_E(x^* \otimes \varphi x) \qquad \varphi \in L(E). \tag{11.9}$$

Next, define linear maps $T^\mu : F \to L(E; F)$ by

$$T^\mu(y)x = \sum_\nu \langle e^{*\nu}, x\rangle R(\varphi_\nu^\mu)y \qquad x \in E,\ y \in F. \tag{11.10}$$

Lemma. *The maps T^μ have the following properties:*

1. $T^\mu(y) \in L_R(E; F)$, $y \in F$.
2. $T^\mu(\alpha x) = \langle e^{*\mu}, x\rangle \alpha$, $\alpha \in L_R(E; F)$.
3. $\sum_\mu T^\mu(y)e_\mu = y$, $y \in F$.

Proof. (1) It has to be shown that

$$T^\mu(y) \circ \varphi = R(\varphi) \circ T^\mu(y) \qquad \psi \in L(E).$$

Fix $y \in F$. Then, by (11.10)

$$R(\varphi)T^\mu(y)x = \sum_\nu \langle e^{*\nu}, x\rangle R(\varphi \circ \varphi_\nu^\mu)y.$$

On the other hand,

$$T^\mu(y)\varphi x = \sum_\nu \langle e^{*\nu}, \varphi x\rangle R(\varphi_\nu^\mu)y.$$

Hence we have to establish the relation

$$\sum_\nu \langle e^{*\nu}, x\rangle R(\varphi \circ \varphi_\nu^\mu) = \sum_\nu \langle e^{*\nu}, \varphi x\rangle R(\varphi_\nu^\mu). \tag{11.11}$$

Formula (11.9) yields, for $x^* = e^{*\mu}$ and $x = e_\nu$,

$$\varphi \circ \varphi_\nu^\mu = T_E(e^{*\mu} \otimes \varphi e_\nu).$$

It follows that

$$\begin{aligned}\sum_\nu \langle e^{*\nu}, x\rangle \varphi \circ \varphi_\nu^\mu &= \sum_\nu \langle e^{*\nu}, x\rangle T_E(e^{*\mu} \otimes \varphi e_\nu) \\ &= T_E(e^{*\mu} \otimes \varphi x) = T_E\left(e^{*\mu} \otimes \sum_\nu \langle e^{*\nu}, \varphi x\rangle e_\nu\right) \\ &= \sum_\nu \langle e^{*\nu}, \varphi x\rangle \varphi_\nu^\mu.\end{aligned}$$

Thus,

$$\sum_\nu \langle e^{*\nu}, x\rangle \varphi \circ \varphi_\nu^\mu = \sum_\nu \langle e^{*\nu}, \varphi x\rangle \varphi_\nu^\mu.$$

Applying R to this formula we obtain (11.11).

(2) Let $a \in E$. Then

$$\begin{aligned}T^\mu(\alpha x)a &= \sum_\nu \langle e^{*\nu}, a\rangle R(\varphi_\nu^\mu)(\alpha x) \\ &= \sum_\nu \langle e^{*\nu}, a\rangle \alpha(\varphi_\nu^\mu x).\end{aligned}$$

Since

$$\varphi_\nu^\mu x = \langle e^{*\mu}, x\rangle e_\nu,$$

it follows that

$$\begin{aligned}T^\mu(\alpha x)a &= \sum_\nu \langle e^{*\nu}, a\rangle \langle e^{*\mu}, x\rangle \alpha(e_\nu) \\ &= \langle e^{*\mu}, x\rangle \cdot \alpha(a)\end{aligned}$$

whence

$$T^\mu(\alpha x) = \langle e^{*\mu}, x\rangle \alpha.$$

(3) In fact, since

$$T_E \sum_\nu (e^{*\nu} \otimes e_\nu) = \iota,$$

it follows that

$$\sum_\mu T^\mu(y) e_\mu = y \qquad y \in F.$$

This completes the proof of the lemma. □

Now let

$$\Psi: F \to L_R(E; F) \otimes E$$

be the linear map given by

$$\Psi(y) = \sum_\mu T^\mu(y) \otimes e_\mu.$$

Then we have for $\alpha \in L_R(E; F)$ and $x \in E$

$$\begin{aligned} \Psi\Phi_R(\alpha \otimes x) = \Psi(\alpha x) &= \sum_\mu T^\mu(\alpha x) \otimes e_\mu \\ &= \sum_\mu \langle e^{*\mu}, x \rangle \alpha \otimes e_\mu = \alpha \otimes x \end{aligned}$$

whence

$$\Psi \circ \Phi_R = \iota.$$

On the other hand, for $y \in F$,

$$\Phi_R \Psi(y) = \Phi_R \sum_\mu T^\mu(y) \otimes e_\mu = \sum_\mu T^\mu(y)(e_\mu) = y$$

whence $\Phi_R \circ \Psi = \iota$. It follows that Φ is an isomorphism. □

Corollary.

$$\dim F = \dim L_R(E; F) \cdot \dim E.$$

Thus $\dim E$ *divides* $\dim F$.

11.17. The Isomorphism θ_R

Observe that the composition map

$$L(F) \times L(E; F) \to L(E; F)$$

restricts to a bilinear mapping

$$L_R(F) \times L_R(E; F) \to L_R(E; F).$$

To simplify notation we set $L_R(E; F) = U$. Then a linear map

$$\theta_R : L_R(F) \to L(U)$$

is defined by

$$\theta_R(\psi)\alpha = \psi \circ \alpha \qquad \psi \in L_R(F), \alpha \in U.$$

Clearly,

$$\theta_R(\psi_2 \circ \psi_1) = \theta_R(\psi_2) \circ \theta_R(\psi_1)$$

and so θ is an algebra homomorphism.

11.18

Theorem 11.18.1. (Wedderburn). *θ_R is an isomorphism.*

Proof. We construct an inverse map

$$\Omega_R : L(U) \to L_R(F).$$

Let $\gamma \in L(U)$ and set

$$\Omega_R(\gamma) = \Phi_R \circ (\gamma \otimes \iota) \circ \Phi_R^{-1}.$$

Then $\Omega_R(\gamma) \in L(F)$. Since, by (11.8), $R(\varphi) \circ \Omega_R(\gamma) = \Omega_R(\gamma) \circ R(\varphi)$, $\varphi \in L(E)$, it follows that $\Omega_R(\gamma) \in L_R(F)$.

Now we show that

$$\theta_R \circ \Omega_R = \iota \quad \text{and} \quad \Omega_R \circ \theta_R = \iota. \tag{11.12}$$

In fact, let $\gamma \in L(U_R)$ and $\alpha \in U$. Then

$$\begin{aligned}[(\theta_R \circ \Omega_R)(\gamma)]\alpha &= [\theta_R(\Phi_R \circ (\gamma \otimes \iota) \circ \Phi_R^{-1})]\alpha \\ &= \Phi_R \circ (\gamma \otimes \iota) \circ \Phi_R^{-1} \circ \alpha.\end{aligned}$$

Now fix $x \in E$. Then, by definition of Φ_R,

$$\Phi_R^{-1}(\alpha x) = \alpha \otimes x.$$

It follows that

$$[\theta_R \Omega_R(\gamma)\alpha]x = \Phi_R(\gamma \otimes \iota)(\alpha \otimes x) = \Phi_R(\gamma(\alpha) \otimes x) = \gamma(\alpha)x.$$

Thus,

$$(\theta_R \circ \Omega_R)\gamma = \gamma \qquad \gamma \in L(U)$$

and so the first relation (11.12) follows.

To establish the second relation, let $\psi \in L_R(F)$. Then

$$(\Omega_R \circ \theta_R)\psi = \Phi_R \circ (\theta_R(\psi) \otimes \iota) \circ \Phi_R^{-1}.$$

From the proof of Theorem 11.16.1 we have for $y \in F$

$$\Phi_R^{-1}(y) = \sum_\mu T^\mu(y) \otimes e_\mu$$

whence

$$(\theta_R(\psi) \otimes \iota)\Phi_R^{-1}(y) = \sum_\mu \theta_R(\psi)T^\mu(y) \otimes e_\mu.$$

It follows that

$$\Phi_R(\theta_R(\psi) \otimes \iota)\Phi_R^{-1}(y) = \sum_\mu \theta_R(\psi)[T^\mu(y)e_\mu] = \psi\left[\sum_\mu T^\mu(y)e_\mu\right] = \psi(y)$$

(see the lemma in Section 11.16). We thus obtain

$$\Phi_R \circ (\theta_R(\psi) \otimes \iota) \circ \Phi_R^{-1} = \psi;$$

i.e.,

$$(\Omega_R \circ \theta_R)\psi = \psi \qquad \psi \in L_R(F).$$

This completes the proof of Theorem 11.18.1. □

11.19

Theorem 11.19.1. *Let A be an associative algebra with unit element e and let R be a representation of the algebra $A \otimes L(E)$ in a vector space F. Then there exists both a representation R_U of A in a vector space U and an isomorphism $\Phi: U \otimes E \xrightarrow{\cong} F$ which makes the diagrams*

$$\begin{array}{ccc} U \otimes E & \xrightarrow{R_U(a)\otimes\varphi} & U \otimes E \\ \Phi\downarrow\cong & & \cong\downarrow\Phi \\ F & \xrightarrow{R(a\otimes\varphi)} & F \end{array} \qquad a \in A,\ \varphi \in L(E) \tag{11.13}$$

commute. Thus R is equivalent to $R_U \otimes \iota$ where ι denotes the standard representation of $L(E)$ in E.

Proof. Define representations P and Q of A and $L(E)$ in F by setting

$$P(a) = R(a \otimes \iota) \qquad a \in A$$

and

$$Q(\varphi) = R(e \otimes \varphi) \qquad \varphi \in L(E).$$

Let $L_Q(F)$ denote the subspace of F which is invariant under Q. We show that

$$P(a) \in L_Q(F) \qquad a \in A.$$

In fact, let $\varphi \in L(E)$. Then

$$\begin{aligned} P(a) \circ Q(\varphi) &= R(a \otimes \iota) \circ R(e \otimes \varphi) \\ &= R(a \otimes \varphi) = R[(e \otimes \varphi) \circ (a \otimes \iota)] \\ &= R(e \otimes \varphi) \circ R(a \otimes \iota) \\ &= Q(\varphi) \circ P(a). \end{aligned}$$

Now set $U = L_R(E; F)$. Then, by Theorem 11.16.1, there is an isomorphism

$$\Phi : U \times E \xrightarrow{\cong} F$$

such that

$$\Phi \circ (\iota \otimes \varphi) = Q(\varphi) \circ \Phi \qquad \varphi \in L(E). \tag{11.14}$$

By Theorem 11.18.1 there is an algebra isomorphism

$$\Omega : L(U) \xrightarrow{\cong} L_Q(F).$$

It is defined by

$$\Omega(\gamma) = \Phi \circ (\gamma \otimes \iota) \circ \Phi^{-1} \qquad \gamma \in L(U). \tag{11.15}$$

Thus a representation R_U of A in U is given by

$$R_U(a) = \Omega^{-1} P(a) \qquad a \in A. \tag{11.16}$$

Relations (11.16) and (11.15) imply that

$$P(a) = \Omega R_U(a) = \Phi \circ (R_U(a) \otimes \iota) \circ \Phi^{-1}.$$

Thus,

$$\Phi \circ (R_U(a) \otimes \iota) = P(a) \circ \Phi \qquad a \in A. \tag{11.17}$$

Relations (11.14) and (11.17) yield

$$\begin{aligned} \Phi \circ (R_U(a) \otimes \varphi) &= \Phi \circ (R_U(a) \otimes \iota) \circ (\iota \otimes \varphi) \\ &= P(a) \circ \Phi \circ (\iota \otimes \varphi) \\ &= P(a) \circ Q(\varphi) \circ \Phi \\ &= R(a \otimes \varphi) \circ \Phi. \end{aligned}$$

Thus Diagram (11.13) commutes and the proof of Theorem 11.19.1 is complete. □

Representations of $C_k(-)$

11.20. The Radon–Hurwitz Number

In this section we shall denote $C_k(-)$ simply by C_k. Let R be a representation of C_k in a Euclidean n-space $\mathbb{R}^n$. We show that then $k \leq n - 1$. In fact, by Proposition 11.3.1 we may assume that R is a (negative) orthogonal representation. Let $\{e_1, \ldots, e_k\}$ be an orthonormal basis of $\mathbb{R}^k$ and set $R(e_i) = \sigma_i$, $(i = 1, \ldots, k)$. Then we have the relations

$$\sigma_i \circ \sigma_j + \sigma_j \circ \sigma_i = -2\delta_{ij} \cdot \iota$$

in particular, $\sigma_i^2 = -\iota\,(i = 1, \ldots, k)$. Moreover, the remark preceding Proposition 11.3.1 shows that σ_i is skew $(i = 1, \ldots, k)$. Now fix a unit vector $a \in \mathbb{R}^n$ and set $\sigma_i(a) = a_i (i = 1, \ldots, k)$. Then

$$(a, a_i) = (a, \sigma_i a) = 0 \qquad (i = 1, \ldots, k)$$

and

$$(a_i, a_j) = (\sigma_i a, \sigma_j a) = (\sigma_j \sigma_i a, \sigma_j^2 a).$$

It follows that

$$2(a_i, a_j) = -((\sigma_j \sigma_i + \sigma_i \sigma_j)a, a) = 2\delta_{ij}(a, a) = 2\delta_{ij}$$

whence

$$(a_i, a_j) = \delta_{ij} \qquad (i, j = 1, \ldots, k).$$

Thus the vectors $a, a_1, \ldots, a_k$ form an orthonormal $(k + 1)$-frame in $\mathbb{R}^n$. This implies that $k + 1 \leq n$. Thus to every $n \geq 1$ there is a largest $k \geq 0$ such that C_k represents in $\mathbb{R}^n$. This number is called the *Radon–Hurwitz number* of $\mathbb{R}^n$ and will be denoted by $K(n)$. It follows from the above that

$$K(n) \leq n - 1. \tag{11.18}$$

Proposition 11.20.1. *The Radon–Hurwitz number satisfies the functional equation*

$$K(16n) = K(n) + 8 \qquad n \geq 1.$$

PROOF. In view of Theorem 10.17.1, Section 10.21, and the table in Section 10.20, we have an isomorphism

$$\Psi: C_{k+8} \xrightarrow{\cong} C_k \otimes L(\mathbb{R}^{16}).$$

Thus, if P is a representation of C_k in $\mathbb{R}^n$, then

$$Q = (P \otimes \iota) \circ \Psi$$

is a representation of C_{k+8} in $\mathbb{R}^n \otimes \mathbb{R}^{16} \cong \mathbb{R}^{16n}$. It follows that

$$K(16n) \geq K(n) + 8.$$

Conversely, let Q be a representation of C_{k+8} in $\mathbb{R}^{16n}$ and set $R = Q \circ \Psi^{-1}$. Then R is a representation of $C_k \otimes C_8$ in $\mathbb{R}^{16n}$. By Theorem 11.19.1 (applied with $A = C_k$ and $E = \mathbb{R}^{16}$) there is a representation of C_k in a vector space U, where

$$U \otimes \mathbb{R}^{16} \cong \mathbb{R}^{16n}.$$

It follows from this relation that $\dim U = n$. Thus,

$$K(n) \geq K(16n) - 8. \qquad \square$$

Theorem 11.20.2. *Write*

$$n = 16^a \cdot 2^b \cdot q \qquad a \geq 0, 0 \leq b \leq 3, q \text{ odd}.$$

Then the Random–Hurwitz number of $\mathbb{R}^n$ *is given by*

$$K(n) = 8a + 2^b - 1 \qquad n \geq 1.$$

In particular, if n is odd, then $K(n) = 0$.

Proof. In view of Proposition 11.20.1 we have only to show that

$$K(2^b \cdot q) = 2^b - 1 \qquad 0 \leq b \leq 3, q \text{ odd}.$$

This will be proved in Lemma II of the next section.

11.21

Lemma I. *Let q be odd and* $0 \leq b \leq 3$. *Then*

$$K(2^b \cdot q) \leq 7.$$

Proof. Assume that C_k represents in $\mathbb{R}^n$ and $k \geq 8$. Write $k = l + 8$, $l \geq 0$. Then $C_k \cong C_l \otimes C_8 \cong C_l \otimes L(\mathbb{R}^{16})$ represents in $\mathbb{R}^n$. Thus, by Theorem 11.19.1, 16 divides n and so n cannot be of the form $2^b \cdot q$ ($0 \leq b \leq 3$, q odd). $\square$

Lemma II. *Let q be odd and* $0 \leq b \leq 3$. *Then*

$$K(2^b \cdot q) = 2^b - 1.$$

Proof. It has to be shown that, for odd q,

1. $K(q) = 0$.
2. $K(2q) = 1$.
3. $K(4q) = 3$.
4. $K(8q) = 7$.

(1) $K(q) = 0$: By Lemma I, $K(q) \leq 7$. Thus it has to be shown that if C_k represents in $\mathbb{R}^q$ and $0 \leq k \leq 7$, then $k = 0$. By the table in Section 10.20

every C_k $(1 \le k \le 7)$ contains $\mathbb{C}$ as a subalgebra. Thus a representation of C_k $(1 \le k \le 7)$ in $\mathbb{R}^q$ determines a representation of $\mathbb{C}$ in $\mathbb{R}^q$. This is impossible, since q is odd. It follows that $q = 0$.

(2) $K(2q) = 1$: We show that

$$K(2q) \le 1. \tag{11.19}$$

Lemma I implies that $K(2q) \le 7$. Now the isomorphisms in Section 10.20 show that C_k contains $\mathbb{H}$ as a subalgebra if $2 \le k \le 7$ and so a representation of C_k $(2 \le k \le 7)$ in $\mathbb{R}^{2q}$ induces a representation of $\mathbb{H}$ in $\mathbb{R}^{2q}$. This is impossible since $2q$ is not divisible by 4 (see Lemma III below). Thus, $k = 1$ and so (11.19) is proved.

On the other hand, $C_1 \cong \mathbb{C}$ represents in $\mathbb{R}^{2q}$ and so

$$K(2q) \ge 1.$$

It follows that $K(2q) = 1$.

(3) $K(4q) = 3$: We show first that

$$K(4q) \le 3. \tag{11.20}$$

By Lemma I, $K(4q) \le 7$. Thus we have to show that for $4 \le k \le 7$ the algebra C_k does not represent in $\mathbb{R}^{4q}$. By the isomorphisms in Section 10.20 every such algebra is of the form

$$C_k = B_k \otimes L(\mathbb{R}^2),$$

where B_k contains $\mathbb{H}$ as a subalgebra. In fact,

$$B_4 = \mathbb{H}, \qquad B_5 = \mathbb{H} \otimes \mathbb{C},$$
$$B_6 = \mathbb{H} \otimes \mathbb{H}, \qquad B_7 = \mathbb{H} \otimes \mathbb{H} \oplus \mathbb{H} \otimes \mathbb{H}.$$

Now let R be a representation of C_k in $\mathbb{R}^{4q}$. Then Theorem 11.19.1 (applied with $A = B_k$ and $E = \mathbb{R}^2$) shows that there is a representation R_U of B_k in a vector space U where

$$\mathbb{R}^{4q} \cong U \otimes \mathbb{R}^2.$$

Since $\mathbb{H} \subset B_k$, R_U determines a representation of $\mathbb{H}$ in U. Thus dim U is divisible by 4 (see Lemma III below) and hence dim $\mathbb{R}^{4q}$ is divisible by 8. This is impossible since q is odd, and so (11.20) follows.

On the other hand,

$$K(4q) \ge 3. \tag{11.21}$$

In fact, write

$$C_3 = \mathbb{H} \oplus \mathbb{H}$$

(see Section 10.20). Then a representation R of C_3 in $\mathbb{R}^4$ $(\cong \mathbb{H})$ is given by

$$R(p \oplus q)x = p \cdot x \qquad x \in \mathbb{H}.$$

Thus

$$\underbrace{R \oplus \cdots \oplus R}_{q}$$

is a representation of C_3 in $\mathbb{R}^{4q}$ and (11.21) follows.

(4) $K(8q) = 7$: By Lemma I,

$$K(8q) \le 7.$$

To show that

$$K(8q) \ge 7,$$

we construct a representation of C_7 in $\mathbb{R}^{8q}$. Write

$$C_7 = L(\mathbb{R}^8) \oplus L(\mathbb{R}^8)$$

(see Section 10.20) and set

$$R(\alpha \oplus \beta) = \alpha \qquad \alpha, \beta \in L(\mathbb{R}^8).$$

Then R is a representation of C_7 in $\mathbb{R}^8$ and so

$$\underbrace{R \oplus \cdots \oplus R}_{q}$$

is a representation of C_7 in $\mathbb{R}^{8q}$. Thus, $K(8q) \ge 7$.

This completes the proof of Lemma II and hence the proof of the theorem. □

EXAMPLES

$$K(2) = 2^1 - 1 = 1, \qquad K(4) = 2^2 - 1 = 3$$

$$K(6) = 2^1 - 1 = 1, \qquad K(8) = 2^3 - 1 = 7$$

$$K(10) = 2^1 - 1 = 1, \qquad K(16) = 8 + 2^0 - 1 = 8.$$

Lemma III. *Let R be a representation of $\mathbb{H}$ in an n-dimensional vector space E. Then n is a multiple of 4.*

PROOF. Write

$$R(p)(x) = p \cdot x \qquad p \in \mathbb{H}, x \in E.$$

We shall say that a family of vectors $x_1, \ldots, x_k \in E$ *generates E over* $\mathbb{H}$, if every $x \in E$ can be written in the form

$$x = \sum_{i=1}^{k} p_i \cdot x_i \qquad p_i \in \mathbb{H}.$$

Let m be the least number such that E is generated by m vectors and let $x_1, \ldots, x_m$ be such a family. Then it is easy to show that the relation

$$\sum_{i=1}^{m} p_i \cdot x_i = 0$$

implies that $p_i = 0\,(i = 1, \ldots, m)$. Now choose a basis $\{e, e_1, e_2, e_3\}$ of $\mathbb{H}$. Then it follows that the $4m$ vectors

$$x_i, e_1 x_i, e_2 x_i, e_3 x_i \qquad (i = 1, \ldots, m)$$

form a basis of E over $\mathbb{R}$. Thus, $n = 4m$. □

11.22. Orthogonal Multiplications

Let E and F be Euclidean spaces. An *orthogonal multiplication* between E and F is a bilinear mapping $E \times F \to F$, denoted by $(x, y) \to x \cdot y$ with the following properties:

i. $|x \cdot y| = |x||y|$, $x \in E$, $y \in F$.
ii. There is an element $e \in E$ such that $e \cdot y = y$, $y \in F$.

Property (i) implies, in view of the symmetry of the inner products, that

$$(x_1 \cdot y, x_2 \cdot y) = (x_1, x_2)|y|^2 \qquad x_1, x_2 \in E, y \in F$$

and

$$(x \cdot y_1, x \cdot y_2) = |x|^2(y_1, y_2) \qquad x \in E, y_1, y_2 \in F.$$

As an example consider the algebra of quaternions ($E = F \cong \mathbb{H}$) (see Section 7.23 of *Linear Algebra*).

Given an orthogonal multiplication denote the orthogonal complement of e by E_1. Every vector $a \in E_1$ determine a linear transformation σ_a of F given by

$$\sigma_a(y) = a \cdot y \qquad y \in F.$$

Relations (i) and (ii) imply that this transformation satisfies

$$(\sigma_a y, \sigma_a y) = |a|^2 \cdot |y|^2$$

and

$$(\sigma_a y, y) = 0 \qquad a \in E_1, y \in F.$$

In particular, if $|a| = 1$, then σ_a is a rotation of F.

11.23. Orthogonal Systems of Skew Transformations

Let F be an n-dimensional Euclidean space and let $\{\sigma_1, \ldots, \sigma_k\}$ be a family of skew linear transformations of F. The family $\{\sigma_1, \ldots, \sigma_k\}$ will be called *orthogonal*, if

$$(\sigma_i y, \sigma_j y) = (y, y) \cdot \delta_{ij} \qquad y \in F. \tag{11.22}$$

This relation is equivalent to the relation

$$\sigma_i \circ \sigma_j + \sigma_j \circ \sigma_i = -2\delta_{ij} \cdot \iota. \tag{11.23}$$

In fact, (11.22) implies that

$$(\sigma_i y, \sigma_j z) + (\sigma_i z, \sigma_j y) = 2(y, z) \cdot \delta_{ij} \qquad y, z \in F.$$

Since the σ_i are skew, it follows that

$$(\sigma_j \sigma_i y, z) + (z, \sigma_i \sigma_j y) = -2(y, z)\delta_{ij}$$

whence

$$(\sigma_j \sigma_i + \sigma_i \sigma_j)y = -2\delta_{ij} y \qquad y \in F.$$

Conversely, assume that (11.23) holds. Then we have

$$\begin{aligned}(\sigma_i y, \sigma_j y) &= -(\sigma_j \sigma_i y, y) = 2\delta_{ij}(y, y) + (\sigma_i \sigma_j y, y)\\ &= 2\delta_{ij}(y, y) - (\sigma_j y, \sigma_i y) \qquad y \in F.\end{aligned}$$

It follows that

$$(\sigma_i y, \sigma_j y) = \delta_{ij}(y, y).$$

Every orthogonal family of k skew transformations of F determines an orthogonal multiplication between $\mathbb{R}^{k+1}$ and F. In fact, choose an orthonormal basis $\{e, e_1, \ldots, e_k\}$ in $\mathbb{R}^{k+1}$ and set

$$x \cdot y = \lambda y + \sum_i \lambda^i \sigma_i(y) \qquad x \in \mathbb{R}^{k+1}, y \in F,$$

where

$$x = \lambda e + \sum_i \lambda^i e_i.$$

Then, clearly, $e \cdot y = y$, $y \in F$. Moreover,

$$\begin{aligned}|x \cdot y|^2 &= \lambda^2 |y|^2 + 2\lambda \sum_i \lambda^i (y, \sigma_i y) + \sum_{i,j} \lambda^i \lambda^j (\sigma_i y, \sigma_j y)\\ &= \lambda^2 |y|^2 + \sum_{i,j} \delta_{ij} \lambda^i \lambda^j |y|^2\\ &= \left(\lambda^2 + \sum_i \lambda^i \lambda^i\right) \cdot |y|^2 = |x|^2 |y|^2\end{aligned}$$

and so we have an orthogonal multiplication $\mathbb{R}^{k+1} \times F \to F$.

Conversely, let $\mathbb{R}^{k+1} \times F \to F$ be an orthogonal multiplication. Denote the orthogonal complement of e by $\mathbb{R}^k$ and choose an orthonormal basis $\{e_1, \ldots, e_k\}$ of $\mathbb{R}^k$. Define σ_i by

$$\sigma_i(y) = e_i \cdot y \qquad y \in F.$$

Then the σ_i form an orthogonal system of skew transformations.

11.24. Orthogonal Multiplications and Representations of C_k

Let $E \times F \to F$ be an orthogonal multiplication and denote the orthogonal complement of e by E_1. Consider the linear map $P: E_1 \to L(F)$ given by

$$P(a)y = a \cdot y \qquad a \in E, y \in F.$$

Then

$$(y, P(a)y) = (e, a)(y, y) = 0$$

and so the transformations $P(a)$ are skew. This implies that

$$(P(a)^2 y, z) = -(P(a)y, P(a)z) = -(a \cdot y, a \cdot z) = -(a, a)(y, z) \qquad y, z \in F,$$

whence

$$P(a)^2 = -(a, a) \cdot \iota.$$

Now introduce a negative definite inner product in E_1 by setting

$$(a, b)^- = -(a, b) \qquad a, b \in E_1.$$

Then the equation above reads

$$P(a)^2 = (a, a)^- \cdot \iota \qquad a \in E_1^-.$$

Thus P extends to a homomorphism from the Clifford algebra C_k, ($\dim E_1 = k$) into $L(F)$ and so it determines a representation of C_k in F.

Conversely, let R be a representation of C_k in F. By Proposition 11.3.1 R is equivalent to a (negative) orthogonal representation P of C_k in F. P satisfies the relations

$$P(a)^2 = P(a^2) = (a, a)^- \cdot \iota = -(a, a) \cdot \iota \qquad a \in E_1$$

and

$$(P(a)y, P(a)y) = -(a, a)^- \cdot (y, y) = (a, a) \cdot (y, y) \qquad a \in E_1, y \in F.$$

In particular, the transformations $P(a)$ are skew.

Now set $E = (e) \oplus \mathbb{R}^k$ and define a bilinear mapping $E \times F \to F$ by setting

$$x \cdot y = \lambda y + P(a)y,$$

where $x \in E$, $y \in F$, $x = \lambda e + a$, $a \in \mathbb{R}^k$.

Then, clearly, $e \cdot y = y$. Moreover,

$$|x \cdot y|^2 = \lambda^2 |y|^2 + 2\lambda(y, P(a)y) + |P(a)y|^2 = \lambda^2 |y|^2 + (a, a) \cdot |y|^2$$
$$= |x|^2 \cdot |y|^2 \qquad x \in E, y \in F,$$

and so this bilinear mapping is an orthogonal multiplication between E and F.

Thus there is a one-to-one correspondence between orthogonal multiplications $E \times F \to F$ and representations of C_k in F ($\dim E = k + 1$).

Theorem 11.24.1. *Assume that there exists an orthogonal multiplication* $\mathbb{R}^n \times \mathbb{R}^n \to \mathbb{R}^n$. *Then* $n = 1, 2, 4,$ *or* 8.

PROOF. The orthogonal multiplication in $\mathbb{R}^n$ determines a representation of C_{n-1} in $\mathbb{R}^n$. Thus, $n - 1 \leq K(n)$. On the other hand, in view of Relation (11.18) $K(n) \leq n - 1$. Thus,

$$K(n) = n - 1. \tag{11.24}$$

Now write

$$n = 16^a \cdot 2^b \cdot q \qquad a \geq 0, 0 \leq b \leq 3, q \text{ odd}$$

Then, by Theorem 11.20.2,

$$K(n) = 8a + 2^b - 1$$

and so Equation (11.24) implies that

$$8a + 2^b = 16^a \cdot 2^b \cdot q; \tag{11.25}$$

that is,

$$8a = 2^b(16^a \cdot q - 1) \qquad 0 \leq b \leq 3.$$

It follows that

$$8a \geq 16^a - 1.$$

Since $8x < 16^x - 1$ for $x > 1$, $x \in \mathbb{R}$, we obtain $a = 0$. Now Formula (11.25) shows that $q = 1$. Thus, $n = 2^b$ $(0 \leq b \leq 3)$; i.e., $n = 1, 2, 4, 8$. □

Remark. If $n = 1, 2, 4$, or 8, there are indeed orthogonal multiplications in $\mathbb{R}^n$. For $n = 1$ and $n = 2$ we have the ordinary multiplication of real (respectively complex) numbers, for $n = 4$ the algebra of quaternions and for $n = 8$ the algebra of Cayley numbers (see Problem 5).

11.25. Orthonormal k-Frames on S^{n-1}

Let S^{n-1} be the unit sphere in $\mathbb{R}^n$. A (continuous) *orthonormal k-frame* on S^{n-1} is a system of k continuous maps $X_i: S^{n-1} \to \mathbb{R}^n$ satisfying

$$(x, X_i(x)) = 0 \quad \text{and} \quad (X_i(x), X_j(x)) = \delta_{ij} \qquad x \in S^{n-1}.$$

Now let $\sigma_1, \ldots, \sigma_k$ be an orthonormal system of k skew transformations of $\mathbb{R}^n$ and set

$$X_i(x) = \sigma_i(x) \qquad x \in S^{n-1}, (i = 1, \ldots, k).$$

Then we have for $x \in S^{n-1}$

$$(x, X_i(x)) = (x, \sigma_i(x)) = 0$$

and

$$(X_i(x), X_j(x)) = \delta_{ij}.$$

Thus every orthogonal system of k skew transformations of $\mathbb{R}^n$ determines an orthonormal k-frame on S^{n-1}. Now, combining the results of Sections 11.23 and 11.24 and Theorem 11.20.1, we see that the $(n-1)$-sphere admits an orthogonal k-frame with $k = K(n)$.

$$\text{Hence there is an orthonormal} \begin{cases} \text{1-frame on } S^1, \\ \text{3-frame on } S^3, \\ \text{1-frame on } S^5, \\ \text{7-frame on } S^7, \\ \text{1-frame on } S^9, \\ \text{8-frame on } S^{15}, \\ \text{9-frame on } S^{31} \end{cases}$$

(see the examples in Section 11.21).

Remark. F. Adams has shown that this is in fact the best possible result; that is, there are no orthonormal k-frames on S^{n-1} if $k > K(n)$, (cf. *Ann. of Math.* **75** (1962) p. 603.).

Problems

1. Suppose that an orthogonal multiplication is defined in a Euclidean space E. Denote the orthogonal complement of e by E_1. Show that the following conditions are equivalent:

   1. $(x \cdot y, e) = -(x, y) \qquad x, y \in E_1.$
   2. $x^2 = -(x, x) \cdot e, \qquad x \in E_1.$

2. *Cross-products in Euclidean spaces.* A *cross-product* in a Euclidean space F is a bilinear mapping $F \times F \to F$, denoted by $\times$, which satisfies the conditions:

   1. $(x, x \times y) = 0$ and $(y, x \times y) = 0$ for all $x, y \in F$.
   2. $|x \times y|^2 = |x|^2 \cdot |y|^2 - (x, y)^2$ for all $x, y \in F$.

   i. Show that a cross-product is skew-symmetric.
   ii. Let E be a Euclidean space with an orthogonal multiplication which satisfies the relation

   $$(x \cdot y, e) = (x, y) \qquad x, y \in E.$$

Let F be the orthogonal complement of e and denote by π the projection $\pi x = x - (x, e)e$, $x \in E$. Define a bilinear mapping $F \times F \to F$ by

$$x \times y = \pi(x \cdot y) \qquad x, y \in F.$$

Show that this map is a cross-product in F.

3. *The complex cross-product.* Let $\mathbb{C}^3$ be a 3-dimensional complex vector space with a positive Hermitian inner product $(\,,)$. Choose a normed determinant function D. Then the *complex cross-product* of two vectors a and b is defined by the equation

$$(a \times b, x) = \overline{D(a, b, x)} \qquad x \in \mathbb{C}^3.$$

Show that the complex cross-product has the following properties:

1. $(\lambda_1 a_1 + \lambda_2 a_2) \times b = \bar{\lambda}_1(a_1 \times b) + \bar{\lambda}_2(a_2 \times b)$ for all $\lambda_1, \lambda_2 \in \mathbb{C}$.
2. $(a \times b, a) = 0$ and $(a \times b, b) = 0$.
3. $a \times b = -b \times a$.
4. $(a_1 \times b_1, a_2 \times b_2) = \overline{(a_1, a_2) \cdot (b_1, b_2) - (a_1, b_2)(b_1, a_2)}$.
5. $|a \times b|^2 = |a|^2 |b|^2 - |(a, b)|^2$.
6. $a \times (b \times c) = (\overline{a, c})b - (\overline{a, b})c$.

4. *Orthogonal multiplications in* $\mathbb{C}^4$. Let $\mathbb{C}^4$ be a complex 4-dimensional vector space with a positive definite Hermitian inner product. Choose a normed determinant function Δ and a unit vector e. Let $\mathbb{C}^3$ denote the orthogonal complement of e. Then a normed determinant function D is defined in $\mathbb{C}^3$ by the equation

$$D(y_1, y_2, y_3) = \Delta(e, y_1, y_2, y_3) \qquad y_i \in \mathbb{C}^3.$$

Define a multiplication in $\mathbb{C}^4$ by the equations

$$x_1 \cdot y_1 = -(x_1, y_1)e + x_1 \times y_1$$
$$\lambda e \cdot y_1 = \lambda y_1, \qquad x_1 \cdot (\lambda e) = \bar{\lambda} x_1 \qquad \lambda \in \mathbb{C}$$
$$(\lambda e) \cdot (\mu e) = \lambda\mu \cdot e \qquad \lambda, \mu \in \mathbb{C}.$$

i. Prove the formula

$$|x \cdot y|^2 = |x|^2 \cdot |y|^2 \qquad x, y \in \mathbb{C}^4.$$

ii. The *conjugate* of an element $x \in \mathbb{C}^4$ is defined by $\bar{x} = \bar{\lambda} e - x_1$, where $x = \lambda e + x_1$, $\lambda \in \mathbb{C}$, $x_1 \in \mathbb{C}^3$. Show that

$$x \cdot \bar{x} = \bar{x} \cdot x = (x, x)_H \cdot e, \qquad x \in \mathbb{C}^4.$$

iii. Verify the formulas $x \cdot y^2 = (x \cdot y) \cdot y$ and $x^2 \cdot y = x \cdot (x \cdot y)$.

5. *The algebra of Cayley numbers.* Let E be an 8-dimensional Euclidean space. Choose a complex structure J in E (that is, a skew transformation J which satisfies $J^2 = -\iota$) and use it to make E into a 4-dimensional complex space H. Define a Hermitian inner product in H by setting

$$(x, y)_H = (x, y) + i(x, Jy) \qquad x, y \in H.$$

Choose a normed determinant function Δ in H.

i. Show that the multiplication defined in Problem 4 makes E into a real (non-associative) division algebra with e as unit element. It is called the *algebra of Cayley numbers*.

ii. Verify the relations

$$(ax, ay) = |a|^2(x, y) \qquad x, y \in E$$

and

$$(xa, ya) = (x, y)|a|^2.$$

6. *The cross-product in* $\mathbb{R}^7$. With the notation and hypotheses of Problem 5, let F denote the (7-dimensional) orthogonal complement of e in E.

   i. Prove the relation

   $$(xy, e) = -(x, y) \qquad x, y \in F.$$

   Conclude that the Cayley multiplication in E determines a cross-product in F.

   ii. Let $x \in E$ and write $x = \alpha e + y$ where $\alpha \in \mathbb{R}$, $y \in F$. Show that the conjugate element of x (see Problem 4, (ii)) is given by $x = \alpha e - y$. Conclude that if x is a non-zero Cayley number, then

   $$x^{-1} = \frac{\bar{x}}{(x, x)}.$$

   iii. Show that the relation $x \times y = 0$ holds if and only if $y = \lambda x$, $\lambda \in \mathbb{R}$.

7. Recall from Section 10.20 that $C_6(-) \cong L(\mathbb{R}^8)$. Construct an explicit isomorphism $\Phi: C_6(-) \xrightarrow{\cong} L(\mathbb{R}^8)$ in the following way: Regard $\mathbb{R}^6$ as the underlying real vector space of $\mathbb{C}^3$ and define a negative inner product in $\mathbb{R}^6$ by setting $(x, y)_- = -(x, y)$, where $(x, y) = \mathrm{Re}(x, y)_{\mathbf{H}}$. Show that the map $\varphi: \mathbb{R}^6 \to L(\mathbb{R}^8)$ given by $\varphi_a(x) = ax$ (Cayley multiplication) $a \in \mathbb{R}^6$, $x \in \mathbb{R}^8$, is a Clifford map and that the homomorphism $\Phi: C_6^- \to L(\mathbb{R}^8)$ extending φ is an isomorphism.
   Consider the linear transformations of $\mathbb{R}^8$ given by

   $$\omega_{\mathbb{C}}(x) = (x, e)_{\mathbf{H}} e \qquad \omega_A(x) = (e, x)_{\mathbf{H}} e,$$

   $$\omega(x) = (x, e)e \quad \text{and} \quad x \mapsto \bar{x}$$

   and describe the corresponding elements in C_6^-.

8. Use Problem 7 to construct explicit isomorphisms

   $$C_7(-) \xrightarrow{\cong} L(\mathbb{R}^8) \oplus L(\mathbb{R}^8)$$

   and

   $$C_8(-) \xrightarrow{\cong} L(\mathbb{R}^{16}).$$

Index

A

B

C

D

E

F

G

H

I

J

K

L

M

N

O

P

R

S

T

Graduate Texts in Mathematics

Soft and hard cover editions are available for each volume up to vol. 14, hard cover only from Vol. 15

1 TAKEUTI/ZARING. Introduction to Axiomatic Set Theory. vii, 250 pages. 1971.
2 OXTOBY. Measure and Category. viii, 95 pages. 1971. (Hard cover edition only.)
3 SCHAEFFER. Topological Vector Spaces. xi, 294 pages. 1971.
4 HILTON/STAMMBACH. A Course in Homological Algebra. ix, 338 pages. 1971. (Hard cover edition only)
5 MACLANE. Categories for the Working Mathematician. ix, 262 pages. 1972.
6 HUGHES/PIPER. Projective Planes. xii, 291 pages. 1973.
7 SERRE. A course in Arithmetic. x, 115 pages. 1973. (Hard cover edition only.)
8 TAKEUTI/ZARING. Axiomatic Set Theory. viii, 238 pages. 1973.
9 HUMPHREYS. Introduction to Lie Algebras and Representation Theory. 2nd printing, revised. xiv, 171 pages. 1978. (Hard cover edition only.)
10 COHEN. A Course in Simple Homotopy Theory. xii, 114 pages. 1973.
11 CONWAY. Functions of One Complex Variable. 2nd ed. approx. 330 pages. 1978. (Hard cover edition only.)
12 BEALS. Advanced Mathematical Analysis. xi, 230 pages. 1973.
13 ANDERSON/FULLER. Rings and Categories of Modules. ix, 339 pages. 1974.
14 GOLUBITSKY/GUILLEMIN. Stable Mappings and Their Singularities. x, 211 pages. 1974.
15 BERBERIAN. Lectures in Functional Analysis and Operator Theory. x, 356 pages. 1974.
16 WINTER. The Structure of Fields. xiii, 205 pages. 1974.
17 ROSENBLATT. Random Processes. 2nd ed. x, 228 pages. 1974.
18 HALMOS. Measure Theory. xi, 304 pages. 1974.
19 HALMOS. A Hilbert Space Problem Book. xvii, 365 pages. 1974.
20 HUSEMOLLER. Fibre Bundles. 2nd ed. xvi, 344 pages. 1975.
21 HUMPHREYS. Linear Algebraic Groups. xiv. 272 pages. 1975.
22 BARNES/MACK. An Algebraic Introduction to Mathematical Logic. x, 137 pages. 1975.
23 GREUB. Linear Algebra. 4th ed. xvii, 451 pages. 1975.
24 HOLMES. Geometric Functional Analysis and Its Applications. x, 246 pages. 1975.
25 HEWITT/STROMBERG. Real and Abstract Analysis. 4th printing. viii, 476 pages. 1978.
26 MANES. Algebraic Theories. x, 356 pages. 1976.
27 KELLEY. General Topology. xiv, 298 pages. 1975.
28 ZARISKI/SAMUEL. Commutative Algebra I. xi, 329 pages. 1975.

29 ZARISKI/SAMUEL. Commutative Algebra II. x, 414 pages. 1976.

30 JACOBSON. Lectures in Abstract Alegbra I: Basic Concepts. xii, 205 pages. 1976.

31 JACOBSON. Lectures in Abstract Algebra II: Linear Algebra. xii, 280 pages. 1975.

32 JACOBSON. Lectures in Abstract Algebra III: Theory of Fields and Galois Theory. ix, 324 pages. 1976.

33 HIRSCH. Differential Topology. x, 222 pages. 1976.

34 SPITZER. Principles of Random Walk. 2nd ed. xiii, 408 pages. 1976.

35 WERMER. Banach Algebras and Several Complex Variables. 2nd ed. xiv, 162 pages. 1976.

36 KELLEY/NAMIOKA. Linear Topological Spaces. xv, 256 pages. 1976.

37 MONK. Mathematical Logic. x, 531 pages. 1976.

38 GRAUERT/FRITZSCHE. Several Complex Variables. viii, 207 pages. 1976

39 ARVESON. An Invitation to C^*-Algebras. x, 106 pages. 1976.

40 KEMENY/SNELL/KNAPP. Denumerable Markov Chains. 2nd ed. xii, 484 pages. 1976.

41 APOSTOL. Modular Functions and Dirichlet Series in Number Theory. x, 198 pages. 1976.

42 SERRE. Linear Representations of Finite Groups. 176 pages. 1977.

43 GILLMAN/JERISON. Rings of Continuous Functions. xiii, 300 pages. 1976.

44 KENDIG. Elementary Algebraic Geometry. viii, 309 pages. 1977.

45 LOÈVE. Probability Theory. 4th ed. Vol. 1. xvii, 425 pages. 1977.

46 LOÈVE. Probability Theory. 4th ed. Vol. 2. xvi, 413 pages. 1978.

47 MOISE. Geometric Topology in Dimensions 2 and 3. x, 262 pages. 1977.

48 SACHS/WU. General Relativity for Mathematicians. xii, 291 pages. 1977.

49 GRUENBERG/WEIR. Linear Geometry. 2nd ed. x, 198 pages. 1977.

50 EDWARDS. Fermat's Last Theorem. xv, 410 pages. 1977.

51 KLINGENBERG. A Course in Differential Geometry. xii, 192 pages. 1978.

52 HARTSHORNE. Algebraic Geometry. xvi, 496 pages. 1977.

53 MANIN. A Course in Mathematical Logic. xiii, 286 pages. 1977.

54 GRAVER/WATKINS. Combinatorics with Emphasis on the Theory of Graphs. xv, 368 pages. 1977.

55 BROWN/PEARCY. Introduction to Operator Theory. Vol. 1: Elements of Functional Analysis. xiv, 474 pages. 1977.

56 MASSEY. Algebraic Topology: An Introduction. xxi, 261 pages. 1977.

57 CROWELL/FOX. Introduction to Knot Theory. x, 182 pages. 1977.

58 KOBLITZ. p-adic Numbers, p-adic Analysis, and Zeta-Functions. x, 122 pages. 1977.

59 LANG. Cyclotomic Fields. xi, 272 pages. 1978.

60 ARNOLD. Mathematical Methods in Classical Mechanics. approx. 350 pages. 1978.

61 WHITEHEAD. Elements of Homotopy Theory. approx. 500 pages. 1978.